Lecture Notes in Mathematics

Edited by A. Dold, Heidelberg and B. Eckmann, Zürich

364

Proceedings on Infinite Dimensional Holomorphy

University of Kentucky 1973

Edited by T. L. Hayden and T. J. Suffridge
University of Kentucky, Lexington, KY/USA

Springer-Verlag
Berlin · Heidelberg · New York 1974

AMS Subject Classifications (1970): 32-xx, 46-xx, 58-xx

ISBN 3-540-06619-5 Springer-Verlag Berlin · Heidelberg · New York
ISBN 0-387-06619-5 Springer-Verlag New York · Heidelberg · Berlin

Offsetdruck: Julius Beltz, Hemsbach/Bergstr.

PREFACE

The first eight authors in this volume gave the invited addresses to the International Conference on Infinite Dimensional Holomorphy held at the University of Kentucky during the week of May 28–June 1, 1973. The second paper of Waelbroeck contains his talk at the conference, while the first paper of Waelbroeck is a result obtained at the conference which was inspired by the address of Siciak. In the last few years the field of infinite dimensional holomorphy has seen a rapid growth in many directions. For those not familiar with the recent work in this area we suggest that the paper by Nachbin be consulted for many of the definitions and notation used by the other authors.

The remaining papers contain original results and were chosen on a refereed basis. The only exception is the paper by Schottenloher which is an invited expository paper.

The conference received generous funding from the National Science Foundation under Grant No. G.U. 2614 and from the University of Kentucky. We would like to express our appreciation for this support.

We thank all the participants for their helpful suggestions and cooperation. In particular we wish to thank Professors Lelong and Nachbin who were involved in the formation of the conference and provided invaluable advice and assistance, particularly in bringing to our attention the names of foreign mathematicians active in infinite dimensional holomorphy. Finally we wish to give special thanks to Wanda Jones who not only attended to many of the details of the conference, but did an excellent job of typing the entire manuscript.

T. L. Hayden

T. J. Suffridge

Lexington, September, 1973

TABLE OF CONTENTS

PARTICIPANTS

Belgium

Waelbroeck, L.

Brazil

Nachbin, L.
Pisanelli, D.

Canada

Beauchamp, J.-P.

France

Coeure, G.
Colombeau, J.-F.
Hervé, M.
Hervier, Y.
Hogbe-Nlend, H.
Lelong, P.
Noverraz, P.
Ramis, J.-P.

Germany

Aurich, V.
Schottenloher, M.

Ireland

Aron, R.
Boland, P.
Dineen, S.
Dwyer, T.
Varilly, J.

Poland

Ligocka, E.
Siciak, J.

Spain

Mª Isidro, J.

Sweden

Josefson, B.

Yugoslavia

Globevnik, J.

U.S.A.

Alves, M.
Basener, R.
Berner, P.
Berenstein, C.
Bogdanowicz, W.
Chae, S. B.
Cima, J.
Dostal, M.
Glickfeld, B.
Greenfield, S.
Gupta, C.
Hahn, K.
Harris, L. A.
Hayden, T. L.
Helton, J.
Helton, W.
Henrich, C.
Kato, T.
Katz, G. I.
Kelleher, J.
Koranyi, A.
Kujala, R.
Laufer, H.
Lorch, E.
Massey, F.
McGehee, O.
Mujica, J.
Patil, D.
Pfaltzgraff, J.
Resnikoff, H.
Rickart, C.
Rubel, L.
Rudin, W.
Selesnick, S.
Simpson, J.
Sommese, A.
Stevenson, J.
Suffridge, T.
Taylor, A. E.
Warren, H.
Weill, G.
Weiss, S.
Zame, W.

HOLOMORPHIC FUNCTIONS AND SURJECTIVE LIMITS

Seán Dineen

Introduction

In infinite dimensional holomorphy we are interested in characterising categories or collections of locally convex spaces[1] for which certain holomorphic properties are true. For example we would like to know in which locally convex spaces do the pseudo-convex and the holomorphically convex open sets coincide or we are interested in determining which locally convex spaces are holomorphically complete. The classical theory of several complex variables is concerned with holomorphy in a finite dimensional setting, in other words it limits its investigation to the simplest collection of locally convex spaces - the locally compact topological vector spaces.

Infinite dimensional holomorphy proceeds by investigating the next simplest and most interesting collection of locally convex spaces, i.e. the locally bounded or normed linear spaces. The next step presents us with many interesting choices of collections of locally convex spaces e.g. Baire, metrizable, barrelled, bornological, nuclear, $\mathcal{DFS}$, etc, spaces. All of these collections have proved themselves - within the context of linear functional analysis - of interest so it is quite natural that we should consider them. However, within the context of holomorphic functional analysis, they have proved to be inadequate so we are forced to find a new method for classifying locally convex spaces.

In this lecture I will describe a method for classifying (or generating) locally convex spaces which lends itself to the study of infinite dimensional holomorphy. The germ of this method is the classical theorem of Liouville concerning bounded entire functions but this fact is not usually apparent.

A study of the literature shows that Hirschowitz ([5]) and Rickart ([10]) first used this method in the study of holomorphic functions over the Cartesian product of complex planes. Nachbin ([7]) essentially uses this method in his study of uniform holomorphy and subsequently further applications to the study of pseudo-convex domains were made by Dineen ([1], [4]) and Noverraz ([8], [9]).

It then occurred to me that here was an essential tool of infinite dimensional holomorphy which needed to be formalized and applied in a systematic fashion. A subsequent investigation turned out to be both longer and more interesting than anticipated and resulted in a lengthy paper which has now gone through several drafts without reaching its final form.[2]

In this lecture I will describe the method (section 1) and give examples and applications (sections 2 and 3) which emphasize both its range of usefulness and its unifying effect in the theory of infinite dimensional holomorphy.

---

(1) Locally convex space means, throughout this work, a locally convex topological vector space over C.

(2) Some preliminary results were announced in [2].

At the conference I learned that E. Ligocka was engaged in a similar study (see [6]) and that some of our results coincided.

We refer to the book of Noverraz ([8]) for any unexplained notation and also for a general background about the concepts in infinite dimensional holomorphy which we use (especially those introduced in section 2). More complete references will be provided with the proofs.

§1. Surjective Systems and Limits

Let A denote a directed system. For each $i \in A$ let E_i denote a locally convex space and for $i,j \in A$, $i \geqslant j$ we let $\pi_j^i : E_i \to E_j$ denote a continuous linear mapping from E_i into E_j. $(E_i, \pi_j^i)_{\substack{i,j \in A \\ i > j}}$ is called a surjective system if

(1) π_j^i is an **onto** mapping for all $i,j \in A$, $i \geqslant j$.

(2) for $i \geqslant j \geqslant k$ the following diagram commutes

i.e. $\pi_k^i = \pi_k^j \circ \pi_j^i$.

A locally convex space E is a surjective limit of the surjective system $(E_i, \pi_j^i)_{\substack{i,j \in A \\ i > j}}$ if there exists a collection of continuous linear mappings $(\pi_i)_{i \in A}$, $\pi_i : E \to E_i$, such that the following conditions hold;

(3) each π_i is an **onto** mapping

(4) if $i,j \in A$, $i \geqslant j$ the following diagram commutes

$$
\begin{array}{ccc}
E & \xrightarrow{\ \pi_i\ } & E_i \\
 & \pi_j \searrow & \downarrow \pi_j^i \\
 & & E_j
\end{array}
$$

i.e. $\pi_j = \pi_j^i \circ \pi_i$.

(5) E has the weakest topology for which the mappings $(\pi_i)_{i \in A}$ are continuous.

We write E as $\lim_{\substack{\leftarrow \\ i \in A}} (E_i, \pi_j^i, \pi_i)$ and when no confusion is possible we write

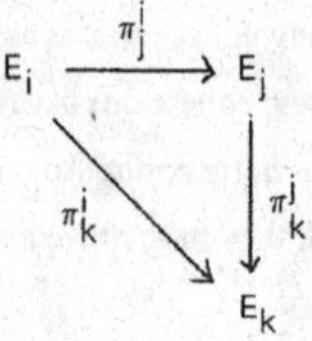

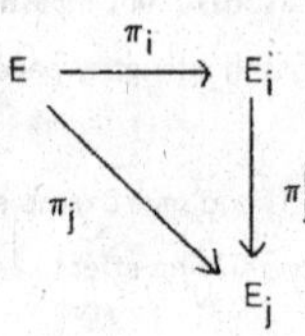

$E = \lim\limits_{\substack{\leftarrow \\ i \in A}} (E_i, \pi_i)$ or $E = \lim\limits_{\substack{\leftarrow \\ i \in A}} E_i$. Ligocka ([6]) calls $(E_i, \pi_i)_{i \in A}$ a basic system for E.

Remark. There may be more than one surjective limit associated with the same surjective system. In fact if $(E_i, \pi_j^i)_{\substack{i,j \in A \\ i \geq j}}$ is a surjective system and each E_i is a complete locally convex space then it does not follow that $E = \lim\limits_{\substack{\leftarrow \\ i \in A}} (E_i, \pi_i)$ is complete, however, the completion of E, $\hat{E}$, is a surjective limit associated with the same surjective system. This situation frequently arises. Our method consists in proving that various infinite dimensional holomorphic properties are preserved under surjective limits. We shall consider various different types of surjective limits and systems.

Definition. $E = \lim\limits_{\substack{\leftarrow \\ i \in A}} (E_i, \pi_i)$ is a complete surjective limit if for any collection, $(x_i)_{i \in A}$, $x_i \in E_i$ all i, for which $\pi_j^i(x_i) = x_j$ there exists an $x \in E$ such that $\pi_i(x) = x_i$ for all $i \in A$.

The complete surjective limit is the usual projective limit.

Let $\mathcal{O}(E)$ (resp. $\mathcal{K}(E)$) denote the set of open (resp. compact) subsets of a locally convex space E.

$E = \lim\limits_{\substack{\leftarrow \\ i \in A}} (E_i, \pi_i)$ is an open (resp. compact) surjective limit if $\pi_i; E \to E_i$ maps $\mathcal{O}(E)$ (resp. $\mathcal{K}(E)$) onto $\mathcal{O}(E_i)$ (resp. $\mathcal{K}(E_i)$) for all $i \in A$. In a similar fashion we define open and compact surjective systems.

The following lemma is useful for recognizing open surjective limits.

Lemma 1. **(a) Any surjective limit associated with an open surjective system is open. (b) A surjective system of Frechet spaces or of $\mathcal{DRF}$ (dual of reflexive Frechet spaces) is an open surjective system.**

The proof of (b) is just an application of the open mapping theorem.

We now give a number of examples.

Example 1. Any locally convex spaces is a surjective limit of normed linear spaces.

2. Every nuclear space is a surjective limit of pre Hilbert or inner product spaces.

3. $\prod_{i \in A} E_i$ is an open and compact surjective limit of $(F_j)_{j \in \mathcal{A}}$ where $\mathcal{A}$ consists of the finite subsets of A for each $j \in \mathcal{A}$, $F_j = \prod_{i \in j} E_i$.

Note. It is important to remember that $E_1 \times E_2$ is not the surjective limit of E_1 and E_2.

4. $\mathcal{C}(X)$, the complex valued continuous functions on the completely regular topological space X endowed with the compact open topology, is an open surjective limit of $\mathcal{C}(K)$ where K ranges over the compact subsets of X.

Note. $\mathcal{C}(X)$ may not be a complete locally convex space but its completion is also a surjective limit of the spaces $\mathcal{C}(K)$, K compact in X.

5. $\mathscr{D}'(\Omega)$, the space of distributions defined on an open subset of R^n, is an open surjective limit of $\mathscr{DFS}$ (dual of Frechet-Schwartz) spaces.

6. $\sum_{n=1}^{\infty} C \times \prod_{n=1}^{\infty} C$ is an open surjective limit of $\sum_{n=1}^{\infty} C \times \prod_{n=1}^{N} C \approx \sum_{n=1}^{\infty} C$.

§2. Applications of Surjective Limits to Infinite Dimensional Holomorphy

In this section we use surjective limits to extend several known results in infinite dimensional holomorphy. The next section is devoted to extending Hartogs' theorem.

(1) An open subset U of a locally convex space E is said to be finitely polynomially convex if $U \cap F$ is polynomially convex for all finite dimensional subspace F of E.

We say a locally convex space has property (α) if every finitely polynomially convex open subset of E is polynomially convex.

Proposition 1. A surjective limit of locally convex spaces with property (α) has property (α).

The crucial portion of the proof consists in using Liouvilles theorem to show the following;

If U is a pseudo convex open subset of E and p is a continuous semi-norm on E then $U_p = \{x \in U, d_p(x, CU) > 0\}$ is an open pseudo convex subset of E ($d_p(x,CU) = \inf_{y \notin U} p(x,y)$).

Example. We know that every pre-Hilbert space has property (α) and hence we can conclude that every nuclear space has property (α).

(2) A locally convex space E is said to have property (β) if every pseudo convex open subset of E is the domain of existence of a holomorphic function.

Proposition 2. A open surjective limit of locally convex spaces with property (β) has property (β).

The essential part of the proof of this proposition consists in showing that every pseudo-convex open subset of $E = \varprojlim_{i \in A} (E_i, \pi_i)$ has the form $\pi_i^{-1}(V)$ for some $i \in A$ and some pseudo-convex open subset V of E_i.
Noverraz ([9]) has shown that $\mathscr{DFS}$ spaces have property (β) and hence we may conclude that $\mathscr{D}'(\Omega)$ also has property (β).

(3) We now apply surjective limits to extend Zorn's theorem concerning the set of points of continuity of G-(Gateaux) holomorphic function.

Definition. Let E and F denote locally convex spaces. E is an F-Zorn space if the set of points of continuity of F-valued G-holomorphic functions defined on open subsets of E is both open and closed. E is a Zorn space if E is an F-Zorn space for any locally convex space F. We note in passing that it suffices to consider only normed linear spaces F in the above statement since every locally convex space is a surjective limit of normed spaces (example 1).

Proposition 3. **The surjective limit of Zorn spaces is a Zorn space.**

The essential part of the proof consists in noting that a G-holomorphic function factors through one of the space E_i at each point of continuity.

Examples (a) $\sum_{n=1}^{\infty} C \times \prod_{n=1}^{\infty} C$ is a Zorn space.

(b) A. Hirschowitz has shown that the space $H(C) \times \sum_{n=1}^{\infty} C$ is not a Zorn space and hence we may conclude that $H(C) \times F'_\beta$ is not a Zorn space where F is a Frechet space which does not admit a continuous norm and F'_β is its strong dual (see §3).

(c) $H(C) \times \sum_{n=1}^{\infty} C$ is a nuclear space and hence, as already noted, is the surjective limit of pre-Hilbert spaces. We may thus conclude that there exist pre-Hilbert spaces which are not Zorn spaces. Furthermore, since any dense subspace of a space which is not Zorn is itself not a Zorn space, we may conclude that any countably infinite dimensional inner-product space is not a Zorn space.

(d) $\mathcal{C}(X)$ is a Zorn space for any completely regular space X.

A number of other applications to "Zorn-type" results are possible. We restrict ourselves to one further application so that we may make some use of the concept of compact surjective limit.

A G-holomorphic function defined on a locally convex space E with values in a locally convex space F is said to be hypoanalytic if its restriction to each compact subset K of E is continuous.

We let $\mathcal{H}_Y(U;F)$ denote the set of all hypoanalytic mappings from the open subset U of E into F. A locally convex space E is called a $\tilde{k}$-space if $\mathcal{H}_Y(U;F) = \mathcal{H}(U;F)$ ($\mathcal{H}(U;F)$ is the set of all (continuous) holomorphic functions from U into F). A locally convex space E which is a topological k-space is a $\tilde{k}$-space but the converse is not true (e.g. $\prod_{w} C$, w uncountable is an example of a $\tilde{k}$-space which is not a k-space).

Proposition 4. **If** $E = \lim\limits_{\substack{\leftarrow \\ i \in A}} (E_i, \pi_i)$ **is a compact surjective limit of $\tilde{k}$-spaces then the set of points of continuity of hypoanalytic functions defined on open subsets of E is both open and closed.**

Proposition 4 is useful in the study of paracomplete locally convex spaces.

(4) Extending Zorn's theorem leads us to the problem of global factorization of holomorphic functions. More precisely, if $E = \lim\limits_{\substack{\leftarrow \\ i \in A}} (E_i, \pi_i)$ and $f \in \mathcal{H}(E;F)$ does there exist an $i \in A$ and $f_i \in \mathcal{H}(E_i,F)$ such that the following diagram commutes

$$\begin{array}{ccc} E & \xrightarrow{\pi_i} & E_i \\ & \searrow^{f} & \downarrow{f_i} \\ & & F \end{array} \qquad ?.$$

If this is true then we say (? write)

$$\mathcal{H}(E;F) = \bigcup_{i\in A} \mathcal{H}(E_i;F).$$

Using a proof similar to that used in extending Zorn's theorem we obtain the following result.

Proposition 5. If $E = \lim\limits_{\substack{\leftarrow \\ i\in A}} (E_i,\pi_i)$ and F is a locally convex space on which there exists a continuous norm then

$$\mathcal{H}(E;F) = \bigcup_{i\in A} \mathcal{H}(E_i;F)$$

in either of the following cases;

(a) each E_i is an F-Zorn space

(b) E is an open surjective limit.

Remark. Since $H(\prod_{n=1}^{\infty} C; \prod_{n=1}^{\infty} C) \neq \bigcup_n H(C^n; \prod_{n=1}^{\infty} C)$ we see that the existence of a continuous norm on F is a necessary condition for proposition 5. This curious condition concerning the existence of a continuous norm appears frequently in infinite dimensional holomorphy and is related to several other conditions which also play a role in holomorphy (see section 3).

(5) Let E and F denote locally convex spaces, $\hat{E}_d(F)$, the F-holomorphic completion of E, is the subspace of E which satisfies the following properties

(a) $E \subset \hat{E}_d(F) \subset \hat{E}$

(b) each holomorphic function from E into F has a unique extension to a holomorphic function from $\hat{E}_d(F)$ into F.

(c) $\hat{E}_d(F)$ is the largest subspace of $\hat{E}$ which satisfies (a) and (b).

$\hat{E}_d(F)$ always exists and may be shown to coincide with the intersection of all domains of existence of F–holomorphic functions defined on open subsets of $\hat{E}$ which contain E ([3]). The next proposition gives a criterion in which we may deduce that $\hat{E}_d(F) = \hat{E}$.

Proposition 6. Let $E = \lim\limits_{\substack{\leftarrow \\ i\in A}} (E_i,\pi_i)$ be an open surjective limit of complete locally convex spaces or a surjective limit of complete F-Zorn spaces then $\hat{E}_d(F) = \hat{E}$ if there exists a continuous norm on F.

We may extend proposition 6 to the situation where the space E_i are not necessarily complete. If $\hat{E}_d(F) = E$ we say E is F-holomorphically complete.

Proposition 7. Let $E = \lim\limits_{\substack{\leftarrow \\ i\in A}} (E_i,\pi_i)$ and F be locally convex spaces. If each E_i is F-holomorphically complete and there exists a continuous norm on F then the F-holomorphic completion of E is the complete surjective limit associated with E in either of the following cases:

(a) E is an open surjective limit.

(b) each E_i is an F-Zorn space.

Example. $\mathcal{C}(X)$ is F-holomorphically complete if and only if it is complete.

(6) Holomorphic completeness concerns finding the largest domain of definition of a family of holomorphic function. We now turn to the dual problem of finding the smallest range of a family of holomorphic functions. Let U and V be domains spread over a locally convex space E. V is called an analytic extension of U if U can be identified with an open subset of V and each C-valued holomorphic function on U can be extended to a C-valued holomorphic function on V.

A locally convex space F is said to be E-paracomplete if for all pairs of domains spread over E, (U,V), V being an analytic extension of U, every F-valued holomorphic function of U has an extension to an F-valued holomorphic function on V.

Proposition 8. The complete surjective limit of E-paracomplete spaces is an E-paracomplete space.

We mention in passing that there appears to be a relationship between paracomplete and holomorphically complete locally convex spaces. In fact if each F-valued holomorphic function f on a domain U, has an extension as a holomorphic function $\tilde{f}$ from V into $\hat{F}$ then $\tilde{f}(V) \subset \hat{F}_d(C)$. Results of this nature will be discussed in a later paper.

§3. Hartogs' theorem.

We have reserved this section for just one application of surjective limits - to extend Hartogs' theorem - because of the number of technical details involved. The conditions we find are sufficient and as we shall see by counterexample they are not unnecessary. Some of the concepts we discuss are of interest because they occur in other infinite dimensional holomorphic settings and others are of interest because they extend our methods of classifying locally convex spaces by means of surjective limits.

Definition. A topological space X is an i-space if for any closed non-compact subset B of X there exists a sequence of disjoint open subsets of X, $(W_n)_{n=1}^{\infty}$, such that

(a) $B \cap W_n \neq \phi$ for each n

(b) if K is a compact subset of X then there exists a positive integer p such that

$$K \cap W_n = \phi \text{ for all } n \geqslant p.$$

The following is one of the main results concerning i-spaces.

Proposition 8. Let X be a completely regular Hausdorf space then the following are equivalent;

(1) X is an i-space

(2) $\mathcal{C}(X)$ is infrabarrelled

(3) for each closed non-compact subset B of X there exists a positive lower semi-continuous function on X which is bounded on each compact subset of X but which is not bounded on B.

We shall now assume that the directed system A, which appears in the definition of surjective limit, is the set of compact subsets of a completely regular Hausdorf i-space. We write

$$\lim_{\substack{\leftarrow \\ K\subset X}} (E_K, \pi_K) \text{ in place of } \lim_{\substack{\leftarrow \\ i\in A}} (E_i, \pi_i).$$

The set of ordinals less than or equal to the first uncountable ordinal with the order topology is an example of a completely regular Hausdorf space which is not an i-space. There also exist i-spaces X such that $\mathcal{C}(X)$ is not barrelled.

We say $x \in E$ is zero on $A \subset X$ and we write $x|_A = 0$ if $\pi_K(x) = 0$ for all compact subsets K of A. Now let f be a function defined on an open subset U of E. We say f factors through $A \subset X$ if for each $x,y \in E$, $y|_A = 0$ implies $f(x+y) = f(x)$. Now if f factors through A and K is a compact subset of X such that $A \subset K$ then there exists a function $\tilde{f}$ on $U_K \subset E_K$ such that $\tilde{f}\circ\pi_K = f$ and $\pi_K^{-1}(U_K) \supset U$. Furthermore if f is a G-holomorphic function on U then $\tilde{f}$ is a G-holomorphic function on U_K.

Definition. (a) Let $E = \lim\limits_{\substack{\leftarrow \\ K\subset X}} (E_K, \pi_K)$ and let f be a function from $U \subset E$ into F. A compact subset K of X is called a unique support of f if it satisfies the following properties;

(1) f factors through K

(2) if W is an open subset of X such that $K \cap W \neq \phi$ then f does not factor through C(W) (= complement of W in X)

(b) E has the **C**-unique support property if each C-valued holomorphic function defined on an open subset of E has a unique support.

We also need the concept of very strongly and very weakly convergent sequences.

Definition. (a) A sequence $(x_n)_{n=1}^{\infty}$, of elements in a locally convex space E, is every **strongly convergent** if the sequence $(\lambda_n x_n)_{n=1}^{\infty}$ converges to 0 for any sequence of scalars $(\lambda_n)_{n=1}^{\infty}$.

(b) E_*, the very strong completion of E, is the subset of $\hat{E}$ consisting of all elements of the form $\sum_{n=1}^{\infty} \lambda_n x_n$ where $(x_n)_{n=1}^{\infty}$ is a very strongly convergent sequence in E and $(\lambda_n)_{n=1}^{\infty}$ is an arbitrary sequence of scalars.

A sequence $(x_n)_{n=1}^{\infty}$ is very strongly convergent to 0 if and only if for each continuous semi-norm p on E there exists an integer N such that $p(x_n) = 0$ for all $n \geq N$. E_* is a vector subspace of $\hat{E}$ which contains E. We note in passing that $E_* \subset \hat{E}_d(F)$ for any complete locally convex space F.

Definition. A sequence, $(x_n)_{n=1}^{\infty}$, of elements of a locally convex space E is **very weakly** convergent to zero if there exists a sequence of non-zero scalars $(\lambda_n)_{n=1}^{\infty}$ such that $\lambda_n x_n \to 0$ as $n \to \infty$.

Examples (a) Every sequence in a metrizable locally convex space is very weakly convergent to 0.

(b) If $E = \sum_{n=1}^{\infty} C$ and $u_n = (0,\ldots,1,0,\ldots)$ (1 in the nth position) then $(u_n)_{n=1}^{\infty}$ is not a very weakly convergent sequence.

(c) $\prod_{(w)} C$ contains a sequence which does not converge to 0 very weakly if and only if (w) is uncountable.

(d) If every sequence in (E,τ) converges to 0 very weakly and τ_1 is a weaker topology on E than τ then every sequence in (E,τ_1) converges to 0 very weakly.

We now state two propositions which show that there is a certain duality between the concepts of very strongly and very weakly convergent sequences and also between both of these concepts and the existence of a continuous norm.

Proposition 9. (a) If there exists a continuous norm on E then every very strongly convergent sequence is eventually identically zero

(b) If E is separable and every sequence of elements of E converges to 0 very weakly then there exists a continuous norm on the strong dual of E

(c) If E contains a non-trivial (i.e. not eventually identically zero) very strongly convergent sequence then E'_β contains a sequence which does not converge to zero very weakly.

Proposition 10. (a) If E is a Frechet space then the following are equivalent:

(1) There does not exist any continuous norm on E.

(2) E contains a non-trivial very strongly convergent sequence.

(3) E contains a subspace isomorphic to $\prod_{n=1}^{\infty} C$.

(4) There exists a continuous linear mapping of E'_β (the strong dual of E) onto $\sum_{n=1}^{\infty} C$.

(5) $\sum_{n=1}^{\infty} C$ is isomorphic to a quotient of E'_β by a closed subspace.

(b) If E is a Frechet space which is not a Banach space then E'_β contains a sequence which does not converge to 0 very weakly

(c) If E is a Frechet space which is not a Banach space then $\prod_{n=1}^{\infty} C$ is isomorphic to a quotient of E by a closed subspace.

Very weakly convergent sequences occur in our requirements for extending Hartogs' theorem and very strongly convergent sequences are used in the proof of this extension.

Definition. $E = \varprojlim_{K\subset X} (E_K,\pi_K)$ has the extension property if for any sequence of elements of E, $(f_n)_{n=1}^{\infty}$, and for any sequence of compact subsets of X, $(K_n)_{n=1}^{\infty}$, there exists a sequence of elements of E, $(g_n)_{n=1}^{\infty}$, such that

(a) $\pi_{K_n}(f_n-g_n) = 0$

(b) $g_n \to 0$ very weakly in E as $n \to \infty$.

We call this property the extension property since it is equivalent to the following property; given a sequence of compact subsets of X, $(K_n)_{n=1}^{\infty}$, and a sequence of elements, $(f_n)_{n=1}^{\infty}$, such that $f_n \in E_{K_n}$ then each f_n can be extended to an element of E (i.e. extended to all of X) and the extended sequence converges to zero very weakly.

A pair of locally convex spaces (E,F) is called a C-Hartogs' pair of every separately continuous G-holomorphic function defined on an open subset of E x F is continuous.

We are now in a position to state an extension of Hartogs' theorem.

Theorem. **Let $E = \varprojlim_{K \subset X} (E_K, \pi_K)$ and $F = \varprojlim_{L \subset Y} (F_L, \rho_L)$ denote open surjective limits which have the extension property where X and Y are i-spaces. Now if E and F have the unique extension property and (E_K, F_L) is a C-Hartogs pair for each compact subset K of X and each compact subset L of Y then (E,F) is a C-Hartogs pair.**

We now define local partitions of unity of a locally convex space. These occur frequently when the locally convex space E has a sheaf like structure. If E admits local partitions of unity then E has the extension property and the unique support property.

Definition. $E = \varprojlim_{K \subset X} (E_K, \pi_K)$ admits local partitions of unity if it satisfies the following conditions;

(a) If $x \in E$, $x|_{\Omega_1} = 0$ and $x|_{\Omega_2} = 0$ for Ω_1 and Ω_2 open subsets of X then $x|_{\Omega_1 \cup \Omega_2} = 0$

(b) If K is a compact subset of the open subset of the open subset Ω of X then there exists an open subset $\tilde{\Omega}$ of X and $T_{\Omega,\tilde{\Omega},K}$ an operator from E into E such that

(b1) $K \subset \tilde{\Omega} \subset \overline{\tilde{\Omega}} \subset \Omega$

(b2) $f - T_{\Omega,\tilde{\Omega},K}(f)|_{\tilde{\Omega}} = 0$ for all $f \in E$

(b3) $T_{\Omega,\tilde{\Omega},K}(f)|_{C(\Omega)} = 0$ for all $f \in E$

(b4) If V is an open subset of X, $f \in E$ and $f|_V = 0$ then $T_{\Omega,\tilde{\Omega},K}(f)|_V = 0$.

We call $T_{\Omega,\tilde{\Omega},K}$ a local partition of unity about K with support in Ω and 1 on $\tilde{\Omega}$.

(c) If K is compact in X and $f \in E$ is such that $\pi_K(f) = 0$ then for each Ω open containing K we can choose a local partition of unity about K with support in Ω, $T_{\Omega,\tilde{\Omega},K}$, such that $T_{\Omega,\tilde{\Omega},K}(f) \to 0$ as $\Omega \to K$ ($\Omega \to K$ in the usual sense of filtered neighborhoods).

Examples. 1. If (E, E'_β) is a C-Hartogs pair then E is a normed linear space. Hence $(\prod_{n=1}^{\infty} C, \sum_{n=1}^{\infty} C)$ is not a C-Hartogs pair. Applying proposition 10 we may deduce that (E,F'_β) is not a C-Hartogs pair when E and F are

Frechet spaces which are not Banach spaces and there does not exist a continuous norm on F.

2. $\prod_{n=1}^{\infty} C$ is an open surjective limit over the integers and satisfies all the conditions of our theorem. $\sum_{n=1}^{\infty} C$ when looked upon as a surjective limit over one point satisfies all the conditions except the extension property (which would imply that every sequence in $\sum_{n=1}^{\infty} C$ converges to 0 very weakly). Hence we see that the extension property cannot be removed from our theorem.

3. If X is a completely regular i-space and E is a Frechet space then $(\mathcal{E}(X), E)$ is a C-Hartogs pair. ($\mathcal{E}(X)$ admits local partitions of unity for any completely regular space X).

4. $(\prod_{i \in w_1} E_i, \prod_{j \in w_2} F_j)$ is a C-Hartogs pair for any collection of Frechet spaces E_i and F_j.

Remark. This result is not true if we replace the Frechet spaces by $\mathcal{DFS}$ spaces and the extension property is not valid in this case. (see example 2).

5. If Ω is the set of all ordinals less than the first uncountable ordinal then $\mathcal{E}(\Omega)$ has the extension property and the unique support property but Ω is not an i-space. $\mathcal{E}'(\Omega)_\beta$ is a Frechet space; hence $(\mathcal{E}(K), \mathcal{E}'(\Omega)_\beta)$ is a C-Hartogs pair for each compact subset K of Ω. However $(\mathcal{E}(\Omega), \mathcal{E}'(\Omega)_\beta)$ is not a C-Hartogs pair and this implies that the i-space requirement is necessary for our extension of Hartogs' theorem.

We conclude our results by giving a simple application of Hartogs' theorem.

Proposition 11. If X and Y are topological i-space then every holomorphic function defined on an open subset of $\mathcal{E}(X)$ with values in $\mathcal{E}'(Y)_\beta$ is locally bounded.

The method used in extending Hartogs' theorem may also be applied to the study of various locally convex topologies on spaces of holomorphic functions.

Final Remarks. The notation used in this paper differs slightly from that which will appear with the proofs. We hope this will cause no confusion.

Bibliography

1. S. Dineen, **Fonctions analytique dans les espaces vectoriels topologiques locelement convexes,** C. R. Acad. Sc., Paris, t274, 1972 , 544–574.

2. _______, **Holomorphically significant properties of topological vector spaces** (to appear in Proceedings of Colloque International du C.N.R.S. sur "Les fonctions analytiques de plusiers variables", Paris, June 14–June 20, 1972)

3. _______, **Holomorphically complete locally convex topological vector spaces,** Lecture Notes in Mathematics, No.332, Springer-Verlag, Berlin 1973, Seminaire Lelong, 1971–1972.

4. _______, **Holomorphic functions on locally convex topological vector spaces II,** Pseudo convex domains (to appear in Ann. de l'Institut Fourier)

5. A. Hirschowitz, **Remarques sur les ouverts d'holomorphie d'un produit dénombrable de droites,** Ann. Inst. Fourier, t19, 1969, 219–229.

6. E. Ligocka, **A local factorization of analytic functions and its applications** (to appear in Studia Mathematica)

7. L. Nachbin, **Uniformite d'holomorphie et type exponential,** Seminaire Lelong, 1969–1970, Springer lecture notes 205.

8. Ph. Noverraz, **Pseudo-convexite, convexite polynomiale et domain d'holomorphie en Dimension Infinie,** Notas de Matematica 3, North Holland 1973.

9. _________, **Sur le pseudo-convexite et la convexite polynomiale en dimension infinie,** C. R. Acad. Sc., Paris, t274, 1972, 313–316.

10. C. E. Rickart, **Analytic functions of an infinite number of complex variables,** Duke Math. Jour., t36, 1969, 581–597.

University College Dublin
School of Mathematics
Belfield, Dublin 4, Ireland

BOUNDED SYMMETRIC HOMOGENEOUS DOMAINS IN INFINITE DIMENSIONAL SPACES

Lawrence A. Harris[1]

In this article, we exhibit a large class of Banach spaces whose open unit balls are bounded symmetric homogeneous domains. These Banach spaces, which we call J*–algebras, are linear spaces of operators mapping one Hilbert space into another and have a kind of Jordan triple product structure. In particular, all Hilbert spaces and all B*–algebras are J*–algebras. Moreover, all four types of the classical Cartan domains and their infinite dimensional analogues are the open unit balls of J*–algebras, and the same holds for any finite or infinite product of these domains. Thus we have a setting in which a large number of bounded symmetric homogeneous domains may be studied simultaneously. A particular advantage of this setting is the interconnection which exists between function-theoretic problems and problems of functional analysis. This leads to a simplified discussion of both types of problems.

We shall see that the open unit balls of J*–algebras are natural generalizations of the open unit disc of the complex plane. In fact, we give an explicit algebraic formula for Möbius transformations of these balls and show that the origin can be mapped to any desired operator in the ball with one of the Möbius transformations. An extremal form of the Schwarz lemma then leads immediately to the representation of each biholomorphic mapping between the open unit balls of two J*–algebras as a composition of a Möbius transformation and a linear isometry of one of the J*–algebras onto the other. Such linear isometries reduce to a multiplication by unitary operators for mappings in the identity component of the group of all biholomorphic mappings of the open unit ball of a C*–algebra with identity. However, in general, linear isometries of one J*–algebra onto another can be complicated. Still, using the mentioned Schwarz lemma and Möbius transformations, we show that all such linear isometries preserve the J*–structure.

A consequence of these results is that the open unit balls of two J*–algebras are holomorphically equivalent if and only if the J*–algebras are isometrically isomorphic under a mapping preserving the J*–structure. Another consequence is that the open unit ball of a J*–algebra is holomorphically equivalent to a product of balls if and only if the J*–algebra is isometrically isomorphic to a product of J*–algebras.

The last result connects the factorization of domains with the factorization of J*–algebras and has a number of interesting applications. For example, using Cartan's classification of bounded symmetric domains in C^n, we classify all J*–algebras of dimension less than 16. Moreover, we reduce the problem of classifying all finite dimensional J*–algebras to the problem of finding some J*–algebras whose open unit balls are holomorphically equivalent to the two exceptional Cartan domains in dimensions 16 and 27, respectively, when such J*–algebras exist. If there are such J*–algebras in both cases, then every bounded symmetric

[1] Research partially supported by N.S.F. grant GP–33117A#1.

domain in $\mathbf{C}^n$ is holomorphically equivalent to the open unit ball of a J*—algebra. Also, for J*—algebras having an isometry, we obtain an algebraic condition which implies that the open unit ball of the J*—algebra is not holomorphically equivalent to a product of balls. As expected, none of the infinite dimensional analogues of the four types of classical Cartan domains is holomorphically equivalent to a product of balls.

The unit spheres of many J*—algebras contain small subsets playing the same role as the distinguished boundary of polydiscs. In fact, any non-empty subset of the unit sphere of a J*—algebra which is stable under the application of the Möbius transformations and under multiplication by complex numbers of unit modulus is a boundary for the algebra of all bounded complex-valued functions holomorphic in the open unit ball of the J*—algebra and continuous in its closure. Three particularly interesting stable subsets of a J*—algebra are (when non-empty) the set of extreme points of the closed unit ball, the set of all isometries, and the set of all unitary operators. For a finite dimensional J*—algebra, the Shilov boundary for the mentioned algebra of functions is the set of extreme points of the closed unit ball of the J*—algebra. We give an algebraic characterization of extreme points and determine these explicitly for the J*—algebras whose open unit balls are classical Cartan domains.

Finally, we show that the open unit balls of a large number of J*—algebras are holomorphically equivalent to an explicitly constructed unbounded affinely homogeneous domain in the J*—algebra which plays the role of the upper half-plane. These upper half-planes are operator-theoretic analogues of Siegel domains of genus 2. Siegel's generalized upper half-plane and tubes whose base is a future light cone are included as special cases.

This article generalizes and expands Chapter IV of the author's thesis, written under the direction of Professor Clifford Earle.

§ 1. Generalities

Let X and Y be complex normed linear spaces and let $\mathcal{D}$ be an open subset of X. (To avoid trivialities, all normed linear spaces considered will be assumed to be different from the zero space.) A function $h: \mathcal{D} \to Y$ is said to be **holomorphic** (in $\mathcal{D}$) if the Fréchet derivative of h at x (denoted by Dh(x)) exists as a bounded complex-linear map of X into Y for each $x \in \mathcal{D}$. If $\mathcal{D}'$ is an open subset of Y, a function $h: \mathcal{D} \to \mathcal{D}'$ is said to be a **biholomorphic mapping** (of $\mathcal{D}$ onto $\mathcal{D}'$) if the inverse function $h^{-1}: \mathcal{D}' \to \mathcal{D}$ exists and both $h: \mathcal{D} \to Y$ and $h^{-1}: \mathcal{D}' \to X$ are holomorphic. The sets $\mathcal{D}$ and $\mathcal{D}'$ are said to be **holomorphically equivalent** if there exists a biholomorphic mapping of $\mathcal{D}$ onto $\mathcal{D}'$. A domain $\mathcal{D}$ is said to be **homogeneous** if for each pair of points $x, y \in \mathcal{D}$ there exists a biholomorphic mapping $h: \mathcal{D} \to \mathcal{D}$ with $h(x) = y$. If in addition the mappings h can be chosen to be affine mappings, $\mathcal{D}$ is said to be **affinely homogeneous**. Further, $\mathcal{D}$ is said to be a **symmetric** domain if for each $x \in \mathcal{D}$ there exists a biholomorphic mapping $h: \mathcal{D} \to \mathcal{D}$ such that h has x as its only fixed point and $h^2 = I$, where I is the identity map on $\mathcal{D}$.

Throughout, the open (resp., closed) unit ball of a complex normed linear space X is denoted by X_0 (resp., X_1). Thus

$$X_0 = \{x \in X: \|x\| < 1\},\ X_1 = \{x \in X: \|x\| \leq 1\}.$$

A point $x \in X_1$ is said to be an **extreme point (resp., a complex extreme point)** of X_1 if the only $y \in X$ satisfying $\|x+\lambda y\| \leq 1$ for all real (resp., complex) numbers λ with $|\lambda| \leq 1$ is $y = 0$. Clearly any extreme point of X_1 is a complex extreme point of X_1.

In a previous paper [14] (see also [15] and [16]), the author proved

Theorem 1. *If h: $X_0 \to Y_0$ is a biholomorphic mapping with $h(0) = 0$, then h is the restriction to X_0 of a linear isometry of X onto Y.*

Now if X_0 is a homogeneous domain, any biholomorphic mapping of X_0 onto Y_0 may be preceeded by a biholomorphic mapping of X_0 onto itself so that the composition takes 0 to 0. Thus we obtain

Corollary 1. *Suppose X_0 is a homogeneous domain. Then the domains X_0 and Y_0 are holomorphically equivalent if and only if the spaces X and Y are isometrically isomorphic.*

For example, suppose that H is a complex Hilbert space and that H_0 is holomorphically equivalent to a domain $X_0 \times Y_0$. Clearly $X_0 \times Y_0 = (X \times Y)_0$ when $X \times Y$ has the norm $\|(x,y)\| = \max\{\|x\|, \|y\|\}$, and we shall see later that H_0 is a homogeneous domain. Hence by Corollary 1, the spaces H and $X \times Y$ are isometrically isomorphic. Then since all unit vectors in H are extreme points of H_1, the same must be true of $X \times Y$. But for each $x \in X$ with $\|x\| = 1$, the point $(x,0)$ is a unit vector in $X \times Y$ which is not an extreme point of $(X \times Y)_1$. Thus we conclude that H_0 is not holomorphically equivalent to any domain of the form $X_0 \times Y_0$. Clearly this result contains Theorem 2.1 of [11]. (A more general result is given in Theorem 8 below.)

Further, note that if X_0 is a homogeneous domain then it is automatically symmetric. Indeed, given $x \in X_0$, let g be a biholomorphic mapping of X_0 onto itself with $g(x) = 0$, and define $h = g^{-1} \circ L \circ g$, where $Lx = -x$ for $x \in X_0$. It is easy to verify that h has the required properties.

It would be interesting to know whether or not Corollary 1 holds without the assumption that X_0 is a homogeneous domain.

§2. J*–algebras and Cartan domains

Let H and K be complex Hilbert spaces and let $\mathcal{L}(H,K)$ denote the Banach space of all bounded linear operators from H to K with the operator norm. For each operator $A \in \mathcal{L}(H,K)$ there is a uniquely determined operator $A^* \in \mathcal{L}(K,H)$ such that $(Ax,y) = (x,A^*y)$ for all $x \in H$ and $y \in K$. It is easily verified that $*$ satisfies the usual laws of an adjoint operation [25, p.105].

Definition 1. A J*-algebra is a closed complex-linear subspace $\mathfrak{A}$ of $\mathcal{L}(H,K)$ such that $AA^*A \in \mathfrak{A}$ whenever $A \in \mathfrak{A}$.

Throughout, unless otherwise specified, $\mathfrak{A}$, $\mathfrak{B}$, and $\mathfrak{C}$ denote arbitrary J*–algebras.

Many familiar spaces are J*–algebras. For example, any Hilbert space H may be thought of as a J*–algebra since H can be identified with the space $\mathcal{L}(C,H)$, where C denotes the complex plane. Also any C*–algebra is obviously a J*–algebra and hence by the Gelfand-Naimark theorem [31, p.244], any B*–algebra may be thought of as a J*–algebra. Further, if C and D are self-adjoint operators in $\mathcal{L}(H)$ and $\mathcal{L}(K)$, respectively, then

$$\mathfrak{A} = \{A \in \mathcal{L}(H,K) : AC = DA\}$$

is easily seen to be a J*–algebra.

Define a set $\mathfrak{A}$ to be a **Cartan factor of**

type I if $\mathfrak{A} = \mathcal{L}(H,K)$,

type II if $\mathfrak{A} = \{A \in \mathcal{L}(H) : A^t = A\}$,

type III if $\mathfrak{A} = \{A \in \mathcal{L}(H) : A^t = -A\}$,

where $A^t = QA^*Q$ and Q is a conjugate-linear map on H with $\|Q\| \leq 1$ and $Q^2 = I$. (Such a map always exists and is called a conjugation.) Polarization shows that Q reverses inner products and it follows that $(A^*)^t = (A^t)^*$ for all $A \in \mathcal{L}(H)$. Consequently, all the Cartan factors of type I-III are J*–algebras. Note that the infinite dimensional domains considered in [11] and [28] are the open unit balls of Cartan factors of types I and II. Moreover, each of the Cartan domains [10, §24] of types I-III is the open unit ball of a finite dimensional Cartan factor of the corresponding type.

A **Cartan factor of type IV** is a closed subspace $\mathfrak{A}$ of $\mathcal{L}(H)$ such that the adjoint of each operator in $\mathfrak{A}$ is in $\mathfrak{A}$ and such that the square of each operator in $\mathfrak{A}$ is a scalar multiple of the identity operator I on H. It is clear from the identities

$$\text{(1)} \qquad \begin{aligned} AB + BA &= (A+B)^2 - A^2 - B^2 \\ AB^*A &= (AB^* + B^*A)A - B^*A^2 \end{aligned}$$

that $\mathfrak{A}$ is a J*–algebra. Note that $\mathfrak{A}$ is also a Hilbert space in an equivalent norm since the equation

$$AB^* + B^*A = 2(A,B)I$$

defines an inner product on $\mathfrak{A}$ and clearly $\frac{1}{2}\|A\|^2 \leq (A,A) \leq \|A\|^2$ for all $A \in \mathfrak{A}$. Conversely, any Hilbert space H can be obtained as the Hilbert space associated with a Cartan factor of type IV. In fact, if $x \to \bar{x}$ is a conjugation on H, there exists a Hilbert space K and a linear map $x \to A_x$ of H into $\mathcal{L}(K)$ such that

$$\text{(2)} \qquad A_x^* = A_{\bar{x}} \text{ and } A_x^2 = (x,\bar{x})I ;$$

and consequently, $(A_x,A_y) = (x,y)$ and $\|x\| \leq \|A_x\| \leq \sqrt{2}\,\|x\|$. Moreover, identifying H with its image $\mathfrak{A}$ under the map $x \to A_x$, we have

(3) $$\mathfrak{A}_0 = \{x \in H: \|x\|^2 + \sqrt{\|x\|^4 - |(x,\bar{x})|^2} < 1\}.$$

Thus each of the Cartan domains of type IV is the open unit ball of a finite dimensional Cartan factor of type IV. (See [19, p.48].) Verification of the unproved assertions above is given at the end of this section.

If $\mathfrak{A}$ and $\mathfrak{B}$ are J*–algebras, the product space $\mathfrak{A} \times \mathfrak{B}$ can be made into a J*–algebra in a natural way. Indeed, if $\mathfrak{A}$ and $\mathfrak{B}$ are subspaces of $\mathcal{L}(H,K)$ and $\mathcal{L}(H',K')$, respectively, define

$$\mathfrak{A} \times \mathfrak{B} = \{(A,B): A \in \mathfrak{A}, B \in \mathfrak{B}\},$$

where (A,B) is the operator from the Hilbert space H x H' to the Hilbert space K x K' given by (A,B)(x,y) = (Ax, By). Clearly $\mathfrak{A} \times \mathfrak{B}$ is a J*–algebra and it is easy to show that

(4) $$\|(A,B)\| = \max\{\|A\|, \|B\|\}.$$

There is no difficulty in extending this definition and equality (4) to the case of finite or infinite products. Thus in particular, any product of open unit balls of J*–algebras is the open unit ball of the corresponding product of the J*–algebras, which is itself a J*–algebra.

Clearly J*–algebras are not algebras in the ordinary sense; however, as the following proposition shows, J*–algebras do contain certain symmetrically formed products of their elements.

Proposition 1. Let A, B, C $\in \mathfrak{A}$ and let p be any polynomial. Then

(a) $AB^*C + CB^*A \in \mathfrak{A}$, (b) $A(B^*A)^n = (AB^*)^nA \in \mathfrak{A}$,

(c) $p(AB^*)C + Cp(B^*A) \in \mathfrak{A}$, (d) $p(AB^*)Cp(B^*A) \in \mathfrak{A}$.

Proof. Part (a) follows from the identities

$$4AB^*A = \sum_{k=0}^{3} (-1)^k (B+i^kA)(B+i^kA)^*(B+i^kA)$$

$$AB^*C + CB^*A = (A+C)B^*(A+C) - AB^*A - CB^*C,$$

and part (b) follows from the identity

$$A(B^*A)^n = A[B(A^*B)^{n-1}]^*A,$$

part (a) and induction. To prove part (c), it suffices to observe that the operator

$$(AB^*)^nC + C(B^*A)^n = A[(BA^*)^{n-1}B]^*C + C[B(A^*B)^{n-1}]^*A$$

is an element of $\mathfrak{A}$ by parts (a) and (b). Finally, part (d) follows from part (c) and the identity

$$
\begin{aligned}
p(AB^*)Cp(B^*A) \\
&= p(AB^*)[p(AB^*)C+Cp(B^*A)] \\
&\quad + [p(AB^*)C + Cp(B^*A)]p(B^*A) \\
&\quad - [p^2(AB^*)C + Cp^2(B^*A)].
\end{aligned}
$$

A map L: $\mathfrak{A} \to \mathfrak{B}$ is said to be a J*–**isomorphism** if L is a bounded linear bijection of $\mathfrak{A}$ onto $\mathfrak{B}$ satisfying

(5) $$L(AA^*A) = L(A)L(A)^*L(A)$$

for all $A \in \mathfrak{A}$. Note that the identities given in the proof of Proposition 1 show that any linear map L: $\mathfrak{A} \to \mathfrak{B}$ satisfying (5) commutes with each of the formulas (a)–(d).

Every J*–algebra is isometrically J*–isomorphic to a J*–algebra of operators on a Hilbert space. In fact, if H and K are Hilbert spaces, the map L: $\mathcal{L}(H,K) \to \mathcal{L}(H \times K)$ defined by $L(A)(x,y) = (0,Ax)$ is a linear isometry satisfying (5).

Certain kinds of operators in $\mathfrak{A}$ and relations between them can be characterized entirely in terms of the J*–structure and thus are preserved under J*–isomorphisms. For example, an operator B is a partial isometry [25, p.111] if and only if $BB^*B = B$. Another example, which will be useful later, is the equivalence of the conditions

(6) $$A^*B = 0 \text{ and } BA^* = 0,$$

(6′) $$BA^*A + AA^*B = 0.$$

To prove this, all we must show is that (6′) ⇒ (6). Suppose (6′) holds. Put

$$P = (BA^*)(BA^*)^*, \ Q = AA^*, \ R = BB^*,$$

and note that each of these operators is positive. Multiplying equation (6′) on the right by B^*, we have $P + QR = 0$; and taking adjoints we obtain $QR = RQ$, so both P and QR are positive. Therefore $P = 0$, and consequently, $BA^* = 0$. Similarly, multiplying equation (6′) on the left by B^*, we obtain $A^*B = 0$. This completes the proof. Another example of interest is the equivalence of the conditions

(7) $$CB^*A = AB^*C = C \text{ for all } C \in \mathfrak{A},$$

(7′) $$AB^*C + CB^*A = 2C \text{ and } AB^*CB^*A = C \text{ for all } C \in \mathfrak{A},$$

which is easily proved. If the underlying Hilbert spaces H and K for $\mathfrak{A}$ are chosen so that they are no larger than necessary, condition (7) asserts that B^* is the inverse of A.

Note that the operator norm satisfies

(8) $$\|AA^*A\| = \|A\|^3$$

for all $A \in \mathcal{L}(H,K)$, since

$$\|AA^*A\|^2 = \|(AA^*A)^*(AA^*A)\| = \|(A^*A)^3\| = \|A^*A\|^3 = \|A\|^6.$$

Hence any bounded linear map $L: \mathfrak{A} \to \mathfrak{B}$ which satisfies (8) also satisfies $\|L\| \leqslant 1$, since

$$\|L(A)\|^3 = \|L(A)L(A)^*L(A)\| = \|L(AA^*A)\| \leqslant \|L\| \; \|AA^*A\| = \|L\| \; \|A\|^3$$

for all $A \in \mathfrak{A}$. Consequently, by the closed graph theorem, any J*–isomorphism is an isometry (cf. [20, p.330]). A converse to this is given in Theorem 4 below.

The rest of this section will be devoted to a further discussion of the Cartan factors of type IV, which are of special interest.

A Cartan factor of type IV can be defined, alternately, as the closed linear space spanned by a spin system [38], i.e., a family $\{U_\alpha\}$ of self-adjoint unitary operators on a Hilbert space such that $U_\alpha U_\beta + U_\beta U_\alpha = 0$ when $\alpha \neq \beta$. For, it is easy to see that any such space is a Cartan factor of type IV. Conversely, as we have seen, any Cartan factor of type IV is a Hilbert space with conjugation and hence has an orthonormal basis $\{U_\alpha\}$ of self-conjugate elements. Thus $\{U_\alpha\}$ is a spin system and its closed span is the Cartan factor. Spin systems with any given finite number of elements can be constructed explicitly by taking certain Kronecker products of the Pauli spin matrices. Indeed, the set of all $2^n \times 2^n$ matrices

$$\underbrace{P \times \cdots \times P}_{k} \times Q \times \underbrace{I \times \cdots \times I}_{n-k-1}, \; k = 0, \ldots, n,$$

where $P = \begin{pmatrix} 1 & 0 \\ 0 & -1 \end{pmatrix}$ and $Q = \begin{pmatrix} 0 & 1 \\ 1 & 0 \end{pmatrix}$ or $\begin{pmatrix} 0 & i \\ -i & 0 \end{pmatrix}$, is a spin system having 2n+1 elements.

Let H be a Hilbert space with conjugation $x \to \bar{x}$, and let $x \to A_x$ be a linear map on H satisfying (2). Given $x,y \in H$, let λ_1 and λ_2 be the roots of the polynomial $p(\lambda) = \lambda^2 - 2(x,\bar{y})\lambda + (x,\bar{x})(y,\bar{y})$. It will be useful to know that

(9) $$\sigma(A_y A_x) = \{\lambda_1, \lambda_2\},$$

where σ denotes the spectrum. To see this, put $A = A_y A_x$ and $B = A_x A_y$, and note that $AB = BA$ and $(\lambda I - A)(\lambda I - B) = p(\lambda)I$. Hence $\sigma(A) \subseteq \{\lambda_1, \lambda_2\}$. Also if $\lambda_1 \notin \sigma(A)$, it follows that $B = \lambda_1 I$, so

$A = 2(x,\bar{y})I - B = \lambda_2 I$. But then $\lambda_1 \neq \lambda_2$ and $\lambda_2 A_x = A_x A = BA_x = \lambda_1 A_x$, a contradiction. Thus (9) holds. In particular, since $\|A_x\|^2$ is the largest number in $\sigma(A_x^* A_x)$, we have

$$\|A_x\|^2 = \|x\|^2 + \sqrt{\|x\|^4 - |(x,\bar{x})|^2},$$

which justifies (3). Also, it follows from the above proof that

$$(10)\qquad (I + A_y^* A_x)^{-1} = \frac{I + A_x A_y^*}{1 + 2(x,y) + (x,\bar{x})(\bar{y},y)} .$$

Next, following Chevalley [5, p.38], we construct a linear map $x \to A_x$ on H satisfying (2). (Such a map is a complex analogue of what Segal [34] has called a Clifford distribution over H, which extends a construction of Cartan [4, §93].) Let E be the exterior algebra of H and note that E is an inner product space [13, p.106]. Given $x \in H$, let δ_x be the antiderivation [13, p.112] of E satisfying $\delta_x y = (x,\bar{y})1$ for $y \in H$ and define $\ell_x(e) = x \wedge e$ for all $e \in E$. Then $\ell_x^2 = 0$, $\ell_x \delta_x + \delta_x \ell_x = (x,\bar{x})I$, and $\ell_x^* = \delta_{\bar{x}}$. Define $A_x = \ell_x + \delta_x$. Thus A_x is a linear map on E and the map $x \to A_x$ is linear. Moreover, the above equalities imply that (2) holds, and A_x is bounded on E since

$$\|A_x e\|^2 \leq ((A_x^* A_x + A_x A_x^*)e,e) = 2\|x\|^2 \, \|e\|^2$$

for all $e \in E$. Consequently, each operator A_x extends uniquely to the completion K of E, and thus the map $x \to A_x$ is as required.

Alternately, suppose we already have a spin system $\{U_\alpha\}$ with the same cardinality as the dimension of H. Then there is a bijection between an orthonormal basis for H of self-conjugate elements and $\{U_\alpha\}$, and it follows from the last inequality that this bijection extends to a linear map $x \to A_x$ on H with the required properties.

§3. Möbius transformations and biholomorphic mappings

Theorem 2. For each $B \in \mathfrak{A}_0$, the Möbius transformation

$$T_B(A) = (I-BB^*)^{-1/2}(A+B)(I+B^*A)^{-1}(I-B^*B)^{1/2}$$

is a biholomorphic mapping of $\mathfrak{A}_0$ onto itself with $T_B(0) = B$. Moreover,

$$T_B^{-1} = T_{-B},\ T_B(A)^* = T_{B^*}(A^*),\ \|T_B(A)\| \leq T_{\|B\|}(\|A\|),$$

and

$$DT_B(A)C = (I-BB^*)^{1/2}(I+AB^*)^{-1}C(I+B^*A)^{-1}(I-B^*B)^{1/2}$$

for $A \in \mathfrak{A}_0$ and $C \in \mathfrak{A}$.

Here positive and negative square roots are defined by the usual power series expansions and I at each occurrence denotes the identity mapping on the appropriate underlying Hilbert space.

Corollary 2. The open unit ball of any J*–algebra is a bounded symmetric homogeneous domain.

Example 1. Let C(S) be the space of all continuous complex-valued functions vanishing at infinity on a locally compact Hausdorff space S. Then with the identification mentioned in §2, the space C(S) is a J*–algebra and we have

$$T_y(x) = \frac{x+y}{1+\bar{y}x}$$

for $x,y \in C(S)_0$. Note that if S is a discrete set with exactly n elements, then $C(S)_0$ is the open unit polydisc in C^n. (One may also view the open unit polydisc in C^n as the open unit ball of the J*–algebra of all nxn diagonal matrices with complex entries.)

Example 2. Let H be a Hilbert space and let $y \in H_0$. Then identifying H with $\mathcal{L}(C,H)$, we have $y^*w = (w,y)$ for $w \in H$. Let E_y be the linear projection of H onto the subspace spanned by y. Then by the power series expansion,

$$(I-yy^*)^{-\frac{1}{2}} = (I-\|y\|^2E_y)^{-\frac{1}{2}} = I + [(1-\|y\|^2)^{-\frac{1}{2}}-1]\,E_y\,,$$

so

$$T_y(x) = \frac{y+E_yx+\sqrt{1-\|y\|^2}\;(I-E_y)x}{1+(x,y)}$$

for $x \in H_0$. When H is finite dimensional, H_0 is sometimes referred to as a hyperball.

Example 3. Let H be a Hilbert space with conjugation $x \to \bar{x}$. Then H can be identified (after renorming) with a Cartan factor $\mathfrak{A}$ of type IV and $\mathfrak{A}_0$ is given by (3). Let $y \in \mathfrak{A}_0$ and let E_y be the projection of H onto the space spanned by y and $\bar{y}$. (Explicitly,

$$E_y = \frac{(yy^*-\bar{y}\bar{y}^*)^2}{\|y\|^4-|(y,\bar{y})|^2}$$

if y and $\bar{y}$ are linearly independent.) Then by (1), (10) and a computation, we have

$$T_y(x) = \frac{[1+(x,y)]\,y + [(x,\bar{x})+(x,\bar{y})]\,\bar{y} + L_yx}{1+2(x,y) + (x,\bar{x})(\bar{y},y)}$$

for $x \in \mathfrak{A}_0$, where

$$L_y = (1-\|y\|^2)E_y + \sqrt{1-2\|y\|^2+|(y,\bar{y})|^2}\;(I-E_y).$$

When H is finite dimensional, $\mathfrak{A}_0$ is sometimes referred to as a Lie ball.

As in the case of one complex variable, one can use the Möbius transformations together with the Schwarz lemma and Cauchy estimates to prove function-theoretic inequalities. In particular, Theorems 7 and 11 and Corollaries 5 and 6 of [18] hold for all J*—algebras. Further, with the aid of Theorems 1 and 2, one can show that

$$d(A,B) = \tanh^{-1}(\|T_{-B}(A)\|)$$

is a complete metric on $\mathfrak{A}_0$ and that the Schwarz-Pick inequality holds for d. See [9] for a generalization.

The formula for the Möbius transformations of Theorem 2 is apparently due to Potapov [29; Ch.1, §1], although he considered these transformations in a much more restrictive setting. The formulas deduced in Examples 2 and 3 appear in a slightly different form in [27] and [24, (16)], respectively. For expositions of the classical theory of biholomorphic mappings and homogeneous domains in $\mathbf{C}^n$, see [1], [10], and [26].

Proof of Theorem 2. Clearly $(I-BB^*)^{1/2}B = B(I-B^*B)^{1/2}$ by comparison of the power series expansions. Hence since

$$(A+B)(I+B^*A)^{-1} = B + (I-BB^*)A(I+B^*A)^{-1},$$

we have

$$T_B(A) = B + (I-BB^*)^{1/2}A(I+B^*A)^{-1}(I-B^*B)^{1/2} \tag{11}$$

for $A \in \mathfrak{A}_0$. Now the power series expansions for $A(I+B^*A)^{-1}$, $(I-BB^*)^{1/2}$ and $(I-B^*B)^{1/2}$ converge in the operator norm. Hence $A(I+B^*A)^{-1} \in \mathfrak{A}$ by part (b) of Proposition 1, and consequently $T_B(\mathfrak{A}_0) \subseteq \mathfrak{A}$ by part (d) of the same proposition. Differentiation of (11) shows that T_B has the derivative asserted, so T_B is holomorphic in $\mathfrak{A}_0$. Also it is clear from (11) that $T_B(0) = B$ and that $T_B(A)^* = T_{B^*}(A^*)$.

We will deduce the inequality $\|T_B(A)\| \leqslant T_{\|B\|}(\|A\|)$ from the formula

$$I-T_B(A)^*T_B(A) = (I-B^*B)^{1/2}(I+A^*B)^{-1}(I-A^*A)(I+B^*A)^{-1}(I-B^*B)^{1/2}, \tag{12}$$

which follows as in [29; Ch.1, §1]. Let H be as in Definition 1 and let $x \in H$. Taking $y = (I+B^*A)^{-1}(I-B^*B)^{1/2}x$ and applying (12), we have that

$$\begin{aligned}
([I-T_B(A)^*T_B(A)]x,x) &= ((I-A^*A)y,y) \geqslant (1-\|A\|^2)\|y\|^2 \\
&\geqslant (1-\|A\|^2)(1+\|B\|\ \|A\|)^{-2}(1-\|B\|^2)\|x\|^2 \\
&= [1-T_{\|B\|}(\|A\|)^2]\ \|x\|^2.
\end{aligned}$$

The desired inequality follows.

Thus $T_B(\mathfrak{A}_0) \subseteq \mathfrak{A}_0$ and T_B is holomorphic in $\mathfrak{A}_0$. Therefore to finish the proof of Proposition 1, it suffices to show that the function $h = T_B \circ T_{-B}$ is the identity map I on $\mathfrak{A}_0$. By what we have already shown, h is a holomorphic mapping of $\mathfrak{A}_0$ into itself and $Dh(0) = DT_B(-B) \circ DT_{-B}(0) = I$. Hence an extension of Cartan's uniqueness theorem [14, Th.1] applies to show that $h = I$, as required.

Theorems 1 and 2 allow us to give a characterization of all biholomorphic mappings between the open unit balls of J*–algebras.

Theorem 3. **Every biholomorphic mapping** $h\colon \mathfrak{A}_0 \to \mathfrak{B}_0$ **is of the form**

$$h = T_{h(0)} \circ L = L \circ T_{-h^{-1}(0)}\,,$$

where $L\colon \mathfrak{A} \to \mathfrak{B}$ **is a surjective linear isometry.**

Proof. By Theorem 2, the composition $T_{h(0)}^{-1} \circ h$ is a biholomorphic mapping of $\mathfrak{A}_0$ onto $\mathfrak{B}_0$ which takes zero to zero. Hence by Theorem 1, there is a surjective linear isometry $L\colon \mathfrak{A} \to \mathfrak{B}$ with $T_{h(0)}^{-1} \circ h = L$, i.e., $h = T_{h(0)} \circ L$. Since

$$L(h^{-1}(0)) = T_{h(0)}^{-1}(0) = -h(0),$$

by (14) below

$$L \circ T_{-h^{-1}(0)} = T_{h(0)} \circ L = h,$$

as asserted.

In particular, Theorem 3 implies that any biholomorphic mapping $h\colon \mathfrak{A}_0 \to \mathfrak{B}_0$ is uniquely determined by the first two terms in its Taylor series expansion about 0. To see this, observe that

$$L = D(T_{-h(0)} \circ h)(0) = DT_{-h(0)}(h(0)) \circ Dh(0).$$

By Theorem 3, to obtain an explicit formula for all biholomorphic mappings between the open unit balls of two J*–algebras, it suffices to obtain an explicit formula for the surjective linear isometries between the J*–algebras. This has been done in certain cases. (See, for example, [8, p.442], [12, §3], [24], and [28].) In general, surjective linear isometries between J*–algebras can be complicated. Even in the case of the finite dimensional Cartan factors, there are surjective linear isometries which are not included in expressions given by Hua [19, §4.3] and Siegel [36, §48] for the biholomorphic mappings of the open unit ball. (For example, the map $A \to A^t$ is a surjective isometry of the Cartan factors of type I with $H = K$. A more subtle example is given by Morita [23].) However, we do have

Theorem 4. **If** $L\colon \mathfrak{A} \to \mathfrak{B}$ **is a surjective linear isometry, then**

$$L(AB^*A) = L(A)L(B)^*L(A)$$

for all $A, B \in \mathfrak{A}$.

Corollary 3. The domains $\mathfrak{A}_0$ and $\mathfrak{B}_0$ are holomorphically equivalent if and only if the J*–algebras $\mathfrak{A}$ and $\mathfrak{B}$ are isometrically J*–isomorphic.

Corollary 4. (Kadison [20]). Let $\mathfrak{A}$ and $\mathfrak{B}$ be C*–algebras each having an identity, and let $L: \mathfrak{A} \to \mathfrak{B}$ be a surjective linear isometry. Then

$$L(A) = U\rho(A)$$

for all $A \in \mathfrak{A}$, where U is a unitary operator in $\mathfrak{B}$ and $\rho: \mathfrak{A} \to \mathfrak{B}$ is a linear isometry satisfying

$$\rho(I) = I, \quad \rho(A^2) = \rho(A)^2, \text{ and } \quad \rho(A^*) = \rho(A)^* \tag{13}$$

for all $A \in \mathfrak{A}$.

Corollary 5. Let H be a Hilbert space with conjugation $x \to \bar{x}$ and let L be an invertible linear operator on H which maps the domain (3) onto itself. Then $L = \lambda O$, where $|\lambda| = 1$ and O is a unitary operator on H satisfying $O\bar{x} = \overline{Ox}$ for all $x \in H$.

Note that Theorem 4, the identity (11) and our remarks after Proposition 1 combine to show that any surjective linear isometry $L: \mathfrak{A} \to \mathfrak{B}$ satisfies

$$L \circ T_B = T_{L(B)} \circ L \tag{14}$$

for $B \in \mathfrak{A}_0$.

Proof of Theorem 4. Let $B \in \mathfrak{A}_0$ and set

$$h = T_{-L(B)} \circ L \circ T_B .$$

By hypothesis and the properties of the Möbius transformations, we have that $h: \mathfrak{A}_0 \to \mathfrak{B}_0$ is a biholomorphic mapping with $h(0) = 0$. Therefore h is linear by Theorem 1, so $Dh(0) = Dh(-B)$. Applying the chain rule and the formula for the derivative of the Möbius transformations given in Theorem 2, we have

$$[I-L(B)L(B)^*]^{-\frac{1}{2}} L((I-BB^*)^{\frac{1}{2}}A(I-B^*B)^{\frac{1}{2}})[I-L(B)^*L(B)]^{-\frac{1}{2}}$$
$$= [I-L(B)L(B)^*]^{\frac{1}{2}} L((I-BB^*)^{-\frac{1}{2}}A(I-B^*B)^{-\frac{1}{2}})[I-L(B)^*L(B)]^{\frac{1}{2}}$$

for any $A \in \mathfrak{A}$. Hence by part (d) of Proposition 1,

$$L((I-BB^*)A(I-B^*B)) = [I-L(B)L(B)^*]\,L(A)\,[I-L(B)^*L(B)]$$

for any $A \in \mathfrak{A}$. Replacing B in the last equation by tB (where $0 < t < 1$) and then equating the coefficients of t^2 on both sides, we have

$$L(BB^*A+AB^*B) = L(B)L(B)^*L(A) + L(A)L(B)^*L(B),$$

and clearly this holds for all $A,B \in \mathfrak{A}$. Theorem 4 now follows from the above with A = B and the first identity given in the proof of Proposition 1.

Proof of Corollaries 3-5. Corollary 3 is an immediate consequence of Theorem 4 and Corollaries 1 and 2. To prove Corollary 4, let $U = L(I)$ and define $\rho(A) = U^*L(A)$. By Theorem 4 and the equivalence of (7) and (7′), we have that U is a unitary operator in $\mathfrak{B}$. Hence $L(A) = U\rho(A)$ and $\rho(I) = I$. Applying Theorem 4 to the products AI^*A and IA^*I, we obtain $\rho(A^2) = \rho(A)^2$ and $\rho(A^*) = \rho(A)^*$, respectively.

Finally, to prove Corollary 5, note that as shown in §2, there is a linear bijection $x \to A_x$ satisfying (2) which maps H onto a Cartan factor $\mathfrak{A}$ of type IV. Hence L induces a linear map $\tilde{L}: \mathfrak{A} \to \mathfrak{A}$, and by hypothesis $\tilde{L}$ is a surjective isometry. Let $u \in H$ with $\|u\| = 1$ and $\bar{u} = u$. Then A_u is a unitary operator in $\mathfrak{A}$, so as above, $\tilde{L}(A_u)$ is also a unitary operator in $\mathfrak{A}$. It follows that $\tilde{L}(A_u) = \lambda A_v$, where $|\lambda| = 1$ and $v \in H$ with $\|v\| = 1$ and $\bar{v} = v$. Hence $Lu = \lambda v$. By dividing L by λ, we may suppose that $Lu = v$.

Now since $A_xA_y^*A_x = 2(x,y)A_x-(x,\bar{x})A_y^*$ by (1), Theorem 4 applied to $\tilde{L}$ shows that

$$2[(x,y)-(Lx,Ly)]\,Lx = (x,\bar{x})L\bar{y} - (Lx,\overline{Lx})\overline{Ly} \tag{15}$$

for all $x,y \in H$. Taking y = u in (15) and using the fact that L is one-to-one, we have

$$2[(x,u)-(Lx,Lu)]\,x = [(x,\bar{x})-(Lx,\overline{Lx})]\,u,$$

and consequently $(Lx, \overline{Lx}) = (x,\bar{x})$ for all $x \in H$. Given $y \in H$, this implies that $(Ly,Lu) = (y,u)$, and hence taking x = u in (15), we obtain

$$L\bar{y} - \overline{Ly} = 2[(u,y) - (Lu, Ly)]\,Lu = 0.$$

Thus L = O.

To exclude complicated surjective linear isometries, following R. S. Phillips [28], we now restrict our attention to the identity component of the group $G(\mathfrak{A}_0)$ of all biholomorphic mappings of $\mathfrak{A}_0$ onto itself, where $G(\mathfrak{A}_0)$ has the topology induced by the metric

$$d(h_1,h_2) = \sup\{\|h_1(A)-h_2(A)\|: A \in \mathfrak{A}_0\}.$$

Theorem 5. **Let $\mathfrak{A}$ be a C*–algebra with identity. Then each function h in the identity component of $G(\mathfrak{A}_0)$ is of the form**

$$h(A) = T_{h(0)}(UAV) = UT_{-h^{-1}(0)}(A)V, \tag{16}$$

where U and V are unitary operators in the weak operator closure of $\mathfrak{A}$.

Corollary 6. **Let $\mathfrak{A}$ be a W*–algebra. Then the identity component of $G(\mathfrak{A}_0)$ is the set of all functions of the form (16) where U and V are unitary operators in $\mathfrak{A}$, or equivalently, the set of all functions of the form**

$$Z \to (AZ + B)(CZ + D)^{-1}, \tag{17}$$

where A, B, C, D are operators in $\mathfrak{A}$ satisfying

$$\begin{pmatrix} A & B \\ C & D \end{pmatrix}^* \begin{pmatrix} I & 0 \\ 0 & -I \end{pmatrix} \begin{pmatrix} A & B \\ C & D \end{pmatrix} = \begin{pmatrix} I & 0 \\ 0 & -I \end{pmatrix} = \begin{pmatrix} A & B \\ C & D \end{pmatrix} \begin{pmatrix} I & 0 \\ 0 & -I \end{pmatrix} \begin{pmatrix} A & B \\ C & D \end{pmatrix}^* .$$

Note that Corollary 6 contains the general symplectic case of Theorem 1 of [28].

Proof of Theorem 5. Let G_e be the identity component of $G(\mathfrak{A}_0)$, and let $\mathcal{L}_e$ be the identity component of the group $\mathcal{L}$ of all surjective linear isometries $\rho: \mathfrak{A} \to \mathfrak{A}$ satisfying (13). We first show that if $h \in G_e$ then

$$h(A) = T_{h(0)}(U\rho(A)), \quad A \in \mathfrak{A}_0, \tag{18}$$

where U is a unitary operator in $\mathfrak{A}$ and $\rho \in \mathcal{L}_e$. Given $h \in G_e$, define $L_h = T_{-h(0)} \circ h$. By Theorem 3, L_h is a surjective linear isometry of $\mathfrak{A}$ onto itself, and it can be verified that the map $h \to L_h$ is continuous on $G(\mathfrak{A}_0)$. Define $\varphi(h) = L_h(I)^* L_h$. Then by Corollary 4, φ is a continuous mapping of $G(\mathfrak{A}_0)$ into $\mathcal{L}$ and $\varphi(I) = I$. Hence φ maps G_e into $\mathcal{L}_e$, and (18) follows.

Now $\mathcal{L}_e$ is generated by any of its neighborhoods of I. Hence Theorem 5 follows from (18) and the following:

Lemma 1. **If $\rho \in \mathcal{L}$ and $\|I-\rho\| < 2/3$, then there is a unitary operator U in the weak operator closure of $\mathfrak{A}$ such that $\rho(A) = UAU^*$ for all $A \in \mathfrak{A}$.**

Proof. By a result of Kadison and Ringrose [21], all we need to show is that ρ is product preserving. By [20], ρ is the sum of a *–isomorphism and a *–anti-isomorphism, say φ. In particular, φ is a product reversing mapping of $\mathfrak{B}$ onto $\mathfrak{B}'$, where $\mathfrak{B}$ and $\mathfrak{B}'$ are C*–subalgebras of $\mathfrak{A}$, and $\varphi = \rho/\mathfrak{B}$. We will show that

$\mathfrak{B}'$ is commutative. It then follows that φ is product preserving, and hence so is ρ.

Now for any A, B $\in \mathfrak{B}$,

$$\|\rho(A)\rho(B)-\rho(B)\rho(A)\|$$

$$= \|\rho(BA)-BA+[B-\rho(B)]A + \rho(B)[A-\rho(A)]\|$$

$$\leqslant 3\,\|I-\rho\|\ \|\rho(A)\|\ \|\rho(B)\|\ ;$$

consequently,

$$\|CD-DC\| < 2\,\|C\|\ \|D\|$$

for all nonzero C,D $\in \mathfrak{B}'$. If $\mathfrak{B}'$ is not commutative, then as remarked in [6] there is an operator C $\in \mathfrak{B}'$ with $C^2 = 0$ and $C \neq 0$. Let $D = CC^*-C^*C$. Clearly $D \neq 0$ and $\|D\| \leqslant \|C\|^2$. Moreover, $\|CD-DC\| = 2\,\|CC^*C\| = 2\,\|C\|^3$ by (8). Hence

$$\|CD-DC\| = 2\,\|C\|\ \|D\|,$$

a contradiction. Thus $\mathfrak{B}'$ is commutative, as we wished to show.

Proof of Corollary 6. An easy argument given in [28, p.17] shows that the set of all transformations (17) is a connected subgroup S of G($\mathfrak{A}_0$), and it is easy to see that S contains all transformations of the form (16). Hence S is precisely the set of transformations (16) and S is the identity component of G($\mathfrak{A}_0$).

It seems reasonable to conjecture that Theorem 5 holds for any J*–algebra $\mathfrak{A}$ where U and V are unitary operators on the underlying Hilbert spaces. In particular, it follows from Theorem 3, [3, Theorem E] and [24] that this holds for all finite dimensional J*–algebras which are products of Cartan factors of types I-IV. (Note that by [4, §97] any rotation of C^n can be represented as a transformation $A \to UAV$ on the n–dimensional Cartan factor of type IV.)

§4. Factorization of algebras and domains

Call a domain $\mathcal{D}$ a product of balls if there exist complex normed linear spaces X and Y such that $\mathcal{D} = X_0 \times Y_0$. The following theorem shows that the ball $\mathfrak{A}_0$ is holomorphically equivalent to a product of balls if and only if $\mathfrak{A}$ is isometrically J*–isomorphic to a product of J*–algebras.

Theorem 6. Let X and Y be complex normed linear spaces and suppose $\mathfrak{A}_0$ is holomorphically equivalent to $X_0 \times Y_0$. Then there exist J*–subalgebras $\mathfrak{B}$ and $\mathfrak{C}$ of $\mathfrak{A}$ which are isometrically isomorphic to X and Y, respectively, such that $\mathfrak{A}$ is isometrically J*–isomorphic to the J*–algebra $\mathfrak{B} \times \mathfrak{C}$.

Corollary 7. Let $\mathfrak{A}$ be a J*–algebra containing an isometry. Suppose that if E is a projection of the form $E = B^*B$ where $B \in \mathfrak{A}$ and if

(19) $$EA^*A = A^*AE$$

for all $A \in \mathfrak{A}$, then $E = 0$ or $AE = A$ for all $A \in \mathfrak{A}$. Then $\mathfrak{A}_0$ is not holomorphically equivalent to a product of balls.

It is easy to show that the converse of Corollary 7 holds when $\mathfrak{A}$ is such that $AB^*C \in \mathfrak{A}$ whenever $A,B,C \in \mathfrak{A}$.

Proof of Theorem 6. Let the space $X \times Y$ have the norm $\|(x,y)\| = \max\{\|x\|, \|y\|\}$. Then by hypothesis, $\mathfrak{A}_0$ is holomorphically equivalent to $(X \times Y)_0$ so $\mathfrak{A}$ is isometrically isomorphic to $X \times Y$ by Corollaries 1 and 2. It follows that there exist complementary projections E and F on $\mathfrak{A}$ such that the spaces $\mathfrak{B} = \text{Rge } E$ and $\mathfrak{C} = \text{Rge } F$ are isometrically isomorphic to X and Y, respectively, and

(20) $$\|A\| = \max\{\|E(A)\|, \|F(A)\|\}$$

for all $A \in \mathfrak{A}$. Given $|\lambda| = 1$, define $L_\lambda = E + \lambda F$ and note that L_λ is a linear isometry of $\mathfrak{A}$ onto itself. Hence by Theorem 4,

(21) $$L_\lambda(AB^*A) = L_\lambda(A)L_\lambda(B)^*L_\lambda(A)$$

for all $A,B \in \mathfrak{A}$. Varying λ in (21), we obtain $E(AB^*A) = E(A)E(B)^*E(A)$ for all $A,B \in \mathfrak{B}$ and similarly $F(AB^*A) = F(A)F(B)^*F(A)$ for all $A,B \in \mathfrak{C}$. Hence $\mathfrak{B}$ and $\mathfrak{C}$ are J^*–subalgebras of $\mathfrak{A}$. Defining $L(A) = (E(A),F(A))$, we see by (4) and (20) that L is a linear isometry of $\mathfrak{A}$ onto the J^*–algebra $\mathfrak{B} \times \mathfrak{C}$. Therefore, $\mathfrak{A}$ and $\mathfrak{B} \times \mathfrak{C}$ are isometrically J^*–isomorphic by Theorem 4.

Proof of Corollary 7. Suppose $\mathfrak{A}_0$ is holomorphically equivalent to a product of balls. Then by Theorem 6, there exist J^*–algebras $\mathfrak{B}$ and $\mathfrak{C}$ such that $\mathfrak{A}$ is isometrically J^*–isomorphic to $\mathfrak{B} \times \mathfrak{C}$. Now given any two operators V and A in $\mathfrak{B} \times \mathfrak{C}$, there exist operators V_1, V_2, A_1, and A_2 in $\mathfrak{B} \times \mathfrak{C}$ such that

$$V = V_1 + V_2, \quad A = A_1 + A_2,$$

$$V_1V_2^* = A_1V_2^* = A_2V_1^* = A_1A_2^* = 0, \; V_2^*V_1 = V_2^*A_1 = V_1^*A_2 = A_2^*A_1 = 0;$$

moreover, there are operators A for which neither A_1 nor A_2 is 0. Hence by the equivalence of (6) and (6'), the same is true for any two operators V and A in $\mathfrak{A}$. Let V be an isometry. Then

$$I = V^*V = V_1^*V_1 + V_2^*V_2,$$

and multiplying both sides of this equation by $E = V_1^*V_1$, we see that E is a projection. Moreover,

$A_1(I-E) = A_2E = 0$, so

(22) $$A_1 = AE, \quad A_2 = A(I-E).$$

Since $A_2^*A_1 = 0$, we have $A^*AE = EA^*AE$, and taking adjoints we obtain (19). Hence by hypothesis, one of the operators (22) is always 0, the desired contradiction.

Clearly one can obtain algebraic identities satisfied by any hermitian projection [2] on a J*–algebra by equating the coefficients of the powers of λ in (21) where $|\lambda|^2$ is replaced by 1. These identities imply in particular that the identity map is the only hermitian projection E on a B*–algebra with identity 1 such that $E(1) = 1$.

The rest of this section is devoted to applications of Theorem 6 and its corollary.

Call a J*–algebra $\mathfrak{A}$ **a finite dimensional Cartan factor** if $\mathfrak{A}_0$ is holomorphically equivalent to one of the Cartan domains of types I-VI. By Corollary 3, we have already determined up to isometric J*–isomorphism all finite dimensional Cartan factors except possibly for a 16 dimensional one and a 27 dimensional one. (See §2.) Clearly no finite dimensional Cartan factor (except for the 2-dimensional one of type IV) is isometrically J*–isomorphic to a product of J*–algebras, otherwise the associated Cartan domain would be reducible.

Theorem 7. Every finite dimensional J*–algebra is isometrically J*–isomorphic to a product of finite dimensional Cartan factors. In particular, every J*–algebra of dimension less than 16 is isometrically J*–isomorphic to a product of finite dimensional Cartan factors of types I–IV.

Proof of Theorem 7. Let $\mathfrak{A}$ be an n-dimensional J*–algebra and identify $\mathfrak{A}$ with C^n. By Corollary 2 and Cartan's classification [3] of bounded symmetric domains, there exist domains $\mathcal{D}_2, \ldots, \mathcal{D}_k$ in the spaces $X_2 = C^{n_2}, \ldots, X_k = C^{n_k}$, respectively, such that $\mathfrak{A}_0 = \mathcal{D}_2 \times \cdots \times \mathcal{D}_k$, and each $\mathcal{D}_j$ is holomorphically equivalent to a Cartan domain of type I–VI. Since $\mathfrak{A}_0$ is a balanced convex open subset of C^n, each $\mathcal{D}_j$ is a balanced convex open subset of X_j so $\mathcal{D}_j$ is the open unit ball of X_j with respect to some norm. Then $\mathfrak{A}$ is the space $X_2 \times \cdots \times X_k$ with the max norm on coordinates since the open unit balls of these spaces agree. By successive application of Theorem 6, it follows that each X_j is isometrically isomorphic to a J*–algebra $\mathfrak{A}_j$ and consequently each $\mathfrak{A}_j$ is a finite dimensional Cartan factor. Hence $\mathfrak{A}$ is isometrically isomorphic to the J*–algebra $\mathfrak{A}_2 \times \cdots \times \mathfrak{A}_k$, and this isomorphism is a J*–isomorphism by Theorem 4.

Theorem 8. The open unit balls of the following J*–algebras are not holomorphically equivalent to a product of balls: all Cartan factors of type I–IV except the 2–dimensional one of type IV; all W*–algebras which are factors in the usual sense; all spaces C(S), where S is compact and connected.

For a description of a large number of W*–algebras which are factors, see [33, §4].

Proof of Theorem 8. It is easy to see from Corollary 7 that Theorem 8 holds for W*–algebras which are factors and for spaces C(S), where S is compact and connected.

Suppose $\mathfrak{A}$ is a Cartan factor of type I, i.e., $\mathfrak{A} = \mathcal{L}(H,K)$. There exist conjugations Q_1 on H and Q_2 on K, and clearly the map $A \to Q_1A^*Q_2$ is a linear isometry of $\mathcal{L}(H,K)$ onto $\mathcal{L}(K,H)$. Hence by the comparability of cardinal numbers, it suffices to consider only the case where $\dim H \leqslant \dim K$. Note that in this case $\mathcal{L}(H,K)$ contains an isometry. Suppose E is a projection in $\mathcal{L}(H)$ satisfying (19). Given $x \in H$ and $y \in K$ with $\|y\| = 1$, take $A = yx^*$. Then $A^*A = xx^*$, so,

$$0 = EA^*A(I-E)x = \|(I-E)x\|^2 Ex.$$

Consequently, $Ex = 0$ or $Ex = x$ for each $x \in H$, and it is easy to show that this implies that $E = 0$ or $E = I$.

Suppose $\mathfrak{A}$ is a Cartan factor of type II. Clearly $I \in \mathfrak{A}$. Suppose E is a projection in $\mathcal{L}(H)$ satisfying (19), and let

$$H_r = \{x \in H: \bar{x} = x\},$$

where $Q(x) = \bar{x}$ is the given conjugation on H. Then given $x \in H_r$, the operator $A = xx^*$ satisfies $A^t = A$ so $A \in \mathfrak{A}$. The argument given for Cartan factors of type I now applies to show that $E = 0$ on H_r or $E = I$ on H_r. But $H = H_r + iH_r$, so $E = 0$ or $E = I$.

Suppose $\mathfrak{A}$ is a Cartan factor of type III. Since Theorem 8 is classical [3] when H is finite dimensional, we may suppose that H is infinite dimensional. Then there is an orthonormal basis for H in H_r and this basis can be partitioned into two equivalent sets $\{e_\alpha\}$ and $\{e'_\alpha\}$. Let U be the operator on H defined by $Ue_\alpha = e'_\alpha$ and $Ue'_\alpha = -e_\alpha$. Then $U = QUQ$, $U^2 = -I$ and $U^* = -U$. Hence U is a unitary operator in $\mathfrak{A}$.

Now suppose E is a projection in $\mathcal{L}(H)$ satisfying (19). Let x be a unit vector in H_r and extend x to an orthonormal basis for H_r. This basis is then an orthonormal basis for H and may be partitioned into sets $\{x\}$, $\{e_\alpha\}$, and $\{e'_\alpha\}$, where the last two sets are equivalent. Let A be the operator on H defined by $Ax = 0$, $Ae_\alpha = e'_\alpha$, and $Ae'_\alpha = -e_\alpha$. Then $A = QAQ$, $A^* = -A$, and $A^2 = xx^* - I$. Hence $A \in \mathfrak{A}$ and $A^*A = I - xx^*$. The argument given for Cartan factors of type II now applies to show that $E = 0$ or $E = I$.

Suppose $\mathfrak{A}$ is a Cartan factor of type IV and that $\dim \mathfrak{A} > 2$. Then $\mathfrak{A}$ is the space spanned by a spin system which we write in the form $\{U_0\} \cup \{U_\alpha\}$, where U_0 is distinct from any of the U_α's. Taking $V_\alpha = iU_0U_\alpha$, we see that $\{V_\alpha\}$ is a spin system with at least two elements and that the J*–algebra $\mathfrak{B} = U_0\mathfrak{A}$ is the closed space spanned by the set $\{I\} \cup \{V_\alpha\}$. Since $\mathfrak{A}$ is isometrically J*-isomorphic to $\mathfrak{B}$, it suffices to show that $\mathfrak{B}$ satisfies the hypotheses of Theorem 6. Suppose (19) holds. Taking $A = I + V_\alpha$ in (19), we have $EV_\alpha = V_\alpha E$ for all α. Then $EB = BE = B$, so

$$E = (BB^* + B^*B) - B^*B^2B^* \in \mathfrak{B}$$

since $\mathfrak{B}$ contains the squares and adjoints of each of its elements. Hence $V_\alpha E + EV_\alpha = \lambda_\alpha I + \mu_\alpha V_\alpha$, where λ_α and μ_α are complex numbers. But since $V_\alpha E = EV_\alpha$, it follows that $2E = \mu_\alpha I + \lambda_\alpha V_\alpha$ for all α, so $E = 0$ or $E = I$.

§5. Function-theoretic boundaries and extreme points

In this section we show that the sets defined below play the role of function-theoretic boundaries for the open unit balls of J*–algebras.

Definition 2. A non-empty subset Γ of the unit sphere of $\mathfrak{A}$ is said to be stable if the transformation $A \to T_B(\lambda A)$ takes Γ to Γ for each $B \in \mathfrak{A}_0$ and each complex number λ with $|\lambda| = 1$.

Clearly any non-empty intersection of stable sets is stable. Also, any component of a stable set is stable; indeed, suppose Γ' is a component of a stable subset Γ of $\mathfrak{A}$ and let $A \in \Gamma'$, $B \in \mathfrak{A}_0$ and $|\lambda| = 1$. Then the map $t \to T_{tB}(\lambda^t A)$ on $[0,1]$ is a continuous curve in Γ connecting A to $T_B(\lambda A)$, so $T_B(\lambda A) \in \Gamma'$. Therefore, the transformation $A \to T_B(\lambda A)$ takes Γ' to Γ', as required. Note that by Theorem 3 and (14), every biholomorphic mapping of $\mathfrak{A}_0$ onto $\mathfrak{B}_0$ extends to a homeomorphism of $\mathfrak{A}_1$ onto $\mathfrak{B}_1$ which takes stable sets to stable sets.

Proposition 2. *Any of the following subsets of* $\mathfrak{A}$ *is stable if non-empty:*

(a) the unitary operators in $\mathfrak{A}$, *(b) the isometries in* $\mathfrak{A}$, *(c) the extreme points of* $\mathfrak{A}_1$.

Proof. By definition or by Theorem 11 below, an operator $B \in \mathfrak{A}$ is in the sets described in (a), (b), or (c) if and only if both $B^*B = I$ and $BB^* = I$, $B^*B = I$, or $(I-BB^*)A(I-B^*B) = 0$ for all $A \in \mathfrak{A}$, respectively. Hence Proposition 2 is an immediate consequence of (12) and the complementary formula

$$(23) \qquad I-T_B(A)T_B(A)^* = (I-BB^*)^{1/2}(I+AB^*)^{-1}(I-AA^*)(I+BA^*)^{-1}(I-BB^*)^{1/2},$$

which follows from (12) and the equality

$$I-T_B(A)T_B(A)^* = I-T_{B^*}(A^*)^*T_{B^*}(A^*).$$

There are many J*–algebras for which one of the sets (a)–(c) is non-empty. For example, if $\mathfrak{A}$ is a B*–algebra with identity, obviously $\mathfrak{A}$ contains a unitary element. Also if $\mathfrak{A} = \mathcal{L}(H,K)$, where H and K are (possibly infinite dimensional) Hilbert spaces with dim H $\leq$ dim K, then $\mathfrak{A}$ contains an isometry. Further, if $\mathfrak{A}$ is any finite dimensional J*–algebra or, more generally, any J*–algebra which is closed in the weak operator topology, then $\mathfrak{A}_1$ has an extreme point. (To see the last assertion, apply the Krein–Milman theorem and the fact that $\mathfrak{A}_1$ is compact in the weak operator topology [8; Ex.6, p.512].)

Let X be a complex normed linear space. The holomorphic hull of a subset Λ of X (which we denote by Co Λ) is defined to be the set of all $x \in X$ such that $|g(x)| \leq 1$ whenever $g: X \to C$ is a holomorphic function satisfying $|g(y)| \leq 1$ for all $y \in \Lambda$. Clearly Co Λ is closed, and by a well-known separation theorem [8; Th.10, p.417], it follows that

(24) $$\mathrm{Co}\,\Lambda \subseteq \overline{\mathrm{co}}\,\Lambda,$$

where $\overline{\mathrm{co}}\,\Lambda$ denotes the closed convex hull of Λ.

We can now state the main result of this section (cf. [17] and [22, Prop. 3.2].)

Theorem 9 (Maximum principle). Let Γ be a stable subset of $\mathfrak{A}$ and let $h: \mathfrak{A}_0 \to X$ be a holomorphic function with a continuous extension to $\mathfrak{A}_0 \cup \Gamma$. Then

(25) $$h(\mathfrak{A}_0) \subseteq \mathrm{Co}\, h(\Gamma).$$

In particular, if h is bounded on Γ,

(26) $$\|h(B)\| \leqslant \sup\{\|h(A)\| : A \in \Gamma\}$$

for all $B \in \mathfrak{A}_0$. Moreover, h is completely determined by its values on Γ.

Note that if S is the unit circle, the components of the set of unitary elements of C(S) are just the homotopy classes of continuous mappings of S into itself. Thus C(S) has infinitely many disjoint closed stable subsets.

Letting h be the identity map on $\mathfrak{A}$ in (25) and applying (24), we obtain a result which considerably extends the Russo-Dye theorem [32, Th.1].

Corollary 8. If Γ is a component of any non-empty one of the sets (a)–(c) of Proposition 2, then $\mathfrak{A}_1 = \overline{\mathrm{co}}\,\Gamma$.

For example, let $\mathfrak{A}$ be a Cartan factor of type IV and note that all normal operators in $\mathfrak{A}$ are scalar multiples of self-adjoint unitary operators in $\mathfrak{A}$. By Corollary 8 above and Lemma 2 of [17], the convex hull of the set of all unitary operators in $\mathfrak{A}$ contains $\mathfrak{A}_0$. Hence if H is any Hilbert space with conjugation $x \to \bar{x}$, the norm whose open unit ball is given by (3) is the largest norm on H which agrees with the Hilbert norm on self-conjugate elements. (This has been observed in the finite dimensional case by Druzkowski [7].)

In finite dimensional J*–algebras there is a smallest closed stable set. Indeed,

Corollary 9. Suppose $\mathfrak{A}$ is finite dimensional. Then the set $\mathcal{E}$ of extreme points of $\mathfrak{A}_1$ is the Bergman-Shilov boundary for $\mathfrak{A}_0$, i.e., $\mathcal{E}$ is the smallest closed subset of the unit sphere of $\mathfrak{A}$ with the property that every complex-valued function holomorphic in $\mathfrak{A}_0$ and continuous in $\mathfrak{A}_1$ assumes its maximum absolute value on $\mathcal{E}$. Moreover, $\mathcal{E}$ is connected and is contained in the closure of every stable subset of $\mathfrak{A}$.

For example, it is known [19, p.6] that the Bergman-Shilov boundary for a classical Cartan domain $\mathcal{D}$ of

(a) type I with $n < m$ is the set of all isometries in Cl $\mathcal{D}$,

(b) type I with $n = m$, type II, type III with n even, or type IV is the set of all unitary operators in Cl $\mathcal{D}$,

(c) type III with n odd is the set of all operators $V \in \mathrm{Cl}\,\mathcal{D}$ satisfying $V^*V = I - xx^*$ for some $x \in \mathbb{C}^n$ with $\|x\| = 1$.

To see this, apply Theorem 11 below to show that each of the sets described is a set of extreme points of Cl $\mathcal{D}$ and observe that each of the sets is closed and stable. Note that the Bergman-Shilov boundary for any finite product of the above domains is the corresponding product of the Bergman-Shilov boundaries for each of the domains.

Proof of Theorem 9. Let $g: X \to C$ be a holomorphic function and suppose $|g(x)| \leqslant 1$ for all $x \in h(\Gamma)$. Given $B \in \mathfrak{A}_0$ and $A \in \Gamma$, by hypothesis $T_B(\lambda A) \in \Gamma$ whenever $|\lambda| = 1$. Hence the function

$$f(\lambda) = g \circ h(T_B(\lambda A))$$

satisfies $|f(\lambda)| \leqslant 1$ for $|\lambda| = 1$, and clearly f is holomorphic in the open unit disc and continuous in its closure. Consequently, by the (classical) maximum principle, $|g(h(B))| = |f(0)| \leqslant 1$. Therefore, $h(B) \in \mathrm{Co}\, h(\Gamma)$, which proves (25). Note that (26) follows from (25), (24), and the convexity of balls. The last part of the theorem is immediate from the formula

$$h(B) = \frac{1}{2\pi} \int_0^{2\pi} h(T_B(e^{i\theta}A))d\theta ,$$

which follows from the mean value property for vector-valued holomorphic functions.

Proof of Corollary 9. Let Λ be the Bergman-Shilov boundary for $\mathfrak{A}_0$. (The existence of Λ is well known [10, p.217].) By Theorem 9 and the compactness of $\mathfrak{A}_1$, it follows that Λ is contained in the closure of every stable subset of $\mathfrak{A}$. In particular, since each component of $\mathcal{E}$ is a stable subset of $\mathfrak{A}$ and $\mathcal{E}$ is closed (see Theorem 11 below), we have that $\Lambda \subseteq \mathcal{E}$ and $\mathcal{E}$ is connected. On the other hand, $\mathfrak{A}_1 \subseteq \mathrm{Co}\,\Lambda \subseteq \mathrm{co}\,\Lambda$ so $\mathcal{E} \subseteq \Lambda$ by a converse to the Krein-Milman Theorem [8; Lemma 5, p.440].

Theorem 10 (Schwarz lemma). Let $h: \mathfrak{A}_0 \to X_1$ be a holomorphic function with $h(0) = 0$ and put $L = Dh(0)$. If L takes a stable subset Γ of $\mathfrak{A}$ into a set of complex extreme points of X_1, then $h = L$.

Proof. Given $A \in \Gamma$, define $f(\lambda) = \frac{1}{\lambda} h(\lambda A)$ for $0 < |\lambda| < 1$ and take $f(0) = L(A)$. Clearly the composition of f with any bounded linear functional on X is holomorphic in the open unit disc Δ, and hence by the (classical) maximum principle and the Hahn-Banach theorem, we have $f(\Delta) \subseteq X_1$. Then since $L(A)$ is a complex extreme point of X_1, it follows that f is constant by the Thorp-Whitley maximum principle [37].

(See [14] for a simple proof which applies to this slightly more general situation.)

Now given $0 < r < 1$, define $g(B) = h(rB) - L(rB)$ for $B \in \mathfrak{A}_1$. Then g is continuous in $\bar{\mathfrak{A}}_1$ and holomorphic in $\mathfrak{A}_0$, and by the above g vanishes on Γ. Hence g vanishes identically in $\bar{\mathfrak{A}}_1$ by Theorem 9. Therefore $h = L$.

Theorem 11. The extreme points of $\mathfrak{A}_1$ coincide with the complex extreme points of $\mathfrak{A}_1$ and are precisely those operators $B \in \mathfrak{A}$ satisfying

$$(I-BB^*)A(I-B^*B) = 0 \tag{27}$$

for all $A \in \mathfrak{A}$. In particular, every extreme point of $\mathfrak{A}_1$ is a partial isometry.

Note that Theorem 11 extends a result of Kadison [20]. For the proof, we first establish an inequality of independent interest.

Proposition 3. If $B \in \mathcal{L}(H,K)_1$, then

$$\|B + (I-BB^*)^{1/2}A(I-B^*B)^{1/2}\| \leqslant 1$$

for all $A \in \mathcal{L}(H,K)$ with $\|A\| \leqslant 1/2$.

Proof. First observe that the functions $B \to (I-B^*B)^{1/2}$ and $B \to (I-BB^*)^{1/2}$ are defined and continuous in $\mathcal{L}(H,K)_1$ by the (commutative) Gelfand-Naimark theorem [31, Th. 4.2.2] and the uniform convergence of the binomial series for $(1-t)^{1/2}$ on $[0,1]$. Hence if $T_B(C)$ is defined by (11) for $C \in \mathcal{L}(H,K)_0$, Theorem 2 implies that

$$\|T_B(C)\| \leqslant 1. \tag{28}$$

Given $A \in \mathcal{L}(H,K)$ with $\|A\| < 1/2$, take $C = A(I-B^*A)^{-1}$. Then $C \in \mathcal{L}(H,K)$, $\|C\| \leqslant \|A\|(1-\|A\|)^{-1} < 1$, and $A = C(I+B^*C)^{-1}$. Hence Proposition 3 follows from (28).

Proof of Theorem 11. Note that Proposition 3 holds without the exponents, and hence any complex extreme point of $\mathfrak{A}_1$ satisfies (27). Thus to finish the proof, it suffices to show that any operator B satisfying (27) is a partial isometry and an extreme point of $\mathfrak{A}_1$. Clearly $B = BB^*B$ since

$$(B-BB^*B)^*(B-BB^*B) = B^*[(I-BB^*)B(I-B^*B)] = 0,$$

and consequently $E = B^*B$ and $F = BB^*$ are projections. Hence B is a partial isometry. Suppose $A \in \mathfrak{A}$ and $\|B \pm A\| \leqslant 1$. Let $x \in H$, where H is as in Definition 1 and set $y = Ex$. Then

$$\|y\|^2 \geq \|By \pm Ay\|^2 = \|By\|^2 \pm 2\,\mathrm{Re}(By,Ay) + \|Ay\|^2.$$

Choosing the appropriate sign, we have that $\|y\|^2 \geq \|By\|^2 + \|Ay\|^2$; but $\|By\|^2 = \|y\|^2$, so $Ay = 0$. Hence $AE = 0$. A similar argument beginning with the inequality $\|B^* \pm A^*\| \leq 1$ shows that $A^*F = 0$, i.e., $FA = 0$. Hence $A = (I-F)A(I-E) = 0$, as required.

§6. A generalization of the upper half-plane

Note that any bounded linear operator A on a Hilbert space H can be written in the form $A = \mathrm{Re}\,A + i\,\mathrm{Im}\,A$, where

$$\mathrm{Re}\,A = \frac{A+A^*}{2} \quad \text{and} \quad \mathrm{Im}\,A = \frac{A-A^*}{2i},$$

and clearly both Re A and Im A are self-adjoint operators. Write $A > 0$ when A is self-adjoint and there is an $\epsilon > 0$ such that $(Ax,x) \geq \epsilon \|x\|^2$ for all $x \in H$.

Theorem 12. Let $\mathfrak{A}$ be a J*–algebra containing an isometry V. Define

$$\mathcal{H} = \{A \in \mathfrak{A} : \mathrm{Im}\, V^*A - A^*(I-VV^*)A > 0\}.$$

Then $\mathcal{H}$ is an unbounded convex domain in $\mathfrak{A}$ and the Cayley transformation

$$S(A) = i(A+V)(I-V^*A)^{-1}$$

is a biholomorphic mapping of $\mathfrak{A}_0$ onto $\mathcal{H}$. Moreover, if V is a unitary operator or if $\mathfrak{A}$ is such that $AB^*C \in \mathfrak{A}$ whenever $A,B,C \in \mathfrak{A}$, then $\mathcal{H}$ is affinely homogeneous.

Clearly Cartan factors of type I with H = K and all Cartan factors of type II contain the identity operator. Also, Cartan factors of type III with the dimension of H even or infinite and all Cartan factors of type IV contain a unitary operator. (See the proof of Theorem 8.) Thus the corresponding upper half-planes $\mathcal{H}$ are affinely homogeneous tubular domains. Further, if H and K are (possibly infinite dimensional) Hilbert spaces with dim H ≤ dim K, then $\mathcal{L}(H,K)$ contains an isometry and products AB^*C whenever it contains A, B and C. Thus, for example, Theorem 12 covers the case considered in Theorem 6.1 of [12].

Note that any product of upper half-planes of J*–algebras which each have an isometry is the upper half-plane of the product of the J*–algebras with respect to the product isometry.

Example 1. View the elements of $\mathcal{L}(C^n)$ as nxn matrices with complex entries and let $\mathfrak{A} = \{A \in \mathcal{L}(C^n): A^t = A\}$. Take V = I. Then $\mathfrak{A}_0$ and $\mathcal{H}$ are Siegel's generalized unit disc and generalized upper half-plane, respectively. (See [35].)

Example 2. Let H be a Hilbert space and let v be a unit vector in H. Then v is an isometry since $v^*v = 1$, and the corresponding upper half-plane is

$$\mathcal{H} = \{x \in H: \operatorname{Im}(x,v) > \|x-(x,v)v\|^2\}.$$

Example 3 (Fuks [10, p.318]). Let $\mathfrak{A} = \mathcal{L}(C^n, C^m)$, where $n < m$ and view $\mathfrak{A}$ as the set of all mxn matrices with complex entries. Each $A \in \mathfrak{A}$ can be written in the form $A = \begin{bmatrix} A_1 \\ A_2 \end{bmatrix}$, where A_1 and A_2 are nxn and (m–n)x n matrices, respectively. Take $V = \begin{bmatrix} I \\ 0 \end{bmatrix}$, where I is the identity nxn matrix. Then V is an isometry and the corresponding upper half-plane is

$$\mathcal{H} = \{\begin{bmatrix} A_1 \\ A_2 \end{bmatrix}: \operatorname{Im} A_1 - A_2^* A_2 > 0\} .$$

Example 4 (cf. [3, p.149]). Let H be a Hilbert space with conjugation $x \to \bar{x}$ and let v be a unit vector in H with $\bar{v} = v$. As shown in §2, there exists a linear transformation $x \to A_x$ satisfying (2) which allows us to identity H with a Cartan factor $\mathfrak{A}$ of type IV. Take $V = A_v$ and note that V is unitary. Since $\operatorname{Im} A_v A_x = i[(\bar{x},v)I - A_v A_{\operatorname{Re} x}]$ for all $x \in H$, it follows from (9) that the corresponding upper half-plane is

$$\mathcal{H} = \{x \in H: (\operatorname{Im} x,v) > \|\operatorname{Re}[x-(x,v)v]\|\}.$$

This domain is mapped by the invertible linear map $Lx = (x,v)v + i[x-(x,v)v]$ onto the domain

$$\tilde{\mathcal{H}} = \{x \in H: (\operatorname{Im} x,v) > \|\operatorname{Im}[x-(x,v)v]\|\},$$

which is a tube whose base is a future light cone. Also, $\tilde{S} = L \circ S$ is a biholomorphic map of $\mathfrak{A}_0$ onto $\tilde{\mathcal{H}}$, and by (1) and (10),

$$\tilde{S}(x) = \frac{i[1-(x,\bar{x})]v - 2[x-(x,v)v]}{1-2(x,v) + (x,\bar{x})} .$$

Proof of Theorem 13. Put $\operatorname{IM} A = \operatorname{Im} V^*A - A^*(I-VV^*)A$ for $A \in \mathfrak{A}$. It follows from Proposition 1 and the power series expansion for S(A) that S maps $\mathfrak{A}_0$ into $\mathfrak{A}$. By a computation,

$$\operatorname{IM} S(A) = (I-A^*V)^{-1}(I-A^*A)(I-V^*A)^{-1},$$

so S maps $\mathfrak{A}_0$ into $\mathcal{H}$. A similar identity (which can be derived from the one above) shows that the transformation S^{-1} given by

$$S^{-1}(A) = (AV^*+i\,I)^{-1}(A-iV)$$

maps $\mathcal{H}$ into $\mathfrak{A}_0$. (Note that if $A \in \mathcal{H}$, the numerical range of V^*A lies in the upper half-plane so $-i$ is not in the spectrum of V^*A.) It is easy to verify that S^{-1} is the inverse of S and that both S and S^{-1} are holomorphic where defined. Hence S is a biholomorphic mapping of $\mathfrak{A}_0$ onto $\mathcal{H}$.

Clearly $\mathcal{H}$ is unbounded since it contains all positive multiples of V. To see that $\mathcal{H}$ is convex, let $A, B \in \mathcal{H}$ and take $C = tA + (1-t)B$, where $0 \leqslant t \leqslant 1$. Then

$$\begin{aligned} \text{IM } C &> tA^*(I-VV^*)A + (1-t)B^*(I-VV^*)B - C^*(I-VV^*)C \\ &= t(1-t)(A-B)^*(I-VV^*)(A-B) \geqslant 0, \end{aligned}$$

so $C \in \mathcal{H}$.

Suppose that $\mathfrak{A}$ contains AB^*C whenever it contains A, B and C. To show that $\mathcal{H}$ is affinely homogeneous, it suffices to show that for each $B \in \mathcal{H}$, the transformation

$$\begin{aligned} R_B(A) = V \text{ Re } V^*B &+ V(\text{ IM } B)^{1/2}V^*A(\text{ IM } B)^{1/2} \\ &+ iVB^*(I-VV^*)[2A(\text{ IM } B)^{1/2} + B] \\ &+ (I-VV^*)[A(\text{ IM } B)^{1/2} + B] \end{aligned}$$

is an invertible mapping of $\mathcal{H}$ onto itself with $R_B(iV) = B$. By hypothesis, the map $A \to A(\text{IM } B)$ takes $\mathfrak{A}$ into $\mathfrak{A}$, and since $(\text{IM } B)^{1/2}$ is the limit in the operator norm of a sequence of polynomials in IM B, it follows that the map $A \to A(\text{IM } B)^{1/2}$ takes $\mathfrak{A}$ into $\mathfrak{A}$. Hence R_B maps $\mathfrak{A}$ into $\mathfrak{A}$. A computation shows that $R_B(iV) = B$ and that

$$\text{IM } R_B(A) = (\text{IM } B)^{1/2}(\text{IM } A)(\text{IM } B)^{1/2}.$$

Hence R_B maps $\mathcal{H}$ into itself. It can be verified that the inverse of R_B is given by

$$R_B^{-1}(A) = V(\text{IM } B)^{-1/2}[V^*A - \text{Re } V^*B - iB^*(I-VV^*)(2A-B)](\text{IM } B)^{-1/2} + (I-VV^*)(A-B)(\text{IM } B)^{-1/2}$$

and that R_B^{-1} maps $\mathfrak{A}$ into $\mathfrak{A}$. It then follows from the identity given above that R_B^{-1} maps $\mathcal{H}$ into itself. Therefore R_B is as required.

Now suppose instead that V is a unitary operator. Since the J^*–algebras $\mathfrak{A}$ and $V^*\mathfrak{A}$ are isometrically J^*–isomorphic, we may assume that $V = I$. Then by part (d) of Proposition 1 and the fact that $(\text{Im } B)^{1/2}$ is the limit of a sequence of polynomials in Im B, the transformation

$$R_B(A) = \text{Re } B + (\text{Im } B)^{1/2}A(\text{Im } B)^{1/2}$$

maps $\mathfrak{A}$ into $\mathfrak{A}$. The argument given above now applies to show that $\mathcal{H}$ is affinely homogeneous.

Our construction and discussion of the domain $\mathcal{H}$ is based on the theory of Siegel domains of genus 2 given in [30, Ch.1].

Bibliography

1. H. Behnke and P. Thullen, **Theorie der Funktionen mehrerer komplexen Veränderlichen**, 2nd edition, Ergebnisse der Math. 51, Springer, Berlin, 1970.

2. E. Berkson, **Hermitian projections and orthogonality in Banach spaces**, Proc. London Math. Soc. 24(1972), 101–118.

3. É. Cartan, **Sur les domaines bornés homogènes de l'espace de n variables complexes**, Abh. Math. Sem. Univ. Hamburg 11(1935), 116–162.

4. ________, **The Theory of Spinors**, M.I.T. Press, Cambridge, Mass., 1966.

5. C. Chevalley, **The Algebraic Theory of Spinors**, Columbia Univ. Press, New York, 1954.

6. R. G. Douglas and D. Topping, **Operators whose squares are zero**, Rev. Romaine Math. Pures Appl. 12(1967), 647–652.

7. L. Druzkowski, **Effective formula for the crossnorm in the complexified unitary spaces**, (to appear).

8. N. Dunford and J. T. Schwartz, **Linear Operators**, part I, Interscience, New York, 1958.

9. C. J. Earle and R. S. Hamilton, **A fixed point theorem for holomorphic mappings**, Global Analysis, Proc. of Symposia in Pure Math. XVI, Amer. Math. Soc., Providence, R.I., 1965.

10. B. A. Fuks, **Special Chapters in the Theory of Analytic Functions of Several Complex Variables**, Transl. of Math. Monographs 14, Amer. Math. Soc., Providence, R.I., 1965.

11. S. Greenfield and N. Wallach, **The Hilbert ball and bi-ball are holomorphically inequivalent**, Bull. Amer. Math. Soc. 77(1971), 261–263.

12. ____________, **Automorphism groups of bounded domains in Banach spaces**, Trans. Amer. Math. Soc. 166(1972), 45–57.

13. W. H. Greub, **Multilinear Algebra**, Grundlehren der math. Wissenshaften 136, Springer, New York, 1967.

14. L. A. Harris, **Schwarz's lemma in normed linear spaces**, Proc. Nat. Acad. Sci. U.S.A. 62(1969), 1014–1017.

15. _________, **Schwarz's lemma and the maximum principle in infinite dimensional spaces**, thesis, Cornell University, Ithaca, N.Y., 1969 (available through University Microfilms, Inc., Ann Arbor, Michigan).

16. _________, **A continuous form of Schwarz's lemma in normed linear spaces**, Pacific J. Math. 38(1971), 635–639.

17. _________, **Banach algebras with involution and Möbius transformations**, J. Functional Anal. 11(1972), 1–16.

18. _________, **Bounds on the derivatives of holomorphic functions of vectors**, (to appear in the proceedings of a conference held in Rio de Janeiro in 1972).

19. L. K. Hua, **Harmonic Analysis of Functions of Several Complex Variables in the Classical Domains,** Transl. of Math. Monographs 6, Amer. Math. Soc., Providence, R.I., 1963.

20. R. V. Kadison, **Isometries of operator algebras,** Ann. of Math. 54(1951), 325–338.

21. R. V. Kadison and J. R. Ringrose, **Derivations and automorphisms of operator algebras,** Comm. Math. Phys. 4(1967), 32–63.

22. A. Korányi and J. A. Wolf, **Realization of hermitian symmetric spaces as generalized half-planes,** Ann. of Math. 81(1965), 265–288.

23. K. Morita, **Schwarz's lemma in a homogeneous space of higher dimensions,** Japanese J. Math. 19(1944), 45–56.

24. ________, **On the kernel functions for symmetric domains,** Sci. Rep. Tokyo Kyoiku Daigaku, Sect. A5(1956), 190–212.

25. M. A. Naimark, **Normed Rings,** P. Noordhoff, Groningen, Netherlands, 1964.

26. R. Narasimhan, **Several Complex Variables,** University of Chicago Press, Chicago, 1971.

27. E. Peschl and F. Erwe, **Über beschränkte Systeme von Funktionen,** Math. Ann. 126(1953), 185–220.

28. R. S. Phillips, **On symplectic mappings of contraction operators,** Studia Math. 31(1968), 15–27.

29. V. P. Potapov, **The multiplicative structure of J-contractive matrix functions,** Amer. Math. Soc. Transl. 15(1960), 131–243.

30. Pyatetskii-Shapiro, **Automorphic Functions and the Geometry of Classical Domains,** Gordon and Breach, New York, 1969.

31. C. E. Rickart, **General Theory of Banach Algebras,** Van Nostrand, Princeton, N.J., 1960.

32. B. Russo and H. A. Dye, **A note on unitary operators in C*–algebras,** Duke Math. J. 33(1966), 413–416.

33. S. Sakai, **C*–algebras and W*–algebras,** Ergebnisse der Math. 60, Springer, Berlin, 1971.

34. I. E. Segal, **Tensor algebras over Hilbert spaces II,** Ann. of Math. 63(1956), 160–175.

35. C. L. Siegel, **Symplectic geometry,** Amer. J. Math. 65(1943), 1–86.

36. ________, **Analytic Functions of Several Complex Variables,** Lecture notes at the Institute for Advanced Study, Princeton, N.J., 1948–1949.

37. E. Thorp and R. Whitley, **The strong maximum modulus theorem for analytic functions into a Banach space,** Proc. Amer. Math. Soc. 18(1967), 640–646.

38. D. Topping, **An isomorphism invariant for spin factors,** J. Math. Mech. 15(1966), 1055–1063.

Department of Mathematics
University of Kentucky
Lexington, Kentucky 40506, USA

LINDELÖF'S PRINCIPLE IN INFINITE DIMENSIONS

Michel Hervé

§ 1. Notations and preliminary results.

Throughout the paper, X and Y will denote Hausdorff topological vector spaces over the complex field; X' and Y' their adjoint spaces; Ω and Γ open connected sets in X and Y respectively. Letters such as U and V will denote open sets in the complex plane **C**: for instance, $U(a;h) = \{u \in \mathbf{C}: a+uh \in \Omega\}$ (a and $h \in X$).

According to the now classical definition (e.g. [12], Def. 2.2), a map f: $\Omega \to Y$ is **Gâteaux-analytic**, $f \in \mathcal{G}(\Omega,Y)$, if, for any $a \in \Omega$, $h \in X$, $\frac{1}{u}[f(a+uh)-f(a)]$ has a limit, in the completed space $\hat{Y}$, as $u \to 0$, $u \in \mathbf{C}\setminus\{0\}$; **f is Fréchet-analytic**, $f \in \mathcal{A}(\Omega,Y)$, if moreover f is continuous.

The following facts 1.1 through 1.4 are known ([4], Chap. III; [12], Chap. II) under the additional assumption that Y is locally convex; therefore they hold in the special case Y = **C**, where $f \in \mathcal{G}(\Omega,\mathbf{C})$ means that, for any a and $h \in X$, the function $u \mapsto f(a+uh)$ is holomorphic on U(a;h).

1.1. Given $a \in \Omega$, let $\omega(a)$ be the biggest balanced subset of the translated set $\Omega - a$; any $f \in \mathcal{G}(\Omega,Y)$ has a generalized homogeneous polynomial expansion (g.h.p.e.) valid for $h \in \omega(a)$: $f(a+h) = f(a) + \sum_{n\geq 1} f_n(a;h)$, with $f_n(a;h) = F_n(a;\underbrace{h,\ldots,h}_{n \text{ times}})$, where $(h_1,\ldots,h_n) \to F_n(a;h_1,\ldots,h_n)$ is an n-linear map $X^n \to \hat{Y}$, and a continuous one if f is continuous at the point a.

1.2. Let $a \in \Omega$ and K be a compact subset of $\omega(a)$. For any $f \in \mathcal{A}(\Omega,Y)$, the g.h.p.e. in 1.1 is uniformly summable for $h \in K$.

1.3. $f \in \mathcal{G}(\Omega,Y)$ iff $y' \circ f \in \mathcal{G}(\Omega,\mathbf{C})$ for any $y' \in Y'$.

1.4. A map $f \in \mathcal{G}(\Omega,Y)$ is continuous iff for any continuous seminorm q on Y, $q \circ f$ is continuous or locally bounded on Ω.

One aim of this paper is to get other properties involving the continuity of a Gâteaux-analytic map, as far as possible without any additional assumption on the spaces X and Y; to begin with, none is made in the following Proposition and Theorem.

Proposition 1.5. If $f \in \mathcal{G}(\Omega,Y)$, **then the map** $u \to f(a+uh)$ **is continuous on the open set** U(a;h).

Proof. Let $a \in \Omega$ and f(a) be the origin in Y. Given a balanced neighborhood B of the origin in Y, since $g(u) = f(a+uh)/u$ has a limit in $\hat{Y}$ as $u \to 0$, there exists $u_0 \neq 0$ such that $|u| \leq |u_0|$ implies $u \in U(a;h)$ and $g(u) - g(u_0) \in B$; let $\lambda \geq 1$ be such that $g(u_0) \in \lambda B$. Then $|u| \leq |u_0|$ implies $g(u) \in \lambda(B+B)$, and $|u| \leq \inf(|u_0|, 1/\lambda)$ implies $f(a+uh) \in B+B$.

Theorem 1.6. Let $f \in \mathcal{G}(\Omega,Y)$, $f(\Omega) \subset \Gamma$. **Then** $\varphi \in \mathcal{A}(\Gamma,\mathbf{C})$ **implies** $\varphi \circ f \in \mathcal{G}(\Omega,\mathbf{C})$.

Proof. Let $a \in \Omega$, $h \in X$. We first consider the special case when $\Gamma = Y$ and φ is a continuous homogeneous polynomial, i.e. $\varphi(y) = \Phi(\underbrace{y,\ldots,y}_{n \text{ times}})$, $n \geq 1$, where Φ is a continuous n-linear map $Y^n \to \mathbf{C}$.

If f(a) is the origin in Y, we again set $g(u) = f(a+uh)/u$. Given a neighborhood B of the origin in Y, such that $y_1,\ldots,y_n \in B$ imply $|\Phi(y_1,\ldots,y_n)| \leq 1$, there exist $u_0 \neq 0$ such that $|u| \leq |u_0|$ implies $u \in U(a;h)$, $g(u) - g(u_0) \in B$, and $\lambda > 0$ such that $g(u_0) \in \lambda B$; then $|u| \leq |u_0|$ implies $|\varphi \circ g(u)| \leq (1+\lambda)^n$ since Φ is n-linear, and $|\varphi \circ f(a+uh)| \leq (1+\lambda)^n |u|^n$. Thus $\varphi \circ f(a+uh)/u \to 0$ as $u \to 0$ if $n \geq 2$. If $n = 1$, $\varphi \circ f(a+uh)/u = \varphi \circ g(u)$. For any $\epsilon > 0$, since $\{y \in Y : |\varphi(y)| \leq \epsilon\}$ is a neighborhood of the origin, there exists $\alpha > 0$ such that $|u|$ and $|u'| \leq \alpha$ imply $|\varphi \circ g(u) - \varphi \circ g(u')| \leq \epsilon$, which proves that $\varphi \circ g(u)$ has a limit in $\mathbf{C}$ as $u \to 0$.

If $b = f(a)$ is not the origin in Y, we set $g = f - b$, $\varphi \circ f = \psi \circ g$, with $\psi(y) = \varphi(y+b) = \Phi(\underbrace{y+b,\ldots,y+b}_{n \text{ times}})$, which is the sum of a constant and n continuous homogeneous polynomials.

Now to the general case. Let $b = f(a) \in \Gamma$ and let $\gamma(b)$ be the biggest balanced (open) subset of the translated set $\Gamma - b$. By Prop. 1.5, $U = \{u \in U(a;h) : f(a+uh) - b \in \gamma(b)\}$ is an open set containing the origin in $\mathbf{C}$ and, for each compact subset K of U, $\{f(a+uh) - b : u \in K\}$ is a compact subset of $\gamma(b)$. Then, by 1.2, the expansion $\varphi \circ f(a+uh) = \varphi(b) + \sum_{n \geq 1} \varphi_n[b; f(a+uh)-b]$ is uniformly summable for $u \in K$; since the sum of a finite number of terms in the right hand member is holomorphic on U, so is the left hand member.

A last notation. $\hat{\mathcal{G}}(\Omega,\Gamma)$ will denote the class of maps $f : \Omega \to \Gamma$ such that $\varphi \in \mathcal{A}(\Gamma,\mathbf{C})$ implies $\varphi \circ f \in \mathcal{G}(\Omega,\mathbf{C})$; $\hat{\mathcal{A}}(\Omega,\Gamma)$ the class of continuous maps $\in \hat{\mathcal{G}}(\Omega,\Gamma)$.

By Th. 1.6; $f \in \mathcal{G}$ (resp. $\mathcal{A}$)(Ω,Y), $f(\Omega) \subset \Gamma$, implies $f \in \hat{\mathcal{G}}$ (resp. $\hat{\mathcal{A}}$) (Ω,Γ); the converse statement follows from 1.3 if Y is locally convex. If $f \in \hat{\mathcal{A}}(\Omega,\Gamma)$, then $\varphi \in \mathcal{A}(\Gamma,\mathbf{C})$ implies $\varphi \circ f \in \mathcal{A}(\Omega,\mathbf{C})$. Therefore the classes $\hat{\mathcal{A}}$ have the composition property.

This notation will be used in §6.

§2. Continuity of a Gateaux-analytic scalar function.

Proposition 2.1. *Let $\varphi \in \mathcal{G}(\Omega,\mathbf{C})$. Unless φ is constant, its image $\varphi(\Omega)$ is an open connected set in $\mathbf{C}$.*

Proof. Let $a \in \Omega$, and first assume that $\varphi(\Omega)$ is not a neighborhood of $\varphi(a)$ in $\mathbf{C}$. For any $h \in X$, since the image of the holomorphic function $U(a;h) \ni u \mapsto \varphi(a+uh)$ is not a neighborhood of $\varphi(a)$, its Taylor expansion $\varphi(a+uh) = \varphi(a) + \sum_{n \geq 1} u^n \varphi_n(a;h)$ reduces to a constant term $\varphi(a)$; thus φ is constant on the translated set $a + \omega(a)$, a neighborhood of a, and hence ([4], § III.1.3) constant on the open connected set Ω.

Since $\{\varphi(a+uh) : u \in \mathbf{C}, |u| \leq 1\}$ is connected for each $h \in \omega(a)$, $\varphi[a + \omega(a)]$ is connected and is therefore contained in a connected component of the open set $\varphi(\Omega)$; so the connected components of $\varphi(\Omega)$ have disjoint open inverse images under φ, and $\varphi(\Omega)$ is connected.

Corollary 2.2. *Let $\varphi \in \mathcal{G}(\Omega, \mathbf{C}^m)$. A sufficient (and obviously necessary) condition for the continuity of φ is the existence of an open set $U \supset \varphi(\Omega)$ in $\mathbf{C}^m$ and a locally injective map $\psi \in \mathcal{A}(U,\mathbf{C}^p)$ such that $\psi \circ \varphi$*

is continuous.

Proof. Let $a \in \Omega$, and norms in C^m, C^p be chosen. For any sufficiently small $\alpha > 0$, $0 < \|z-\varphi(a)\| \leqslant \alpha$ implies $z \in U$ and $\psi(z) \neq \psi[\varphi(a)]$, and hence there exists $\beta > 0$ such that $\|z-\varphi(a)\| = \alpha$ implies $\|\psi(z) - \psi[\varphi(a)]\| \geqslant \beta$. Since $\psi \circ \varphi$ is continuous, $\Omega_0 = \{x \in \Omega : \|\psi\circ\varphi(x) - \psi\circ\varphi(a)\| < \beta\}$ is an open set containing a. If $\omega_0(a)$ is the biggest balanced subset of the translated set $\Omega_0 - a$, $\varphi[a + \omega_0(a)]$ is connected (cf. Proof of Prop. 2.1), contains $\varphi(a)$ and does not meet the sphere with center $\varphi(a)$, radius α, and hence is contained in the open ball with same center and radius.

Example 2.3. Let $\varphi \in \mathcal{G}(\Omega,C)$. Then e^{φ} is continuous iff φ is continuous.

Thus there exist Gateaux-analytic scalar functions which omit the value 0 and are not continuous. In the following two Theorems, two values cannot be replaced by one.

Theorem 2.4. Let $\varphi \in \mathcal{G}(\Omega,C)$. **A sufficient (and obviously necessary) condition for the continuity of** φ **at a point** $a \in \Omega$ **is the existence of a neighborhood of a in** Ω **where** φ **omits at least two values.**

Proof. Let A be a balanced neighborhood of the origin in X such that $a + A \subset \Omega$ and $\varphi \neq \alpha$, $\varphi \neq \beta$ on $a + A$. On the disc $\{u \in C : |u| < 1\}$, the holomorphic functions $u \mapsto \varphi(a+uh)$, $h \in A$, omitting both values α and β, are a normal family [10]. Since they assume the same value $\varphi(a)$ for $u = 0$, they are bounded together for $u = ½$, which means that φ is bounded on $a + ½A$.

Corollary 2.5. Let X be a Baire space and let $\varphi \in \mathcal{G}(\Omega,C)$. **If** φ **omits one value** α**, either** φ **is continuous, or the inverse image under** φ **of any other value is dense in** Ω.

Proof. If the latter is not the case, then φ is continuous on some open nonempty subset of Ω, and hence continuous on Ω by a theorem of Zorn [13]. •

Remark 2.6. An example was recently given [5] of a normed, but not Baire, space X and a function $\varphi \in \mathcal{G}(X,C)$ which is continuous on some open set A, but not on X, so e^{φ} is not continuous on X (Ex. 2.3), omits the value 0, and the inverse image under e^{φ} of any other value is nowhere dense in A.

Theorem 2.7. Let $\varphi \in \mathcal{G}(\Omega,C)$. **A sufficient (and obviously necessary) condition for the continuity of** φ **is the existence of at least two values** α, β **having closed inverse images in** Ω **under** φ.

Proof. Let φ be nonconstant. Then (Prop. 2.1) the set E of zeroes of $\psi = (\varphi-\alpha)(\varphi-\beta)$ is closed in Ω and $\mathring{E} = \phi$. By Th. 2.4, φ is continuous on the open set $\Omega \setminus E$, and we only have to show that each point $a \in E$ has a neighborhood where φ is bounded.

Since $\psi \not\equiv 0$ on $a + \omega(a)$, we can choose $h \in \omega(a)$ so that $u = 0$ is an isolated zero of $\psi(a+uh)$, then $r > 0$ so that Ω and $\Omega \setminus E$ contain the compact sets $K = \{a + uh: |u| \leqslant r\}$ and $L = \{a + uh: |u| = r\}$, respectively. Since each point in L has a neighborhood where φ is bounded, the origin in X has a neighborhood

A such that $\Omega \supset K+A$ and $|\varphi| \leqslant M$ on $L + A$. Hence the last inequality holds on $a + A$ by the maximum principle applied to the holomorphic functions $u \mapsto \varphi(x+uh)$, $x \in a + A$.

§3. A generalization of the Green function.

Definition 3.1. An open connected set U in C is a Green open set if it has the equivalent properties: $\complement U$ is nonpolar ([1], chap. III); there exists a nonconstant superharmonic function > 0 on U; there exists a nonharmonic superharmonic function > 0 on U.

For instance, by the Riemann conformal mapping theorem, any simply connected open set, except C itself, is Green.

If U is Green, there exists a continuous and symmetric Green function $U^2 \ni (\alpha,\beta) \mapsto g_U(\alpha,\beta) \in]0,+\infty]$, which is finite iff $\alpha \neq \beta$. If U is not Green, we set $g_U(\alpha,\beta) = +\infty$ for any α and $\beta \in U$.

In the former case, g_U can be characterized as follows: for each $\beta \in U$, $\zeta \mapsto g_U(\zeta,\beta)$ is the smallest positive superharmonic function on U such that $\zeta \mapsto g_U(\zeta,\beta) + \log|\zeta-\beta|$ is harmonic on U. Then this superharmonic function has the same harmonic (or subharmonic) minorants on U as the constant 0. Such a superharmonic function is a potential on U ([2], part IV, chap. 4). Moreover, $\limsup g_U(\zeta,\beta) < +\infty$ as ζ tends to any boundary point of U, or to infinity if U is unbounded.

Proposition 3.2. Let U and U' be open connected sets in C, $U \subset U'$. Then: a) $g_U \leqslant g_{U'}$ on U^2; b) $U' \backslash U$ polar implies $g_U \equiv g_{U'}$ on U^2 and, conversely, $g_U(\alpha,\beta) = g_{U'}(\alpha,\beta)$, for one couple of distinct points $\alpha, \beta \in U$, implies $U' \backslash U$ polar.

Proof of b) when U and U' are Green: Let $U' \backslash U$ be polar. If $\beta \in U$ and $g_U(\zeta,\beta) = h(\zeta) - \log|\zeta-\beta|$, h is harmonic on U with a finite $\limsup |h(\zeta)|$ as ζ tends to any point in $U' \backslash U$. Hence a harmonic extension [1] h' of h to U' exists with $h'(\zeta) - \log|\zeta-\beta| > 0$ for any $\zeta \in U'$.

Now let r be the reduced function [1] on $U' \backslash U$ of the superharmonic function $U' \ni \zeta \mapsto g_{U'}(\zeta,\beta)$. Since r is harmonic on U, we have $g_U(\zeta,\beta) \leqslant g_{U'}(\zeta,\beta) - r(\zeta)$, and $U' \backslash U$ nonpolar implies $r > 0$.

Theorem 3.3 (Lindelöf's principle). Let U and V be open connected sets in C and let $f: U \to V$ be a holomorphic map. Then: a) $g_U(\alpha,\beta) \leqslant g_V[f(\alpha), f(\beta)]$ for any α, $\beta \in U$; b) f injective and $V \backslash f(U)$ polar imply the equality in a) for any α, $\beta \in U$; conversely, $g_U(\alpha,\beta) = g_V[f(\alpha), f(\beta)] < +\infty$, for one couple of points α, $\beta \in U$, implies f injective, $V \backslash f(U)$ polar, and $V = f(U)$ if U contains any open set in $\overline{C} = C \cup \{\infty\}$ which has a polar intersection with $\overline{C} \backslash U$.

A consequence of a): If f is nonconstant and U is not Green, then f(U) is not Green.

Proof. a) Let f be nonconstant and let V be a Green open set. For each $\beta \in U$, $\zeta \mapsto g_V[f(\zeta), f(\beta)]$ is a nonharmonic superharmonic function > 0 on U. Then U too is Green, and the superharmonic function $U \ni \zeta \mapsto g_V[f(\zeta), f(\beta)] - g_U(\zeta,\beta)$ is positive since $g_V > 0$ and $\zeta \mapsto g_U(\zeta,\beta)$ is a potential on U [2].

b) If f is a 1-1 map of U onto V, f^{-1} is holomorphic too, and the equality follows from a). By this remark and Prop. 3.2, both inequalities $g_U(\alpha,\beta) \leq g_{f(U)}[f(\alpha),f(\beta)]$ and $g_{f(U)}[f(\alpha),f(\beta)] \leq g_V[f(\alpha),f(\beta)]$ are equalities if f is injective and $V\backslash f(U)$ polar.

Conversely, let $g_U(\alpha,\beta) = g_{f(U)}[f(\alpha),f(\beta)] = g_V[f(\alpha),f(\beta)] < +\infty$. Then $f(\alpha) \neq f(\beta)$, U and f(U) are Green, and $V\backslash f(U)$ polar by Prop. 3.2. The positive superharmonic function $U \ni \zeta \mapsto g_{f(U)}[f(\alpha),f(\zeta)] - g_U(\alpha,\zeta)$ vanishes for $\zeta = \beta$, hence for any $\zeta \in U$, which yields the first equality for the given point α and an arbitrary point β; but then, for each $\beta \in U$, the positive superharmonic function $U \ni \zeta \mapsto g_{f(U)}[f(\zeta),f(\beta)] - g_U(\zeta,\beta)$ vanishes for $\zeta = \alpha$, hence for any $\zeta \in U$, which yields the first equality for arbitrary α and β; finally $\alpha \neq \beta$ implies $g_{f(U)}[f(\alpha),f(\beta)] < +\infty$ and $f(\alpha) \neq f(\beta)$. Now f^{-1}, a holomorphic function on f(U) omitting a nonpolar set of values, has a meromorphic extension φ to V since $V\backslash f(U)$ is polar ([11], Chap.II); $\varphi(V)$ is an open set in $\bar{C}$ such that $\varphi(V)\backslash U$, equal to $\varphi[V\backslash f(U)]$ because f^{-1} is injective, is again polar ([1], chap. III).

Definition 3.4. Given a and $b \in \Omega$ (Ω open connected in the space X), $e_\Omega(a,b)$ is the infimum of $g_{\varphi(\Omega)}[\varphi(a),\varphi(b)]$ either for all nonconstant $\varphi \in \mathcal{G}(\Omega, C)$, or for all nonconstant $\varphi \in \mathcal{A}(\Omega, C)$.

Both infima coincide. In fact, for any nonconstant $\varphi \in \mathcal{G}(\Omega, C)$, the image $\varphi(\Omega)$ is an open connected set in C (Prop. 2.1). If this set is Green, $C\backslash\varphi(\Omega)$ is infinite and $\varphi \in \mathcal{A}(\Omega, C)$ by Th. 2.4.

Obvious properties of the function e_Ω are: its symmetry on Ω^2; $e_\Omega(a,a) = +\infty$ and $0 \leq e_\Omega(a,b) \leq +\infty$ for $a \neq b$; $e_\Omega \leq e_{\Omega'}$ on Ω^2 if $\Omega \subset \Omega'$; e_Ω generalizes the Green function since, by Lindelöf's principle, $e_U \equiv g_U$ if U is an open connected set in C.

More precisely, if U is a Green open set in C, for each $a \in U$, the sets $\{x \in U: g_U(a,x) \geq \alpha\}$ are a fundamental system of neighborhoods of a; so, in order to make the generalization closer, we shall be led to consider (in Corollary 5.5 below) open sets Ω where e_Ω has the same property. Such an Ω cannot contain any complex affine manifold, by the following

Proposition 3.5. Let $a \in \Omega$. a) $e_\Omega(a,a+uh) \geq g_{U_0}(0,u)$ for $h \in X$, $u \in U_0$, the connected component containing 0 of $U(a;h)$. b) $e_\Omega(a,b) \to +\infty$ as $b \to a$, and $e_\Omega(a,a+uh) = +\infty$ for any $u \in C$ if $U(a;h) = C$.

Proof. a) follows from Th. 3.3.a for $u \mapsto \varphi(a+uh)$, a holomorphic map $U_0 \mapsto \varphi(\Omega)$, $\varphi \in \mathcal{G}(\Omega,C)$, $\varphi \neq$ const.

b) Let $h \in \omega(a)$. Since $U_0 \supset \{u \in C : |u| < 1\}$, by Prop. 3.2.a, $b-a \in r\,\omega(a)$, $0 < r < 1$, implies $e_\Omega(a,b) \geq \log 1/r$.

Theorem 3.6. Let $\varphi \in \mathcal{G}(\Omega,C)$, $\varphi \neq$ const. If $\mathcal{G}[\varphi(\Omega), X] \ni \rho$ such that $\rho \circ \varphi(\Omega) \subset \Omega$ and $\varphi \circ \rho$ is the identity $\varphi(\Omega) \to \varphi(\Omega)$, then $e_\Omega(a,b) = g_{\varphi(\Omega)}[\varphi(a),\varphi(b)]$ for any $a,b \in \rho \circ \varphi(\Omega)$.

Proof. Let $\alpha,\beta \in \varphi(\Omega)$ be such that $a = \rho(\alpha)$, $b = \rho(\beta)$, so $\alpha = \varphi(a)$, $\beta = \varphi(b)$. For any nonconstant $\varphi' \in \mathcal{A}(\Omega,C)$, by Th. 1.6 $\varphi' \circ \rho$ is a holomorphic map $\varphi(\Omega) \to \varphi'(\Omega)$, for which Th. 3.3.a yields $g_{\varphi'(\Omega)}[\varphi'(a),\varphi'(b)] = g_{\varphi'(\Omega)}[\varphi'\circ\rho(\alpha), \varphi'\circ\rho(\beta)] \geq g_{\varphi(\Omega)}(\alpha,\beta)$.

Example 3.7. Let φ be linear $X \to C$, $\varphi(a) \neq \varphi(b)$, $\rho(\zeta) = a + \frac{\zeta - \varphi(a)}{\varphi(b-a)}(b-a)$. If there exists an open connected set U in **C** such that $\rho(U) \subset \Omega \subset \varphi^{-1}(U)$, then $e_\Omega(a,b) = g_U[\varphi(a), \varphi(b)]$.

Since the gap is wide between $\rho(U)$ and $\varphi^{-1}(U)$, no analogue to the second statement in Prop. 3.2.b can be expected. As a substitute for the first statement we have

Theorem 3.8. **Let $\Omega \subset \Omega'$. The equality $e_\Omega \equiv e_{\Omega'}$ on Ω^2 holds in the following two cases: a) any function $\in \mathcal{G}(\Omega,C)$ admits an extension $\in \mathcal{G}(\Omega',C)$; b) $\Omega' \setminus \Omega$ has an empty interior; for any a and $h \in X$, for each connected component U of $U'(a;h) = \{u \in C\colon a+uh \in \Omega'\}$, the set $\{u \in U\colon a+uh \in \Omega' \setminus \Omega\}$ is either polar or U itself.**

Proof. a) Let $\varphi \in \mathcal{G}(\Omega',C)$. For each $\alpha \in C \setminus \varphi(\Omega)$, $1/(\varphi-\alpha)|_\Omega$ has an extension $\psi \in \mathcal{G}(\Omega',C)$ such that $(\varphi-\alpha)\psi \equiv 1$ ([4], § III.1.3). Hence $\alpha \in C \setminus \varphi(\Omega')$ and $\varphi(\Omega') = \varphi(\Omega)$.

b) Let $\varphi \in \mathcal{A}(\Omega,C)$ be such that $\varphi(\Omega)$ is a Green open set. Since the closed set $C \setminus \varphi(\Omega)$ is not locally polar ([1], chap. III), there exists $\alpha \notin \varphi(\Omega)$ such that any neighborhood of α has a nonpolar intersection with $C \setminus \varphi(\Omega)$. Since φ may be replaced by $1/\varphi - \alpha$ if α is finite, we assume that $\alpha = \infty$ and prove that φ has an extension $\varphi' \in \mathcal{A}(\Omega',C)$ such that $\varphi'(\Omega') \setminus \varphi(\Omega)$ is polar, which entails $g_{\varphi'(\Omega')}[\varphi'(a), \varphi'(b)] = g_{\varphi(\Omega)}[\varphi(a), \varphi(b)]$, for any $a,b \in \Omega$, by Prop. 3.2.b.

Let $a \in \Omega' \setminus \Omega$ and $h \in \omega'(a) \cap (\Omega - a)$, with the notation $\omega'(a)$ for the biggest balanced subset of $\Omega' - a$. Since $\{u \in C : |u| \leqslant 1, a+uh \in \Omega' \setminus \Omega\}$ is polar, its complementary set contains arbitrarily small circumferences with center 0 ([1], chap. VII), i.e. there exist $r > 0$ and $r' > r$ such that Ω (resp.: Ω') contains the compact set L (resp.: L') $= \{a+uh\colon r \leqslant |u| \leqslant r'$ (resp.: $|u| \leqslant r')\}$, and hence contains L + A (resp.: L' + A) for a suitably chosen open neighborhood A of the origin in X.

Now let $D' = \{u \in C\colon |u| < r'\}$. For each $x \in a + A$, $K(x) = \{u \in C\colon |u| \leqslant r, x+uh \in \Omega' \setminus \Omega\}$ is a compact polar set and $u \mapsto \varphi(x+uh)$ a holomorphic function on $D' \setminus K(x)$ omitting a nonpolar set of values. Such a function has ([11], chap. II) a meromorphic extension $u \mapsto \psi_{x,h}(x+uh)$ to D', and the set of values $\{\psi_{x,h}(x+uh)\colon u \in K(x)\}$ is again polar [1]. Then each value α in the latter set is such that some neighborhood of α in **C** has a polar intersection with $C \setminus \varphi(\Omega)$, and by the assumption above α is finite. Thus $u \mapsto \psi_{x,h}(x+uh)$ is holomorphic on D', and its set of values either reduces to one point $\in \varphi(\Omega)$ or is open and has a polar intersection with $C \setminus \varphi(\Omega)$.

Finally the integral formula $\varphi'(x) = \frac{1}{2\pi}\int_0^{2\pi} \varphi(x + re^{i\theta}h)d\theta$ defines ([4], § III.2.2) $\varphi' \in \mathcal{A}(a+A,C)$, and $\varphi'(x) = \psi_{x,h}(x) = \varphi(x)$ if $x \in (a+A) \cap \Omega$. This proves the existence of an extension $\varphi' \in \mathcal{A}(\Omega',C)$ such that $\varphi'(\Omega')$ is the union of open sets each of which has a polar intersection with $\varphi'(\Omega') \setminus \varphi(\Omega)$.

Remarks 3.9. a) If $\Omega' \setminus \Omega$ is the set of zeroes of a nonconstant function $\in \mathcal{A}(\Omega', C)$, the assumptions of Th. 3.8.b are satisfied. b) If $|\varphi| < 1$, then $|\varphi'| < 1$ too.

§4. A generalization of the Caratheodory metric.

Let $D = \{\zeta \in C : |\zeta| < 1\}$.

Definition 4.1. Given a and $b \in \Omega$, $E_\Omega(a,b)$ is the infimum of $g_D[\varphi(a), \varphi(b)]$ for all $\varphi \in \mathcal{A}(\Omega, C)$ such that $\varphi(\Omega) \subset D$, or the infimum of $\log 1/|\varphi(b)|$ for all $\varphi \in \mathcal{A}(\Omega, C)$ such that $\varphi(\Omega) \subset D$ and $\varphi(a) = 0$.

Both infima coincide since, by Th. 3.3.b, composing φ with a 1–1 holomorphic map $D \to D$ does not alter $g_D[\varphi(a), \varphi(b)]$.

Obvious properties of the function E_Ω are: its symmetry on Ω^2; $E_\Omega(a,a) = +\infty$ and $0 \leqslant E_\Omega(a,b) \leqslant +\infty$ for $a \neq b$; $E_\Omega \leqslant E_{\Omega'}$ on Ω^2 if $\Omega \subset \Omega'$; $E_\Omega \geqslant e_\Omega$ follows from Prop. 3.2.a and allows the substitution of E_Ω to e_Ω in Prop. 3.5. Finally Th. 3.8 also holds for E_Ω, with a simpler proof, by Remark 3.9.b.

Proposition 4.2. **a)** $(a,b) \to \log \coth \frac{1}{2} E_\Omega(a,b)$ **is a finite pseudo-distance on** Ω. **b)** $E_\Omega > 0$, E_Ω **is continuous on** Ω^2.

Proof. a) We have a pseudodistance because $D^2 \ni (\alpha,\beta) \to \log \coth \frac{1}{2} g_D(\alpha,\beta)$ satisfies the triangle inequality. By Prop. 3.5.b, the pseudodistance is finite for a given a and b in a suitable neighborhood of a, and hence also for a given b and a in a suitable neighborhood of b. Then, given $a \in \Omega$, the set of points $b \in \Omega$ at a finite pseudodistance from a is a closed open subset of Ω.

b) The pseudodistance of a and b tends to 0 as $b \to a \in \Omega$ (Prop. 3.5.b), and therefore is a continuous function of (a,b) by the triangle inequality.

Proposition 4.3. **If the space X is separable (eg., if it has finite dimension), then for each couple** $(a,b) \in \Omega^2$, **there exists** $\varphi \in \mathcal{A}(\Omega, C)$ **such that** $\varphi(\Omega) \subset D$ **and** $E_\Omega(a,b) = g_D[\varphi(a), \varphi(b)]$.

Proof. Let $E_\Omega(a,b) = \lim \log 1/|\varphi_n(b)|$ with $\varphi_n \in \mathcal{A}(\Omega, C)$, $\varphi_n(\Omega) \subset D$, $\varphi_n(a) = 0$. The sequence (φ_n) is equicontinuous, and a suitably chosen subsequence converges, uniformly on any compact subset of Ω, to the desired function φ ([4], § III.2.2).

Theorem 4.4. **Assume that homotopy in** Ω **with fixed extremities holds between any two (continuous) paths in** Ω **with the same end points. Then, for each nonconstant** $\varphi \in \mathcal{A}(\Omega, C)$ **such that** $\varphi(\Omega)$ **is Green, there exist** $\psi \in \mathcal{A}(\Omega, C)$, $\psi(\Omega) \subset D$, **and a holomorphic map** $f: D \to \varphi(\Omega)$, **such that** $\varphi = f \circ \psi$; **hence** $g_{\varphi(\Omega)}[\varphi(a), \varphi(b)] \geqslant g_D[\psi(a), \psi(b)]$ **by Th. 3.3.a.**

A consequence. Under the assumption of Th. 4.4, e_Ω and E_Ω coincide. This allows the substitution of e_Ω to E_Ω in Prop. 4.2 and 4.3 and of E_Ω to e_Ω in § 5, where Ω is assumed to be convex.

Proof. Since $C \setminus \varphi(\Omega)$ is infinite, there exist a holomorphic map f of D onto $\varphi(\Omega)$, and a family $\mathcal{U}$ of open connected sets U in C such that $\varphi(\Omega) = \bigcup_{U \in \mathcal{U}} U$ and the restrictions of f to the connected components of

$f^{-1}(U)$ are 1–1 maps onto U ([3], chap. II). By this property of f, given $a \in \Omega$ and $\alpha \in D$ such that $\varphi(a) = f(\alpha)$, for each path c: $[0,1] \to \Omega$ such that $c(0) = a$, there exists a unique path γ: $[0,1] \to D$ such that $\gamma(0) = \alpha$ and $f \circ \gamma = \varphi \circ c$. The argument is a classical one and may be omitted.

From the homotopy assumption and the local injectiveness of f follows that $\gamma(1)$ depends only on $x = c(1)$, say $\gamma(1) = \psi(x)$, and only $\psi \in \mathcal{H}(\Omega,C)$ remains to be proved. Given $x_0 \in \Omega$, let the paths c_0: $[0,1] \to \Omega$ and γ_0: $[0,1] \to D$ be such that $f \circ \gamma_0 = \varphi \circ c_0$, $c_0(0) = a$, $\gamma_0(0) = \alpha$, $c_0(1) = x_0$, $\gamma_0(1) = \psi(x_0)$. Let $\varphi(x_0) \in U_0$, $U_0 \in \mathcal{U}$, and f_0 be the restriction of f to the connected component containing $\psi(x_0)$ of $f^{-1}(U_0)$. If c is a path in $\varphi^{-1}(U_0)$ starting from x_0, $f_0^{-1} \circ \varphi \circ c$ is a path in D starting from $\psi(x_0)$, which proves $\psi = f_0^{-1} \circ \varphi$ on the connected component containing x_0 of $\varphi^{-1}(U_0)$.

Example 4.5. Showing to what extent the assumption of homotopy is necessary in Th. 4.4. Given an open bounded connected set U in **C**, let U' be the union of U and all $z \in C$ such that some neighborhood of z has a polar intersection with $\complement U$. Again U' is open bounded and connected, and U'\U is polar. Then U' simply connected implies $g_U \equiv E_U$ on U^2 and, conversely, $g_U(\alpha,\beta) = E_U(\alpha,\beta)$, for one couple of distinct points $\alpha, \beta \in U$, implies U' simply connected.

Proof. By Prop. 4.3 (with U' instead of Ω), Remark 3.9.b and Prop. 3.2.b, we have $g_U(\alpha,\beta) = E_U(\alpha,\beta)$ iff there exists a holomorphic map φ': $U' \to D$ such that $g_{U'}(\alpha,\beta) = g_D(\varphi'(\alpha), \varphi'(\beta))$; but, by Th. 3.3, the latter equality occurs iff φ' is a 1–1 map of U' onto D, since $\alpha \neq \beta$ and U' is Green.

§5. Convex open sets.

Proposition 5.1. Let $p_1,\ldots,p_m$ be continuous seminorms on X and let $\alpha_1,\ldots,\alpha_m > 0$. **If** $\Omega = \{x \in X: p_j(x-a) < \alpha_j, j=1,\ldots,m\}$, **then for each** $b \in \Omega$, $e_\Omega(a,b) = \inf\{\log \frac{\alpha_j}{p_j(b-a)} : j=1,\ldots,m\}$.

Proof for $m = 1$, $\alpha_1 = 1$. If $p(b-a) = 0$ although $b-a \neq 0$, $\Omega \supset \{a+u(b-a): u \in C\}$ and $e_\Omega(a,b) = +\infty$ by Prop. 3.5.b.

Let $0 < p(b-a) < 1$. By the Hahn-Banach theorem, the adjoint space X' contains φ such that $|\varphi| \leqslant p$, $|\varphi(b-a)| = p(b-a)$. With $U = \{\zeta \in C : |\zeta - \varphi(a)| < 1\}$, Ex. 3.7 yields $e_\Omega(a,b) = g_U[\varphi(a),\varphi(b)] = \log 1/p(b\text{-}a)$.

Corollary 5.2. If X is locally convex, then $e_\Omega(a,b) \to +\infty$ **as a and b tend to a same point** $c \in \Omega$.

Proof. Let $\Omega \supset \Omega' = \{x \in X: p_j(x-c) < \beta_j, j=1,\ldots,m\}$. For a and $b \in \Omega'$, we have $e_\Omega(a,b) \geqslant e_{\Omega'}(a,b) \geqslant$
$\geqslant \inf\{\log \frac{\beta_j - p_j(a-c)}{p_j(b-a)} : j=1,\ldots,m\}$.

Theorem 5.3. Let Ω be a convex open neighborhood of the origin in X, $\omega = \omega(0)$ **its biggest balanced subset, P and p the gauges of Ω and ω, i.e.**

$$P(x) = \inf\{\lambda > 0 : x \in \lambda\Omega\}, \quad p(x) = \sup_{\theta \in R} P(e^{i\theta}x), \quad x \in X.$$

Then, for any $h \in \Omega$,

$$e_\Omega(0,h) \leqslant \log\frac{2-P(h)}{P(h)}, \quad \log\frac{1}{p(h)} \leqslant e_\Omega(0,h) \leqslant \log[1+\frac{2}{p(h)}].$$

Proof. Both gauges are finite and continuous, and $P \leqslant p$; p is a seminorm, but P is only convex and positively homogeneous. The inequality $e_\Omega(0,h) \geqslant \log 1/p(h)$ is obvious if $p(h) \geqslant 1$; if $p(h) < 1$ or $h \in \omega$, it follows from $e_\Omega(0,h) \geqslant e_\omega(0,h)$ and Prop. 5.1.

Now let $D' = \{\zeta \in C : \operatorname{Re}\zeta < 1\}$. If $\varphi \in X'$, $\operatorname{Re}\varphi \leqslant P$ and $\varphi(h) \neq 0$, then Ex. 3.7 can be used with $\Omega' = \varphi^{-1}(D') \supset \Omega$ instead of Ω and $U = D'$; hence

$$e_\Omega(0,h) \leqslant e_{\Omega'}(0,h) = g_{D'}[0,\varphi(h)] = \log\left|\frac{2-\varphi(h)}{\varphi(h)}\right|.$$

When $\operatorname{Re}\zeta$ remains equal to a given number $\xi \in]0,1[$, $\log\left|\frac{2-\zeta}{\zeta}\right|$ is maximum for $\zeta = \xi$; so the first inequality in the theorem is obtained, for $0 < P(h) < 1$, by choosing such a φ that $\operatorname{Re}\varphi(h) = P(h)$.

When $|\zeta|$ remains bigger than a given number $\rho > 0$, $\log\left|\frac{2-\zeta}{\zeta}\right|$ is maximum for $\zeta = -\rho$; so the last inequality in the theorem is obtained, for $p(h) > 0$, by choosing such θ and φ that $\operatorname{Re}\varphi(e^{i\theta}h) = P(e^{i\theta}h) = p(h)$, which implies $|\varphi(h)| \geqslant p(h)$.

Corollary 5.4. Let Ω be convex, a and $a+h \in \Omega$, $h \neq 0$, $e_\Omega(a,a+h) = +\infty$. Then: a) $\Omega \supset \{a+uh : u \in C\}$ (a converse statement to Prop. 3.5.b); b) for any $b \in \Omega$, $\Omega \supset \{b+uh : u \in C\}$; hence $e_\Omega(b,b+uh) = +\infty$ for any $u \in C$, by Prop. 3.5.b.

Proof. Let a be the origin in X.

a) $e_\Omega(0,h) = +\infty$ iff $p(h) = 0$, or $P(e^{i\theta}h) = 0$ for any $\theta \in R$, or $re^{i\theta}h \in \Omega$ for any $r > 0$ and $\theta \in R$.

b) Since Ω is convex, $rt\,e^{i\theta}h \in \Omega$ $(r > 0, t > 2)$ implies $(1-\frac{2}{t})b + \frac{2}{t}rt\,e^{i\theta}h \in \Omega$. By letting $t \to +\infty$ we get $b+2re^{i\theta}h \in \overline{\Omega}$ and $b+re^{i\theta}h \in \Omega$.

Corollary 5.5. Let Ω be convex, and let $a \in \Omega$ be given. a) Assume that, for each neighborhood A of a in Ω, there exists $\alpha > 0$ such that $e_\Omega(a,x) \geqslant \alpha$, $x \in \Omega$, implies $x \in A$. Then X is a normed space and $\omega(a)$ is bounded, which does not depend on a. b) Conversely, if X is a normed space and $\omega(a)$ contained in the open ball $\{x \in X : \|x\| < R(a)\}$, then $e_\Omega(a,x) \leqslant \log[1+\frac{2R(a)}{\|x-a\|}]$ for any $x \in \Omega$, and the sets $\{x \in \Omega : e_\Omega(a,x) \geqslant \alpha\}$ are a fundamental system of neighborhoods of a.

Proof. Let p_a be the gauge of $\omega(a)$, which is a continuous seminorm on X.

a) By the assumption and Th. 5.3, $p_a(x-a) < e^{-\alpha}$ or $e^\alpha(x-a) \in \omega(a)$ implies $x-a \in A-a$; thus $\omega(a)$ is absorbed by $A-a$, an arbitrary neighborhood of the origin.

Since $\omega(a)$ is bounded, p_a is a norm, and the sets $\{x \in X: p_a(x) < \lambda\} = \lambda\omega(a), \lambda > 0$, are a fundamental system of neighborhoods of the origin.

Finally $a-b \in \lambda(\Omega-a)$, $\lambda > 0$, implies $\Omega-b = (\Omega-a)+(a-b) \subset (1+\lambda)(\Omega-a)$ since $\Omega-a$ is convex; hence $\omega(b) \subset (1+\lambda)\,\omega(a)$.

b) The first statement follows from Th. 5.3 and $p_a(x) \geq \frac{\|x\|}{R(a)}$ for any $x \in X$, and the second statement follows from the first one and Prop. 3.5.b.

Corollary 5.6. **Let Ω be convex in a normed space X. The boundedness of the set $\bigcup_{a\in\Omega} \omega(a)$ is equivalent to the existence, for each neighborhood B of the origin in X, of a $\beta > 0$ such that $e_\Omega(a,b) \geq \beta$, a and $b \in \Omega$, implies $b-a \in B$.**

Proof. By Th. 5.3, the existence of such a β is equivalent to the existence of a $\lambda > 1$ such that $\lambda x \in \omega(a)$ for some $a \in \Omega$ implies $x \in B$; thus $\bigcup_{a\in\Omega} \omega(a)$ is absorbed by B.

Examples 5.7. Showing that the three conditions Ω bounded, $\bigcup_{a\in\Omega}\omega(a)$ bounded, $\omega(a)$ bounded for each $a \in \Omega$, are actually different. For a finite or infinite **m**, let X be the Hilbert space whose elements are all sequences **x** of **m** complex numbers x_n, $0 \leq n < m$, such that $\sum_n |x_n|^2 = \|x\|^2 < +\infty$. For such an **x**, let Re x, Im x, Re^+x denote the sequences of **m** real numbers $Re\, x_n$, $Im\, x_n$, $Re^+x_n = \sup(0, Re\, x_n)$, respectively.

a) Since $\|Re\, y\| \leq \|Re\, x\| + \|y-x\|$, $\Omega = \{x \in X: \|Re\, x\| < 1\}$ is an open unbounded convex set in X; but $h \in \omega(a)$ implies $a \pm h$ and $a \pm ih \in \Omega$; hence $\|Re\, h\|$ and $\|Im\, h\| < 2, \|h\| < 2\sqrt{2}$.

b) Since ξ^+ is a convex function of the real variable ξ, we also have $Re^+[(1-t)x_n+ty_n] \leq (1-t)\,Re^+x_n + t\,Re^+y_n$ for each n; therefore $\|Re^+[(1-t)x+ty]\| \leq \|(1-t)Re^+x + t\,Re^+y\|$, $0 < t < 1$, and $\|Re^+y\| \leq \|Re^+x\| + \|Re^+(y-x)\| \leq \|y-x\|$ by taking $t = \frac{1}{2}$. Again $\Omega' = \{x \in X: \|Re^+x\| < 1\}$ is an open convex set in X, $\Omega' \supset \Omega$; $\bigcup_{a\in\Omega'} \omega'(a)$ is no longer bounded since $Re^+a = Im\, a = 0$ implies $a \in \omega'(a)$, but each $\omega'(a)$ is still bounded since $h \in \omega'(a)$ implies $\|Re^+(\pm h)\|$ and $\|Re^+(\pm ih)\| < 1 + \|Re^+(-a)\|$; hence $\|h\| < 2\sqrt{2}[1 + \|Re^+(-a)\|]$.

§6. A generalization of the Lindelöf principle.

Theorem 6.1. **Let $f \in \hat{\mathcal{G}}(\Omega,\Gamma)$ (cf. end of §1). Then $e_\Gamma[f(a), f(b)] \geq e_\Omega(a,b)$, $E_\Gamma[f(a),f(b)] \geq E_\Omega(a,b)$, for any $a, b \in \Omega$.**

Proof. $\varphi \in \mathcal{A}(\Gamma, C)$ implies $\varphi\circ f \in \mathcal{G}(\Omega, C)$, and $\varphi(\Gamma) \subset D$ implies $\varphi\circ f(\Omega) \subset D$; for a nonconstant $\varphi\circ f$, we have $g_{\varphi(\Gamma)}[\varphi\circ f(a), \varphi\circ f(b)] \geq g_{\varphi\circ f(\Omega)}[\varphi\circ f(a), \varphi\circ f(b)] \geq e_\Omega(a,b)$.

Corollary 6.2. **Both functions e_Ω, E_Ω are unaltered by a 1–1 map f such that f and f^{-1} are analytic.**

Example 6.3. Let X be a Hilbert space, with the notation $(\mid)$ for the scalar product, and let $\Omega = \{x \in X: \|x\| < 1\}$. From the formula $e_\Omega(0,h) = \log 1/\|h\|$, $h \in \Omega$, yielded by Prop. 5.1, $e_\Omega(a,b)$ can be computed for arbitrary $a, b \in \Omega$ through the analytic map $f_a: \Omega \to \Omega$ defined by

$$[1-(x|a)]\, f_a(x) = \lambda_a x - a + \frac{(x|a)}{1+\lambda_a}\, a, \quad \lambda_a = \sqrt{1-\|a\|^2}.$$

In fact $|1-(x|a)|^2[1-\|f_a(x)\|^2] = (1-\|a\|^2)(1-\|x\|^2)$, $f_a \circ f_{-a}$ is the identity, and $f_a(a) = 0$.

Corollary 6.4. **Let Γ be convex in a normed space Y; for $b \in \Gamma$, let $\gamma(b)$ be the biggest balanced subset of $\Gamma - b$. a) If each set $\gamma(b)$ is bounded, then any map $f \in \hat{\mathcal{G}}(\Omega,\Gamma)$ is continuous. b) If $\bigcup_{b\in\Gamma} \gamma(b)$ is bounded, then the family of maps $\hat{\mathcal{G}}(\Omega,\Gamma)$ is equicontinuous, and $\{f(x)-f(a): f \in \hat{\mathcal{G}}(\Omega,\Gamma)\}$ is bounded for given $a, x \in \Omega$.**

Proof. a) and the first statement in b) follow from $e_\Gamma[f(a), f(x)] \geq e_\Omega(a,x)$, Prop. 3.5.b and Corollary 5.5.b. Now let the open ball $\{y \in Y: \|y\| < R\}$ contain $\bigcup_{b\in\Gamma} \gamma(b)$. The inequality in Corollary 5.5.b yields

$$\|f(x)-f(a)\| \leq \frac{2R}{\exp[E_\Omega(a,x)]-1}, \text{ where } E_\Omega(a,x) > 0 \text{ by Prop. 4.2.b.}$$

Corollary 6.5. **Assume that any bounded function $\in \mathcal{A}(\Omega, C)$ is constant; let $f \in \mathcal{G}(\Omega, Y)$ and M be a one dimensional complex affine manifold in Y. If $M \cap f(\Omega)$ contains at least two points, then any open convex set Γ containing $f(\Omega)$ also contains M.**

Proof. The assumption means $E_\Omega(a,b) = +\infty$, which implies $E_\Gamma[f(a), f(b)] = +\infty$, for any $a, b \in \Omega$; the conclusion follows from Corollary 5.4.a.

Corollary 6.6. **Let U be an open connected Green set in C, α and $\beta \in U$, $\alpha \neq \beta$; let Γ be an open convex neighborhood of the origin in Y, $f \in \mathcal{A}(U,Y)$. If $f(U) \subset \Gamma$ and $f(\beta) = 0$, then $\frac{1+\exp[g_U(\alpha,\beta)]}{2}\, f(\alpha) \in \bar{\Gamma}$.**

Proof. If Q is the gauge of Γ, $\frac{2-Q\circ f(\alpha)}{Q\circ f(\alpha)} \geq \exp[g_U(\alpha,\beta)]$ is an equivalent inequality, which follows from Th. 5.3 and 6.1. Th. 7.1 will answer the problem of equality raised by this result, but for a simply connected U only.

Corollary 6.7. **Let Ω and Γ be open convex neighborhoods of the origins in X and Y, p and q the gauges of their biggest balanced subsets, $f \in \mathcal{G}(\Omega, Y)$, $f(\Omega) \subset \Gamma$, $f(0) = 0$. Then: a) f maps the biggest vector subspace of X contained in Ω into the biggest vector subspace of Y contained in Γ, i.e. $p(x) = 0$ implies $q \circ f(x) = 0$; b) for any $h \in X$, $\lim_{u\to 0} q[\frac{f(uh)}{u}] \leq 2\,p(h)$.**

Proof. Th. 5.3 and 6.1 yield $q[f(uh)] \leq 2\frac{p(uh)}{1-p(uh)}$ or $q[\frac{f(uh)}{u}] \leq \frac{2\,p(h)}{1-|u|p(h)}$ for $uh \in \Omega$. For any $\epsilon > 0$, since $\{y \in Y: q(y) \leq \epsilon\}$ is a neighborhood of the origin, there exists $\alpha > 0$ such that $|u|$ and $|u'| \leq \alpha$ imply $q[\frac{f(uh)}{u} - \frac{f(u'h)}{u'}] \leq \epsilon$, which proves that $q[\frac{f(uh)}{u}]$ has a limit as $u \to 0$. Theorem 7.3 will answer the problem of equality raised by this result.

Corollary 6.8 (a generalization of Schwarz's lemma). Let Ω and Γ be convex and balanced, and let p and q be their gauges. Any $f \in \mathcal{G}(\Omega, Y)$, with $f(\Omega) \subset \Gamma$, $f(0) = 0$, satisfies $q \circ f(x) \leqslant p(x)$ for any $x \in \Omega$ and $\lim_{u \to 0} q[\frac{f(ux)}{u}] \leqslant p(x)$ for any $x \in X$.

Proof. Prop. 5.1 is used here instead of Th. 5.3. See Th. 7.4 for the problem of equality.

Theorem 6.9. Let Ω and $\Omega' \supset \Omega$ satisfy the assumptions of Th. 3.8.b, namely: $\Omega' \setminus \Omega$ has an empty interior; for any a and $h \in X$ and for each connected component U of $U'(a;h) = \{u \in C :$ $a + uh \in \Omega'\}$, the set $\{u \in U : a+uh \in \Omega' \setminus \Omega\}$ is either polar or U itself. Then: a) any bounded function $\varphi \in \mathcal{A}(\Omega, C)$ has an extension $\varphi' \in \mathcal{A}(\Omega', C)$; b) if Y is a Banach space, and Γ a convex open set such that $\bigcup_{b \in \Gamma} \gamma(b)$ is bounded, any map $f \in \mathcal{A}(\Omega, Y)$, such that $f(\Omega) \subset \Gamma$, has an extension $f' \in \mathcal{A}(\Omega', Y)$.

Proof. a) was proved with Th. 3.8.b, and can be used to show that the connectedness of Ω' involves the connectedness of Ω.

b) Let x and $x' \in \Omega$ tend to a same point $a \in \Omega' \setminus \Omega$. By Th. 3.8.b (for E_Ω) and Prop. 4.2.b, $E_\Omega(x,x') = E_{\Omega'}(x,x') \to +\infty$; on the other hand, by the proof of Corollary 6.4.b,

$$\|f(x) - f(x')\| \leqslant \frac{2R}{\exp[E_\Omega(x,x')] - 1},$$

which proves the existence of $\lim_{\Omega \ni x \to a} f(x)$ since Y is complete, so f has a continuous extension f' to Ω'.

Now, for each $y' \in Y'$, $y' \circ f'$ is a continuous extension of $y' \circ f$; therefore each point $a \in \Omega' \setminus \Omega$ has an open neighborhood ω' in Ω' such that $y' \circ f$ has a bounded restriction to $\omega = \omega' \cap \Omega$, so $y' \circ f \in \mathcal{A}(\omega, C)$ by Th. 1.6, and $y' \circ f' \in \mathcal{A}(\omega', C)$ by Th. 6.9.a for ω and ω'. This implies $f' \in \mathcal{A}(\Omega', Y)$ ([4], § III.1.3) since Y is a Banach space.

Remark 6.10. Th. 6.9 is to be compared with other extension theorems obtained by Huber [6] for a pointed disc Ω and a Riemann surface Y, by Kwack [8] for a pointed disc Ω and a complex space Y, by Kobayashi [7] for finite dimensional analytic manifolds X and Y, $\Omega' \setminus \Omega$ nowhere dense in an analytic subset of X. All these assume that Y is a hyperbolic space, which is less restrictive than our assumption on Γ but , we make no assumptions on X.

§7. Problems of equality.

We again use the notations $D = \{\zeta \in C : |\zeta| < 1\}$, $D' = \{\zeta \in C : \mathrm{Re}\,\zeta < 1\}$.

Theorem 7.1. Let $U = rD$ be an open disc centered at the origin in C, Γ an open convex neighborhood of the origin in Y, $f \in \mathcal{A}(U,Y)$, $f(U) \subset \Gamma$, and $f(0) = 0$. We say that the equality in Corollary 6.6 is reached at a point $\alpha \in U\setminus\{0\}$ if $\frac{1+\exp[g_U(\alpha,0)]}{2} f(\alpha) = \frac{r+|\alpha|}{2|\alpha|} f(\alpha)$ is on the boundary of Γ. a) If this equality is reached at $\alpha \in U\setminus\{0\}$, then the same equality is reached at all points ζ on the same radius of U as α, and these ζ also satisfy $e_\Gamma[f(\zeta), f(0)] = g_U(\zeta,0)$. b) If Y has a finite dimension m, then the equality can be reached on at most m radii of U; but it can be reached on the whole of U if Y has an infinite dimension, even if Y is separable. c) Nevertheless, if Γ is strictly convex, i.e. y and $y' \in \overline{\Gamma}$, $y \neq y'$, imply $\frac{y+y'}{2} \in \Gamma$, then the equality can be reached on one radius of U only, which f maps onto a real line segment.

Proof. a) Let $h = f(\alpha)$, Q be the gauge of Γ, $\varphi \in Y'$ be such that $\mathrm{Re}\,\varphi \leqslant Q$, $\mathrm{Re}\,\varphi(h) = Q(h) \in]0,1[$ by the assumption. Since $\varphi\circ f$ is a holomorphic map $U \to D'$, bearing in mind the proof of Th. 5.3 we have the inequalities

$$\log\frac{2-Q(h)}{Q(h)} \geqslant \log\left|\frac{2-\varphi(h)}{\varphi(h)}\right| = g_{D'}[\varphi\circ f(\alpha), \varphi\circ f(0)] \geqslant g_U(\alpha,0),$$

which have to be equalities by the assumption; thus $\varphi(h) = Q(h)$ and (Th. 3.3.b) $\varphi\circ f$ is a 1–1 map of U onto D'.

Then $\varphi\circ f$ maps the radius of U which goes through α onto the interval $]0,1[$; if ζ lies on this radius and $h' = f(\zeta)$, since $Q(h') \geqslant \varphi(h')$, we have

$$\log\frac{2-Q(h')}{Q(h')} \leqslant \log\frac{2-\varphi(h')}{\varphi(h')} = g_{D'}[\varphi\circ f(\zeta), \varphi\circ f(0)] = g_U(\zeta,0),$$

but also $\log\frac{2-Q(h')}{Q(h')} \geqslant g_U(\zeta,0)$ by Corollary 6.6. Finally $g_{D'}[\varphi\circ f(\zeta), \varphi\circ f(0)] \geqslant e_\Gamma[f(\zeta), f(0)] \geqslant g_U(\zeta,0)$ by Th. 6.1.

b) From the proof of a) follows the formula $\varphi\circ f(\zeta) = \frac{2\zeta}{re^{i\theta}+\zeta}$ showing that $\varphi\circ f$ is uniquely determined by $e^{i\theta} = \frac{\alpha}{|\alpha|}$, and to distinct values of $e^{i\theta}$ are associated linearly independent functions $\varphi\circ f$ and hence linearly independent $\varphi \in Y'$.

If Y has a finite dimension m, the equality is reached on m distinct radii of U, with arbitrarily given arguments θ_j, in the example $Y = \mathbb{C}^m$, $\Gamma = D'^m$, $f(\zeta) = \left(\frac{2\zeta}{re^{i\theta_1}+\zeta}, \ldots, \frac{2\zeta}{re^{i\theta_m}+\zeta}\right)$.

On the contrary, let Y be the Hilbert space whose elements are all infinite sequences y of complex numbers y_n, $n \geqslant 1$, such that $\sum_{n\geqslant 1} |y_n|^2 = \|y\|^2 < +\infty$; in particular, let $f(\zeta)$ be the sequence of complex numbers $n\left(\frac{\zeta}{r}\right)^n$, $\zeta \in U$, and, for each $\theta \in \mathbb{R}$, y_θ the sequence of complex numbers $\frac{2}{n}(-1)^{n-1}e^{in\theta}$. With the notation $(\,|\,)$ for the scalar product in Y, $\Gamma = \{y \in Y : \sup_{\theta\in\mathbb{R}} \mathrm{Re}(y|y_\theta) < 1\}$ is an open convex neighborhood of the origin in Y. Since $(f(\zeta)|y_\theta) = \frac{2\zeta}{re^{i\theta}+\zeta}$, the equality $\sup_{\theta\in\mathbb{R}} \mathrm{Re}(f(\zeta)|y_\theta) = \frac{2}{1+\exp[g_U(\zeta,0)]}$ holds for any $\zeta \in U$.

c) If f maps a radius of U onto a real line segment, so does $\varphi\circ f$, which by the expression obtained above can happen for one radius of U only. So it is enough to check that a radius of U on which the equality is reached is mapped by f onto a real line segment.

In fact, with the notation used in the proof of a), φ equals 1 at both points $\frac{h}{Q(h)}$, $\frac{h'}{Q(h')}$, which lie on the boundary of Γ; were they different, then φ would also equal 1 at the point $\frac{h+h'}{Q(h)+Q(h')}$ lying in Γ, contradicting $\varphi(\Gamma) \subset D'$.

Now let ζ run from the center to the boundary of U along the radius going through α. Then $Q(h') = \varphi(h') = \varphi \circ f(\zeta)$ runs from 0 to 1, and $f(\zeta) = Q(h') \frac{h}{Q(h)}$ from the origin to $\frac{h}{Q(h)}$ along a line segment.

Remark 7.2. By a conformal mapping, the results in Th. 7.1 are easily transposed to a simply connected $U \neq \mathbf{C}$. If U is multiply connected, it may happen that $g_U < E_U$ (Ex. 4.5), and then the problem of equality should be considered after the substitution of E_U to g_U in Corollary 6.6.

Theorem 7.3. Let Ω and Γ be open convex neighborhoods of the origins in X and Y, p and q the gauges of their biggest balanced subsets, $f \in \mathcal{H}(\Omega,Y)$, $f(\Omega) \subset \Gamma$, and $f(0) = 0$; finally let $h \in X$ be such that $p(h) > 0$. Then the equality is reached in Corollary 6.7.b iff the following two conditions hold: (i) $U = U(0;h) = \{u \in \mathbf{C} : uh \in \Omega\}$ is an open disc centered at the origin, namely $\frac{1}{p(h)} D$. (ii) For the analytic map $u \mapsto f(uh)$ of U into Γ, the equality in Corollary 6.6 is reached at some points in $U \setminus \{0\}$ (for the location of which see Th. 7.1).

Proof. Let Q be the gauge of Γ and let $q(y) = \sup_{\theta \in \mathbf{R}} Q(e^{i\theta}y)$, $y \in Y$. If (i) and (ii) hold, then, by Th. 7.1, for some value of arg u we have $q \circ f(uh) \geq Q \circ f(uh) = \frac{2|u|p(h)}{1+|u|p(h)}$ and hence the equality in Corollary 6.7.b.

Now assume that this equality holds. By the proof of Corollary 6.7.b, for any $\epsilon > 0$ there exists $\alpha > 0$ such that $|u|$ and $|u'| \leq \alpha$ imply $Q[e^{i\theta} \frac{f(uh)}{u} - e^{i\theta} \frac{f(u'h)}{u'}] \leq \epsilon$, a fortiori $|Q[e^{i\theta} \frac{f(uh)}{u}] - Q[e^{i\theta} \frac{f(u'h)}{u'}]| \leq \epsilon$, for all real θ. Then $Q[e^{i\theta} \frac{f(uh)}{u}]$ has a uniform limit as $u \to 0$, and $\sup_{\theta \in \mathbf{R}} \lim_{u \to 0} Q[e^{i\theta} \frac{f(uh)}{u}] = \lim_{u \to 0} q[\frac{f(uh)}{u}]$.

Using the Hahn-Banach theorem in the completed space $\hat{Y}$, we get $\varphi \in Y'$ such that Re $\varphi \leq Q$ and, for a suitable θ, $\mathrm{Re} \lim_{u \to 0} \varphi[e^{i\theta} \frac{f(uh)}{u}] = \lim_{u \to 0} Q[e^{i\theta} \frac{f(uh)}{u}] = \lim_{u \to 0} q[\frac{f(uh)}{u}] = 2\,p(h)$ by the assumption. Then (Th. 1.6) $u \mapsto \psi(u) = \frac{\varphi \circ f(uh)}{2 - \varphi \circ f(uh)}$ is a holomorphic map $U \to D$; in particular, it is a map $\frac{1}{p(h)} D \to D$, with $\psi(0) = 0$, $\psi'(0) = \frac{1}{2} \lim_{u \to 0} \varphi[\frac{f(uh)}{u}]$, $|\psi'(0)| \geq p(h)$. By the classical Schwarz lemma, $\psi'(0) = e^{-i\theta} p(h)$, $\psi(u) = e^{-i\theta} p(h) u$, $U = \frac{1}{p(h)} D$ [i.e. (i)]; finally, $\varphi \circ f(uh) = \frac{2u}{\frac{e^{i\theta}}{p(h)} + u}$, which implies $\frac{1 + \exp[g_U(u,0)]}{2} f(uh)$ on the boundary of Γ for $e^{-i\theta} u \in \left]0, \frac{1}{p(h)}\right[$ [i.e. (ii)], since $Q \geq \mathrm{Re}\, \varphi$.

Theorem 7.4. Let Ω and Γ be convex and balanced, p and q their gauges, $f \in \mathcal{H}(\Omega,Y)$, $f(\Omega) \subset \Gamma$, and $f(0) = 0$; finally let $x \in \Omega$ be such that $p(x) > 0$. a) If either equality in Corollary 6.8, i.e., $q \circ f(x) = p(x)$, $\lim_{u \to 0} q[\frac{f(ux)}{u}] = p(x)$, is reached at the point x, then both equalities are reached at all points $ux \in \Omega$

[i.e. $|u| < \frac{1}{p(x)}$]. **b) If Γ is strictly convex, either equality implies that $f(ux) = uf(x)$ for $|u| < \frac{1}{p(x)}$.**

Proof. a) If $q\circ f(x) = p(x)$, we choose $\varphi \in Y'$ such that $|\varphi| \leqslant q$ and $\varphi\circ f(x) = q\circ f(x) = p(x)$. Then (Th. 1.6) $u \mapsto \psi(u) = \varphi\circ f(ux)$ is a holomorphic map $\frac{1}{p(x)} D \to D$, with $\psi(0) = 0$, $\psi(1) = p(x)$. By the classical Schwarz lemma, $\psi(u) = p(x)u$ and $\lim_{u\to 0} q[\frac{f(ux)}{u}] \geqslant \lim_{u\to 0} |\frac{\varphi\circ f(ux)}{u}| = p(x)$.

If $\lim_{u\to 0} q[\frac{f(ux)}{u}] = p(x)$, we choose $\varphi \in Y'$ such that $|\varphi| \leqslant q$, $\lim_{u\to 0} \varphi[\frac{f(ux)}{u}] = \lim_{u\to 0} q[\frac{f(ux)}{u}] = p(x)$, and a similar argument yields $q\circ f(x) \geqslant p(x)$. The second equality is unaltered if x is changed into tx, $|t| < \frac{1}{p(x)}$.

b) In the first part of the proof of a), φ equals 1 at all points $\frac{f(x)}{p(x)}$ and $\frac{f(ux)}{up(x)}$, $0 < |u| < \frac{1}{p(x)}$, which lie on the boundary of Γ; were these points different, then φ would also equal 1 at the middle point lying in Γ, contradicting $\varphi(\Gamma) \subset D$.

Remark 7.5. A balanced strictly convex open set in a topological vector space X, either is X itself, or its gauge p vanishes only at the origin.

Proof. Let $p(a) = 1$, $p(b) = 0$, $b \neq 0$. Then all points a, a+b, a−b, lie on the boundary of the set.

Corollary 7.6. Let Ω be strictly convex and balanced in X, $\Omega \neq X$, $f \in \mathcal{G}(\Omega,X)$, $f(\Omega) \subset \Omega$, and $f(0) = 0$. Then $E = \{x \in \Omega : f(x) = x\}$ is the intersection of Ω with a linear subspace of X.

Proof. Let p be the gauge of Ω. If $x \in E\setminus\{0\}$, then $p(x) > 0$ by Remark 7.5, and $p\circ f(x) = p(x)$ entails $f(ux) = ux$ for $|u| < \frac{1}{p(x)}$ by Th. 7.4.b.

Now let x_1 and $x_2 \in E\setminus\{0\}$, $x_1 + x_2 \in \Omega\setminus\{0\}$, $\varphi \in X'$ be such that $|\varphi| \leqslant p$, $\varphi(x_1+x_2) = p(x_1+x_2) > 0$. By the classical Hartogs theorem, $(u_1,u_2) \to \varphi\circ f(u_1x_1+u_2x_2)$ is a holomorphic function on some neighborhood of the origin in C^2, with first order derivatives $\varphi(x_1)$, $\varphi(x_2)$ at the origin. Then $\varphi[\frac{f(u(x_1+x_2))}{u}] \to \varphi(x_1+x_2)$ as $u \to 0$, and $\lim_{u\to 0} p[\frac{f(u(x_1+x_2))}{u}] \geqslant p(x_1+x_2)$; hence, by Th. 7.4.b, $f(u(x_1+x_2)) = u\, f(x_1+x_2)$ for $|u| < \frac{1}{p(x_1+x_2)}$.

Then, for any $\varphi \in X'$, we have $\varphi\circ f(x_1+x_2) = \lim_{u\to 0} \varphi[\frac{f(u(x_1+x_2))}{u}] = \varphi(x_1+x_2)$, implying that $y = f(x_1+x_2) - (x_1+x_2)$ vanishes, since otherwise $p(y) > 0$.

Remark 7.7. The assumption $\Omega \neq X$ is obviously necessary. The counterexample $D^2 \ni (x_1,x_2) \to (x_1, \frac{x_1^2+x_2}{2}) \in D^2$ shows that Corollary 7.6 is false, therefore Th. 7.4.b is false too, without the strict convexity.

Corollary 7.8. Let Ω be convex and balanced in X, with a gauge p vanishing only at the origin. Let Γ be strictly convex and balanced in the sequentially complete space Y, and let $f \in \mathcal{G}(\Omega, Y)$, $f(\Omega) \subset \Gamma$, and $f(0) = 0$. If the equalities in Corollary 6.8 are reached at all points on an open nonempty subset of Ω, especially if f is a 1-1 map of Ω onto Γ and $f^{-1} \in \mathcal{G}(\Gamma, X)$, then f coincides with the linear map $x \to \ell(x) = \lim\limits_{u\to 0} \frac{f(ux)}{u}$.

Proof. By our assumptions, the gauge q of Γ cannot vanish identically, i.e. $\Gamma \neq Y$, and q vanishes only at the origin by Remark 7.5.

We first prove that ℓ is linear: for any $\varphi \in Y'$, $(u_1,u_2) \to \varphi\circ f(u_1x_1+u_2x_2)$ is a holomorphic function on some neighborhood of the origin in C^2, with first order derivatives $\varphi\circ\ell(x_1)$, $\varphi\circ\ell(x_2)$ at the origin, therefore $\varphi\circ\ell(x_1+x_2) = \lim\limits_{u\to 0} \varphi[\frac{f(u(x_1+x_2))}{u}] = \varphi[\ell(x_1) + \ell(x_2)]$, implying that $y = \ell(x_1+x_2)-[\ell(x_1)+\ell(x_2)]$ vanishes, since otherwise $q(y) > 0$.

Now, for any $\varphi \in Y'$, $\varphi\circ(f-\ell) \in \mathcal{G}(\Omega, C)$ (Th. 1.6), $\varphi\circ(f-\ell) \equiv 0$ (th. 7.4.b), again implying that $y = f(x)-\ell(x)$ vanishes for any $x \in \Omega$, since otherwise $q(y) > 0$.

Remark 7.9. In the special case when f is a 1-1 map of Ω onto Γ and $f^{-1} \in \mathcal{G}(\Gamma, X)$: for any $x \in \Omega$ and $y = f(x)$, since $q(y) = p(x)$, f is a 1-1 map of $\{ux \in \Omega\}$ onto $\{vy \in \Gamma\}$; therefore, for any $y \in Y$ and $0 < |v| < \frac{1}{q(y)}$, $\frac{f^{-1}(vy)}{v}$ depends only on y, and f^{-1} coincides with the linear map $y \to \frac{f^{-1}(vy)}{v}$, $0 < |v| < \frac{1}{q(y)}$, which is inverse to ℓ.

Corollary 7.10. Let Ω_1, Ω_2 be convex and balanced in X_1, X_2, with gauges p_1, p_2 vanishing only at the origins, and let Γ be strictly convex and balanced in the locally convex and sequentially complete space Y. Then a 1-1 map f of $\Omega = \Omega_1 \times \Omega_2$ onto Γ such that $f \in \mathcal{A}(\Omega, Y)$, $f^{-1} \in \mathcal{A}(\Gamma, X_1 \times X_2)$ and $f(0) = 0$, cannot exist.

Proof. The gauge p of Ω, given by $p(x_1,x_2) = \sup[p_1(x_1), p_2(x_2)]$, vanishes only at the origin in $X = X_1 \times X_2$. Assume that f exists. By Corollary 7.8 and Remark 7.9, $\Gamma \neq Y$, and f and f^{-1} coincide with linear maps ℓ and ℓ^{-1}, respectively.

Now choose $a \in \Omega_1 \setminus \{0\}$ and $\varphi \in X_1'$, $\varphi \not\equiv 0$, $|\varphi| \leq p_1$. $g(x_1,x_2) = (\varphi^2(x_1)a, x_2)$ defines $g \in \mathcal{A}(X,X)$, $g(\Omega) \subset \Omega$, $g(0) = 0$, and $p\circ g(x_1,x_2) = \sup[|\varphi(x_1)|^2 p_1(a), p_2(x_2)] = p(x_1,x_2)$ if $p_2(x_2) > p_1(x_1)$, i.e. on an open nonempty set $A \subset \Omega$.

Now $f\circ g\circ f^{-1}$ is a map $\Gamma \to \Gamma$, with the property that $\varphi \in \mathcal{A}(\Gamma, C)$ implies $\varphi\circ f\circ g\circ f^{-1} \in \mathcal{A}(\Gamma, C)$; hence (by 1.3) $f\circ g\circ f^{-1} \in \mathcal{A}(\Gamma, Y)$, and $f\circ g\circ f^{-1}(0) = 0$. On the other hand, let q be the gauge of Γ. Since $q\circ f(x) = p(x)$ for any $x \in \Omega$, $q\circ f\circ g\circ f^{-1}(y) = q(y)$ iff $p\circ g\circ f^{-1}(y) = p\circ f^{-1}(y)$, in particular if $y \in f(A)$, an open nonempty subset of Γ. Then (Corollary 7.8) $f\circ g\circ f^{-1}$ coincides with a linear map, and so does g, an obvious contradiction.

Bibliography

1. M. Brelot, **Eléments de la théorie classique du potentiel** (Centre de documentation universitaire, Paris, 1965).

2. M. Brelot, **Lectures on potential theory** (Tata Institute of fundamental research, Bombay, 1960).

3. M. Hervé, **Fonctions périodiques d'une on plusieurs variables** (Institut Henri Poincaré, Paris, 1967).

4. M. Hervé, **Analytic and plurisubharmonic functions** (Lecture notes n° 198, 1971).

5. A. Hirschowitz, **Sur un théorème de ZORN** (Prépublication de la Faculté des Sciences, Nice, 1970).

6. H. Huber, Comm. math. Helvet. 27, 1953, p.1.

7. S. Kobayashi, **Hyperbolic manifolds and holomorphic mappings** (New York, 1970).

8. M. Kwack, Annals of Math., 90, 1969, p.9.

9. P. Lelong, **Séminaire d'analyse** (Lecture notes, no 71, exposé 17; no 116, exposé 1).

10. P. Montel, **Legons sur les familles normales de fonctions analytiques et leurs applications** (Paris, 1927).

11. K. Noshiro, **Cluster sets** (Ergebnisse der Math., no 28, 1960).

12. P. Noverraz, Annales Inst. Fourier, 19, 1969/2, p. 419.

13. M. Zorn, Annals of Math., 46, 1945, p.585.

directeur adjoint de l'Ecole
Normale Superieure
et professeur a l'Universite
de Paris VI
45 rue d'Ulm
75230 Paris 05, France

PLURISUBHARMONIC FUNCTIONS IN TOPOLOGICAL VECTOR SPACES: POLAR SETS AND PROBLEMS OF MEASURE

Pierre Lelong

§1. **Introduction.** In complex analysis we meet classes of "small" sets; for instance, polar sets and negligible sets (see below for the definition). In the finite dimensional case, i.e., in C or in C^n, such sets are null sets in the classical sense; that is, they are sets of measure zero in R^2 or R^{2n} for the invariant Lebesgue measure of the space. This property is often precise enough for classical problems (for instance to prove the Hartogs lemma). Only for specific problems do we need more precise properties of those sets (for example, to be also null sets on the real subspace R^n of C^n). In infinite dimensional complex analysis the situation seems to be different. There exists no invariant measure which remains locally bounded, and therefore we have to work with the classes of polar sets and negligible sets.

In this lecture I first recall properties of the polar sets with some improvements of the results I have given provioulsy in [5b] and [5d].

In the second part, I give a generalization to a Fréchet space E of a previous result concerning the family of continuous polynomial mappings E → F which are bounded at each point of a set A ⊂ E which is not polar. In my previous paper [5e] the proof was given for locally bounded spaces. A result (which can be seen as a generalization of the classical Banach-Steinhaus theorem) is given here to some classes of spaces which are projective and surjective limits of Banach spaces E_α.

The last part gives the definition and some properties of the number ν_x, which is the so-called Lelong number (see [5f] and [5g]) of a plurisubharmonic function V. Recall that ν_x was defined in the finite dimensional case using properties of the positive Laplacian measure ΔV; if E is a topological vector space, ν_x can be defined using properties which are true except for a polar set of complex linear subspaces. In the following, the vector spaces are Hausdorf and quasi-complete.

§2. **Plurisubharmonic functions and polar sets.** In this paragraph E is a linear vector space on the complex field C. We do not suppose that E is a topological vector space and we consider topologies T on E such that the space (E,T) has the following properties and we call (E,T) a semi-topological vector space:

1. Addition (x,y) → x+y is a mapping E x E → E which is continuous if x (or y) remains in a finite dimensional subspace (i.e., addition is nct continuous).

2. Multiplication (λ,x) → λx is a mapping C x E → E which is continuous.

We denote by τ the family of topologies T on E with the preceeding properties. In the following we consider only topologies T ∈ τ which are Hausdorf.

Each T ∈ τ is invariant by translation; T induces on each finite dimensional space M^n (n = dim M) the topology of C^n; T ∈ τ is weaker than the finite topology T_f. (In T_f the open sets are all the sets whose intersection which each M^n, for all n, are open sets for the C^n topology). We have T_f ∈ τ and T_f is the upper bound of the family τ (see [3], and [4b], [5h]).

If T is a τ-topology, the filter $\mathfrak{F}_0$ of the neighborhoods of the origin 0 has a basis of balanced neighborhoods $\{V_\alpha\}$; V_α is called balanced if for $x \in V_\alpha$ we have $\lambda x \in V_\alpha$ for each $\lambda \in C$, $|\lambda| \leqslant 1$. A disc $\Delta_{x,y} \subset E$ is the linear image in E of the disc D: $|u| \leqslant 1$ in C given by

$$\Delta_{x,y} = [x' = x + uy, \text{ for } |u| \leqslant 1] = [x + Dy]$$

$x \in E$, $y \in E-\{0\}$.

The star S(x,G) of $x \in G$, where G is a set in (E,T), is by definition the union of the discs $\Delta_{x,y}$ with center x which are contained in G.

The following statement (see [5,h]) is an improvement of a lemma given in [1,b].

Proposition 1. **a/ If G is an open set in (E,T) and if T is a τ-topology, then the star S(x,G) is T-open.**
b/ Moreover, if G is a domain (open and connected set) in (E,T) and if A is a subset of G such that for each $x \in A$ the star S(x,G) belongs to A, then A is empty or A is G.

Proof. Obviously the space (E,T) has the property studied by C.O. Kiselman [4a]: the mapping $C \times E \to E$ defined by $x' = x + uy$, $(u \times x) \to x'$ is continuous for each given y. Then if x_0 belongs to S(x,G), S(x,G) contains a compact disc $\Delta = \Delta_{x,y_0} = x + Dy_0$ containing x_0. By the property 1, there exists a balanced neighborhood W of the origin such that $W_1 = x + W + Dy_0$ is a neighborhood of Δ and is contained in G. Writing $W = \bigcup_{|v|\leqslant 1} vW = D \cdot W$, we get $W_1 = x + D(y_0 + W) \subset G$. Then W_1 is a union of discs of center x contained in G and the inclusions $x_0 + W \subset x + Dy + W = W_1 \subset S(x,G)$ make the proof of a/ complete.

We consider x_1 and x_2 in G with $x_1 \neq x_2$ and we define

$$S'(x_1,G) = \cup\Delta_{x_1,y} \text{ for } \Delta_{x_1,2y} \subset S(x_1,G).$$

Then for $x_2 \in S'(x_1,G)$ we obtain $x_1 \in S(x_2,G)$, which is an elementary consequence obtained by considering the intersection of the complex line determined by x_1, x_2 with G.

For the proof of b/, first note that A is T-open by the hypothesis and a/, and if $x' \in \bar{A} \cap G$, there exists $x \in A$ with $x \in S'(x',G)$. Then $x' \in S(x,G)$ by the preceeding remark; consequently, we obtain $x' \in A$ and A is closed in G, which completes the proof.

Plurisubharmonic functions. We recall the definition: in an open set G of (E,T), f(x) is called plurisubharmonic if f is upper semi-continuous and if for each disc $\Delta_{x,y} \subset G$ the "mean value inequality" holds, i.e.,

$$f(x) \leqslant \frac{1}{2\pi}\int_0^{2\pi} f(x+ye^{i\theta})d\theta.$$

If G is a domain we denote by P(G) the class of all plurisubharmonic functions in G different from the constant function $-\infty$. We say that $f \in P(G)$ is continuous if e^f is continuous.

Polar sets. A set A on a domain G of (E,T) is called **polar** if there exists $f \in P(G)$ such that A is contained in $A' = [x \in G; f(x) = -\infty]$. If we suppose in addition that $f \leq 0$, then A is called **strictly polar**. A conic set S is called polar if it is polar in some neighborhood of its vertex.

Proposition 2. If A is a polar set in a domain G of (E,T), its interior $\mathring{A}$ is empty.

Proof: $\mathring{A}$ is open and $x \in \mathring{A}$ gives $S(x,G) \subset A$. By Proposition 1a), S(x,G) is T-open and we obtain $S(x,G) \subset \mathring{A}$. Applying Proposition 1b), we obtain $\mathring{A} = \phi$.

The following proposition, though obvious, is important.

Proposition 3. a/ Let us consider a linear subspace M of E, a domain G in (E,T) and a topology T' on M which is finer than the topology induced on M by T. If $G_1 \subset G \cap M$ is a domain in (M,T') then the restriction to $G_1 \subset (M,T')$ of a function f plurisubharmonic in $G \subset (E,T)$ is plurisubharmonic or is the constant $-\infty$.

b/ With the same assumptions on G and G_1, if A is a polar set in $G \subset (E,T)$ then $A_1 = A \cap G_1$ is G_1 or is a polar set in G_1.

For the applications we recall the following result [5b], [5d]:

Proposition 4. Given two complex semi-topological vector spaces (E,T) and (F,T') with (F,T') quasi-complete, we consider a mapping φ: $E \to F$ which maps the domain $G \subset (E,T)$ into the domain $G' \subset (F,T')$, and we suppose that φ is continuous for the given T and T' and that $\varphi(x + uy) = \varphi_{x,y}(u)$ is analytic as a function of $u \in \mathbb{C}$ for $|u| \leq 1$ on each compact disc $\Delta_{x,y} \subset G$. Then given a function f plurisubharmonic in G', $f_1 = f \circ \varphi$ is plurisubharmonic in G or is the constant $-\infty$.

One can obtain properties of null measure for the polar sets on a special class of compact sets by using Proposition 4 and the Jessen (or cylindrical) measure and following a construction given by G. Coeuré [2].

To do this, we suppose that E is a topological vector space which is quasi-complete with locally convex topology. We consider a sequence $\{a_n\}$ in E, which is a summable family in E, and define a mapping $\ell^\infty(N) \to E$ by

$$x(y) = x_o + \Sigma y_k a_k \,, \quad \|y\| = \sup |y_k| < \infty.$$

If we give $\ell^\infty(N)$ the topology of the convergence of the y_k, then x(y) is continuous on the bounded set of $\ell^\infty(N)$ and the image of the ball $\|y\| \leq 1$ is a compact polycylinder $P(x_o,a_k)$ with center x_o in E. Let us consider (see [2], Theorem 5.3) a domain G in E containing $P(x_o,a_k)$, and denote by $\mathring{P}(x_o,a_k)$ the set of the points $x \in P(x_o,a_k)$ such that there exists a $y \in u^{-1}(x)$ with $\|y\| < 1$. Then if f is plurisubharmonic in G, we have only two possibilities for its restriction on the affine Banach space $M = x_o + E(a_k)$ of the x such that $x - x_o = \Sigma y_k a_k$, $\|y\| < \infty$:

a/ f_M is the constant $-\infty$ on the connected component of $E(a_k) \cap G$ which contains x_o,

b/ f_M is plurisubharmonic and integrable for the Jessen measure $\prod_1^\infty d\theta_k$ on the set $x = x_o + \Sigma a_k e^{i\theta_k}$.

As a consequence we obtain

Proposition 5. A polar set in a domain G of a quasi-complete topological vector space E is of measure zero on the distinguished boundary of a polycylinder $P(x_o,a_k)$ in E or it contains the connected component of $M = x_o + E(a_k)$ in G which contains x_o.

Proposition 5 is a generalization to infinite dimensional spaces of the classical conditions for plurisubharmonic functions to be integrable on the distinguished boundary of a polycylinder. As an application to a classical problem we give

Proposition 6. Given a domain Ω in C^n and a sequence f_q of functions holomorphic in Ω such that $\sum_1^\infty |f_k|$ converges uniformly on the compact sets in Ω, then the functions

$$f_\theta = \sum_1^\infty e^{i\theta_q} f_q, \quad 0 \leqslant \theta_q \leqslant 2\pi$$

that have a domain of holomorphy Ω_f such that $\Omega_f \neq \Omega$ are given by a set A of measure zero for the Jessen measure $\prod_1^\infty \frac{d\theta_q}{2\pi}$ or are all the f_θ.

Proof: Let us consider the Fréchet algebra $A(\Omega)$ of the holomorphic functions in Ω with compact convergence. We have proved in [5,d] that if M is a Banach subspace of $A(\Omega)$ and has a topology finer than the topology induced by $A(\Omega)$ on M, then the set η of the functions $f \in M$ with $\Omega_f \neq \Omega$ is all of M or is a polar set in M. Let us consider for M the Banach space of the $f \in A(\Omega)$ given by $f = \Sigma y_k f_k$, $\|y\| = \sup |y_k|$. Define $\|f\| = \|y\|$. Then M has a topology finer than $A(\Omega)$ and we finish the proof using Proposition 5.

§3. Polar sets and negligible sets. A general result which holds for the τ-topologies in a semi-topological space (E,T) (see [2], [5,h])is the following:

a/ Given a family $f_\alpha(x)$ of plurisubharmonic functions in a domain G of (E,T) which is locally bounded above in G, then

$$g = \sup f_i$$

has an upper regularizing $g^* = \limsup_{y \to x} g(y)$ which is plurisubharmonic.

b/ The same result holds for

$$h = \limsup f_i$$

in $i \in \mathcal{I}$ and $\mathcal{I}$ is a directed set (see [5,h]).

Definition. A set A in a domain G of (E,T) is called negligible if there exists a sequence $\{f_n\} \subset P(G)$ locally bounded above such that $g = \sup f_n$ and

(1) $A \subset A' = [x \in G; g(x) < g^*(x)]$.

A polar set in G is negligible in G. Very little is known about this class of sets in the infinite dimensional case. For results in C^n see [5,a]. However the following result makes it possible to get polar sets in (1) (for the proof see [5,d]).

Proposition 7. We suppose (E,T) quasi-complete. If in (1) we obtain for g^* a constant or a pluriharmonic function, then A (and A') are strictly polar sets or we have $A' = G$. Moreover if (E,T) is a Baire space and the f_n are continuous, then A and A' are strictly polar.

Proposition 8. Given a sequence $\{f_n\}$ of continuous plurisubharmonic functions in a domain G of (E,T) with E quasi-complete, we suppose $f_n \leq 0$ on G and $g(x_o) = \lim\sup f_n(x_o) = 0$ at a point $x_o \in G$. Then $\eta = [x \in G; g(x) = \lim\sup f_n(x) < 0]$ is strictly polar and is a countable union of closed sets with empty interior.

Proposition 9. Given a space (E,T), a sequence A_q of strictly polar sets in a domain G and $A_q \subset A'_q = [x \in G; f_q(x) = -\infty]$, then $A = \cup A'_q$ is strictly polar in G or $G = \cup A'_q$. If (E,T) is a Baire space and the f_q are continuous, then A is strictly polar.

For the proof see [5,d].

§4. **A "Banach-Steinhaus" theorem for polynomial mappings.** The polar sets allow us to give a more precise theorem about equicontinuity for a family $P_\alpha(x)$ of continuous polynomials mapping $E \to F$.

A plurisubharmonic function q(y) defined in a topological vector space will be called a pseudo-norm (in [5,e] such functions were called quasi-norms) if it has the property that $q(ux) = |u|q(x)$ for each $u \in C$ and $x \in F$. Then $\log q(x)$ is plurisubharmonic [5,b]. If the filter $\mathfrak{F}_o$ of the neighborhoods of the origin in F has a basis $\{V_\alpha\}$ such that $V_\alpha = [x \in F; q_\alpha(x) < 1]$, where q_α is a continuous pseudo-norm, F will be said to be endowed with a pseudo-convex topology and F will be called a pseudo-convex space. Locally convex topologies on complex spaces given by semi-norms $\{p_\alpha\}$ are pseudo-convex topologies because continuous convex functions are plurisubharmonic. In [5,e, Théorème 4] the following result was proved.

If E is a complex Banach space and $q_i(x)$ is a family of continuous pseudo-norms on E, we have the following alternatives for the set A defined on E by

$$\pi(x) = \sup_i q_i(x) < \infty.$$

a/ A is a conic set with vertex at the origin and is meager and strictly polar,

b/ Or $A = E$, $\pi(x)$ is bounded in a neighborhood U of the origin and we have $\pi(x) \leq C\, p_U(x)$ where p_U is the gauge of U.

A consequence of this result is, cf. [5,e, see p.41]:

Proposition 10. Given a complex Banach space E and a pseudo-convex topological vector space F, we consider a family $P_i(x)$ of continuous, homogeneous polynomial mappings, n_i = degree P_i, and we suppose that there exists for each $x \in A$, $A \subset E$, a bounded set $B_x \subset F$, and $\sigma_x > 0$ such that

(2) $$P_i(x) \subset \sigma_x^{n_i} B_x \quad \text{for } x \in A.$$

Then we have the following alternatives:

a/ A is meager, polar, and contained in $[x \in E, S(x) = -\infty]$, $S \in P(E)$.

b/ Or for each pseudo-norm q_W continuous on F, there exists a neighborhood U of the origin in E with gauge p_U, such that

$$\log q_W \circ P_i(x) \leqslant \log p_U(x).$$

Plurisubharmonic functions on $E = \varprojlim E_i$. We consider a directed set of indices $\mathcal{I}$, a family of topological vector spaces E_i, $i \in \mathcal{I}$ and a given set of continuous linear mappings π_j^i from E_i onto E_j given for each $i,j \in \mathcal{I}$, $i \geqslant j$, and such that $\pi_k^j \circ \pi_j^i = \pi_k^i$, for all $i, j, k \in \mathcal{I}$, $i \geqslant j \geqslant k$. A space E is called the projective limit

$$E = \varprojlim E_i$$

if there exist continuous linear mappings π_i: $E \to E_i$ such the following conditions hold: each π_i is onto; for each $i \geqslant j \in \mathcal{I}$, $\pi_j^i \circ \pi_i = \pi_j$; the topology on E is the weakest topology for which the π_i are all continuous.

Example: If E is a Fréchet space, we have

$$E = \varprojlim E_n, \; E_n = E/p_n^{-1}(0)$$

for a fundamental and increasing sequence p_n of continuous semi-norms. If f(x) is plurisubharmonic at a point $x_o \in E$, there exists a neighborhood of the origin $p_n(x) < c$ such that $f(x_o + h) - f(x_o) < 1$; then $f(x_o + uh)$, $u \in C$ is a bounded subharmonic function for $u \in C$; consequently $f(x_o + uh) = f(x_o)$ for all $u \in C$ and $h \in p_n^{-1}(0)$. Therefore $f(x_o + N_n) = f(x_o)$ if $N_n = p_n^{-1}(0)$. The subspace N_n depends generally on x_o. It does not depend on x_o if we suppose that f(x) is upper semicontinuous relative to the same p_n in all the points of E.

Given a polynomial mapping P(x): $E \to F$, where F is a quasi-complete pseudo-convex topological vector space, we consider a continuous pseudo-norm $q_W(x)$ on F. The function

$$q_W \circ P(x)$$

is plurisubharmonic (Proposition 4) and the same is true for $L_W(x) = [q_W \circ P(x)]^{1/n}$ where n = deg P. Also, $L_W(x)$ is a continuous pseudo-norm on E; for, given $W = [y \in F, q_W(y) < 1]$ in F, $L_W(x)$ is continuous for the same set $\{p_\lambda\}$ of semi-norms at all the points of E. Then $L_W(x)$ takes the same value at x and x + y for $y \in N_\lambda$ if P is p_λ-continuous. We obtain:

Proposition 11. Let us consider a family $P_\alpha(x)$ of continuous, homogeneous polynomial mappings $E \to F$. We suppose that $E = \varprojlim E_n$ is a Fréchet space it is both a projective and surjective limit of the Banach spaces $E_n = E/p_n^{-1}(0))$ and that F is quasi-complete with topology given by the pseudo-norms q_W. Then the $L_{W,\alpha}(x) = [q_W \circ P(x)]^{1/n_\alpha}$ are continuous pseudo-norms on E. Moreover, for each $L_{W,\alpha}$, there exists an $m \in N$ such that $L_{W,\alpha}(x + N_n) = L_{W,\alpha}(x)$ for all $n \geq m$, and $N_n = p_n^{-1}(0)$; for then, there exist continuous pseudo-norms $\tilde{L}_{W,\alpha}$ on E_n such that

$$(3) \qquad L_{W,\alpha}(x) = \tilde{L}_{W,\alpha} \circ \pi_n(x) \text{ for } n \geq m.$$

Then we obtain:

Proposition 12. If $E = \varprojlim E_n$ is a Fréchet space, if the P_α satisfy (2) for $x \in A$, and if A is not meager, then (3) holds and there exists a Banach space E_m in the sequence $\{E_n\}$ and $\tilde{P}_\alpha$: $E_m \to F$ such that for each α

$$P_\alpha(x) = \tilde{P}_\alpha \circ \pi_m(x),$$

where $\tilde{P}_\alpha$ is a family of continuous homogeneous mappings $E_m \to F$. For each q_W on F there exists a constant $C_W > 0$ such that

$$(4) \qquad q_W \circ \tilde{P}_\alpha(y) \leq C_W \|y\|_m$$

$y \in E_m$ with norm $\| \ \|_m$.

The following problem is open. Consider the Fréchet space $E = \varprojlim E_n$, $E_n = E/N_n$, $N_n = p_n^{-1}(0)$. Does there exist a sequence $P_\alpha(x)$ of homogeneous continuous mappings $E \to F$ such that

a/ condition (2) holds for $x \in A$ where A is a meager conic set which is not a polar set in E.

b/ there exists no $n \in N$ such that all the P_α are continuous for the semi-norm p_n on the Fréchet space E?

§5. **The number $\nu(a)$ for plurisubharmonic functions.**

In this last section we use the given properties of the polar sets to define the number $\nu(a)$ for a plurisubharmonic function f. In C^n, $\nu(a)$ is defined more generally for a closed positive current, t; $\nu(a)$, which is sometimes called the Lelong number of t at a, is a positive measure of value $\nu(a)$ of support a [5,f and g]. We consider a positive and closed current t of type (p,p) in C^n, and the fundamental form $\beta = \frac{i}{2} \Sigma dz_k \wedge d\bar{z}_k$, $\beta_p = \frac{\beta^p}{p!}$. By the "multiplication theorem" for positive forms and currents, we obtain two positive measures

$$\sigma = t \wedge \beta_{n-p}, \quad \nu_a = t \wedge \alpha_a^{n-p}$$

where $\alpha_a = i\pi^{-1} d_z d_{\bar{z}} \log \|z-a\|$ is a positive form in $C^n - \{a\}$; ν_a is a positive measure defined in $C^n - \{a\}$ which can be extended to all of C^n by addition of a positive Radon measure in a of value

$\nu(a) = \lim_{r=o} [\tau_{2p}(1)]^{-1} r^{-2p} \sigma(r)$, where $\sigma(r)$ is the measure σ carried by $\|z\| \leq r$. For this result see [5,g].
Of course, $\nu(a) = 0$ if $a \notin \operatorname{supp} t$.

Now we consider $p = 1$ and a ball B: $\|z\| < R$ in C^n; given a positive closed current in B there exists f plurisubharmonic in B such that

$$t = 2i d_z d_{\bar z} f,$$

and the number $\nu(0)$ at the origin can be computed directly from the mean values of f on the spheres $\|z\| = r$ in C^n

$$\lambda(f, o, r) = \omega_{2n-1}^{-1} \int f(r\alpha) d\omega_{2n-1}(\alpha), \quad \|\alpha\| = 1.$$

This method was used before general positive currents were defined: $\lambda(f, 0, r)$ is a convex increasing function of $\log r$, $0 \leq r < R$, and we obtain if $f(0) = -\infty$,

(4)
$$\nu(0) = \lim_{r \to o} \frac{\partial \lambda(f,0,r)}{\partial \log r} = \lim_{r \to o} \frac{\lambda(f,o,r)}{\log r}.$$

(see [5,f].)

For $p = 1$, $\sigma = 2i d_z d_{\bar z} f \wedge \beta_{n-1} = \Delta f \cdot \beta_n$, where Δf is the Laplacian of f. (4) is then an elementary consequence of the Gauss theorem.

Now we consider the same problem in a topological vector space E. We suppose that f is plurisubharmonic in a balanced neighborhood W of the origin, and for $z \in E$ and $r > 0$ sufficiently small, we define

(5)
$$\ell_r(z) = \frac{1}{2\pi} \int_o^{2\pi} f(z\, re^{i\theta}) d\theta = I(z).$$

Proposition 13. $\ell_r(z)$ **is a plurisubharmonic function of** $z \in W$ **for** $r < 1$**, and has the property**

(6)
$$\ell_{tr}(z) = \ell_r(tz) \qquad t > 0.$$

Proof. We prove that $I(z)$ is an upper semi-continuous function of z; for given $\epsilon > 0$ there exists a finite covering of the circle $u = e^{i\theta}$, $0 \leq \theta \leq 2\pi$, by compact intervals A_q of length $\Delta\theta_q$ such that

$$I(z_o) \geq \Sigma \Delta\theta_q [\sup_{\theta \in A_q} f(z_o re^{i\theta})] - \epsilon.$$

But by the continuity of $(z,\theta) \to z\, re^{i\theta}$, the function $g_q(z) = \sup_{\theta \in A_q} f(z\, re^{i\theta})$ is upper semi-continuous in z, and we obtain for h in a neighborhood U of the origin,

$$I(z_o + h) \leq \Sigma \Delta\theta_q g_q(z_o + h) \leq \Sigma \Delta\theta_q g_q(z_o) + \epsilon \leq I(z_o) + 2\epsilon.$$

Hence (6) is an obvious consequence of the definition of ℓ_r.

Proposition 14. If f is plurisubharmonic in a balanced neighborhood W of the origin in a topological space E with E quasi-complete and $f(0) = -\infty$, then

a/ the following limits exist for $r > 0$, $r \to 0$:

$$\lim_{r\to 0} \frac{\ell_r(z)}{-\log r} = \lim_{r\to 0} -\frac{d\ell_r(z)}{d\log r} = \psi(z) = -\nu(0, z) \tag{7}$$

b/ $\psi(z)$ is defined in E and has an upper regularization $\psi^*(z)$ which is a constant $-\nu(0)$.

c/ For all $0 < r < 1$ and $z \in W$,

$$\ell_r(z) \leqslant \nu(0) \log r + \ell_1(z).$$

d/ The set of the $z \in E$ such that $\psi(z) > \nu(0)$ is either a polar conic set with vertex 0 or it is the entire space E.

Proof. If $f(uz)$ is not the constant $-\infty$ for $u \in \mathbf{C}$, we write for $z \in W$ and $0 < r < 1$,

$$U_r(z) = \frac{\ell_r(z) - \ell_1(z)}{-(\log r - \log 1)} = \frac{\ell_r(z)}{-\log r} + \frac{\ell_1(z)}{\log r}\,. \tag{8}$$

$\ell_r(z)$ is an increasing and convex function of $\log r$, so $U_r(z) \leqslant 0$. $U_r(z)$ is a decreasing function of r, or if $r \searrow 0$, $U_r(z)$ is an increasing family. Moreover $U_r(z)$ is a plurisubharmonic function of z (because $-\log r > 0$). Then $\lim_{r\to 0} U_r(z) = \psi(z)$ exists, and the two limits in (7) exist and have the same value $\psi(z)$; $\psi(z)$ is the limit of a sequence of plurisubharmonic functions $U_r(z)$ which are negative (if z belongs to the polar conic set for which $f(uz) \equiv -\infty$, we put $U_r(z) = -\infty$). By the regularization theorem (see §3), $\psi^*(z)$ is plurisubharmonic; using (6) we obtain

$$\psi^*(tz) = \psi^*(z)$$

for all $t > 0$; then ψ^* is bounded above on the space E; ψ^* is a constant and the parts a/ and b/ of the Proposition 14 are proved. Moreover ψ^* is negative and $\nu(0) = -\psi^*$ is positive.

The property c/ is a consequence of

$$U_r(z) \leqslant \psi(z) \leqslant -\nu(0)$$

and d/ is a consequence of Proposition 7.

The number $\nu(a)$ for plurisubharmonic functions in a Fréchet space. If E is a Fréchet space, there exists in each neighborhood of the origin compact Coeuré polycylinders P such that if $d\mu_o$ is the image of the Jessen-measure $\prod_1^\infty \frac{d\theta_k}{2\pi}$ (cf [2, p.382]) we obtain

$$\psi^*(z_o) \leqslant \int_{P^*} \psi(z_o + z)\,d\mu_o(z),$$

where P^* is the image of the distinguished boundary $\sup_k |x_k| = 1$ in P (See §2.). Then if ψ^* is a constant γ and $\psi(z_o + z) \leqslant \gamma$, the set $\psi(z) = \gamma = \psi^*(z_o)$ is of positive measure on P^* and we obtain

Proposition 15. **If E is a Fréchet space, the negligible set $g(x) < g^*(x)$ as defined in Proposition 7 is a strictly polar set if g^* is a constant. Consequently, in Proposition 14, if E is a Fréchet space, $\psi(z) = \nu(0)$ except for z in a conic polar set A. Of course the same conclusion holds in the general case for E (E quasi-complete) if there exists an open set $\omega \subset E$ and $z_0 \in \omega$ such that $\inf_{z\in\omega} \nu(0,z) = \nu(0,z_0)$.**

Continuous plurisubharmonic functions. If F is an analytic function on E, $f = \log |F|$ is a continuous plurisubharmonic function. To obtain the existence of a $z_0 \in E$ such that $\psi(z_0) = \nu(0,z_0) = \nu(0)$, it is sufficient by Proposition 7 to suppose that E is a Baire space.

Proposition 16. **If A is an analytic set locally defined in a Baire space E by F = 0, where F is holomorphic in a neighborhood of the origin and F(0) = 0, then the number $\nu(0)$ for $f = \frac{1}{2\pi} \log |F|$ is a positive integer. On the projective space of the complex lines through 0 we have $\nu(0,z) = \nu(0)$ except for a conic polar set η for z.**

To extend analytic geometry to such vector spaces (Baire and quasi-complete), it seems desirable to get more information on the set η. We have a decomposition $\eta = \cup \eta_n$; η_n is the conic set of the $z \in E$ such that $\psi(z) = \nu(0,z) = \nu(0) + n$, $n = 1,2,\ldots$ We make following conjectures which are not independent: There exist holomorphic functions F such that $\eta_n \neq \phi$ for each n. The set η_n is an analytic cone (with vertex 0) in E of finite codimension. The set η is a countable union of analytic sets.

Bibliography

1. J. Bochnak and J. Siciak, a/ **Polynomials and multilinear mappings in topological vector spaces.**

b/ **Analytic functions in topological vector spaces**, Studia Math., v. 39 (1971), 59–76.

2. G. Coeuré, **Fonctions plurisousharmoniques sur les espaces vectoriels topologiques et applications à l'étude des fonctions analytiques**, Ann. Inst. Fourier, Grenoble, v. 20(1970), 361–432.

3. S. Kakutani and V. Klee, **The finite topology of a linear space**, Arch. Math. Bâle, v.11(1963), 55–58.

4. C.O. Kiselman, a/ **On entire functions of exponential type and indicators of analytic functionals**, Acta Math., v.117(1967), 1.

b/ **Plurisubharmonic functions in vector spaces** (in preparation).

5. P. Lelong, a/ **Fonctions entières et fonctionnelles analytiques** (Cours professé à Montréal, publié en 1968 aux Presses de Montréal).

b/ **Fonctions plurisousharmoniques dans les espaces vectoriels topologiques**, Séminaire d'Analyse, Lecture-Notes Springer, No.71, exposé 17, 1968.

c/ **Petits ensembles dans les espaces vectoriels topologiques et les algebres complexes et probabilité nulle**, Colloque International du C.N.R.S. "Probabilités sur les structures algébriques", Clermont-Ferrand, édité par le C.N.R.S.

d/ **Fonctions plurisousharmoniques et ensembles polaires sur une algèbre de fonctions holomorphes**, Séminaire d'Analyse, Lecture-Notes Springer No.116, exposé 1, 1969.

e/ **Sur les fonctions plurisousharmoniques dans les espaces vectoriels topologiques et une extension du théorème de Banach-Steinhaus aux familles d'applications polynomiales**, Colloque d'Analyse fonctionnelle, Liège, September 1970.

f/ **Propriétés métriques des variétés analytiques complexes définies par une équation**, Ann. Ec. Norm., v.67(1950), 393–419.

g/ **Intégration sur un ensemble analytique complexe**, Bull. Soc. Math. France, v.85(1957), 239–262.

h/ **Fonctions plurisoursharmoniques dans les espaces vectoriels topologiques et les algèbres de fonctions analytiques**, Colloque International du C.N.R.S. sur l'Analyse complexe, Juin 1972. Revue Agora Mathematica, Gauthier-Villars No.1.

6. Ph. Noverraz, **Fonctions plurisousharmoniques et analytiques dans les espaces vectoriels topologiques complexes**, Ann. Inst. Fourier, Grenoble, v.19(1969), 419–493.

Faculte des Sciences
Universite de Paris
95 Blvd Jourdan
75-Paris 14e
France

A GLIMPSE AT INFINITE DIMENSIONAL HOLOMORPHY

Leopoldo Nachbin

§1. Introduction

Holomorphic mappings between locally convex spaces have been the object of research by analysts and geometers with an increasing emphasis on quite a variety of interesting aspects, mostly from the viewpoints of foundations, functional analysis, holomorphic continuation, holomorphic sets and differential equations. In this one-hour lecture, we shall try to describe some of the basic aspects of such a study.

Unless the contrary is said explicitly, E and F will represent two complex locally convex spaces, and U will denote a non-void open subset of E.

Let $\mathcal{L}_s(^mE;F)$ be the vector space of all continuous symmetric m-linear mappings $A: E^m \to F$ for $m = 1,2,\ldots$; and $\mathcal{L}_s(^0E;F) = F$. For simplicity, we write $\mathcal{L}_s(^1E;F) = \mathcal{L}(E;F)$.

If $A \in \mathcal{L}_s(^mE;F)$ and $x \in E$, we write $Ax^m = A(x,\ldots,x)$ where x is repeated m times in the right hand side for $m = 1,2,\ldots$; and $Ax^0 = A$ if $m = 0$.

Let $\mathcal{P}(^mE;F)$ be the vector space of all continuous m-homogeneous polynomials $P: E \to F$. We have the natural linear bijection

$$A \in \mathcal{L}_s(^mE;F) \to \hat{A} \in \mathcal{P}(^mE;F)$$

defined by $\hat{A}(x) = Ax^m$ for every $x \in E$, where $m \in N$.

A mapping $f: U \to F$ is said to be amply bounded if, for every $\xi \in U$ and every continuous seminorm β on F, there is a neighborhood V of ξ in U such that $\beta \circ f$ is bounded on V. We denote by $\mathcal{AB}(U;F)$ the vector space of all amply bounded mappings from U into F. More generally, a set $\mathfrak{X}$ of mappings from U into F is said to be amply bounded if, for every $\xi \in U$ and every continuous seminorm β on F, there is a neighborhood V of ξ in U such that

$$\sup\{\beta[f(x)]; f \in \mathfrak{X}, x \in V\} < +\infty.$$

By wF we shall denote F endowed with the weak topology $\sigma(F,F')$; and $\hat{F}$ will represent a completion of F.

§2. The Weierstrass Viewpoint

A mapping $f: U \to F$ is said to be holomorphic on U if, for every $\xi \in U$, there is a sequence $A_m \in \mathcal{L}_s(^mE;F)$ $(m \in N)$ such that, given any continuous seminorm β on F, we can find a neighborhood V of ξ in U for which

$$\lim_{m\to\infty} \beta[f(x) - \sum_{k=0}^{m} A_k(x-\xi)^k] = 0$$

uniformly for $x \in V$. Let $\mathcal{H}(U;F)$ be the vector space of all holomorphic mappings from U into F.

Assume that F is separated. For each $f \in \mathcal{H}(U;F)$ and $\xi \in U$, the above sequence (A_m) is unique. We set

$$d^m f(\xi) = m!\, A_m,$$

$$\hat{d}^m f(\xi) = m!\, \hat{A}_m,$$

and call each of these mappings the differential of order m of f at ξ. Moreover, each of the mappings

$$d^m f\colon x \in U \mapsto d^m f(x) \in \mathcal{L}_s(^m E;F),$$

$$\hat{d}^m f\colon x \in U \mapsto \hat{d}^m f(x) \in \mathcal{P}(^m E;F),$$

is called the differential of order m of f on U.

If $E = \mathbf{C}$, then $d^m f(\xi)$ and $\hat{d}^m f(\xi)$ are identified in a natural way to an element $f^{(m)}(\xi)$ of F called the derivative of order m of f at ξ. Then we have the derivative of order m of f on U given by

$$f^{(m)}\colon x \in U \mapsto f^{(m)}(x) \in F.$$

Besides the above attitude in defining holomorphic mappings, there is another one that we want to mention right away. The two attitudes differ only slightly; in fact, they are essentially the same thing and any one of the concepts in question may be then defined in terms of the other.

We introduce the vector space H(U;F) formed by every $f\colon U \to F$ such that $f \in \mathcal{H}(U;\hat{F})$ when f is considered as having its values in $\hat{F}$; it is clear that H(U;F) is independent of the choice of $\hat{F}$.

Conversely, $\mathcal{H}(U;F)$ is defined in terms of the preceding concept as follows. By assuming that F is separated, for simplicity, then $\mathcal{H}(U;F)$ consists of every $f \in H(U;F)$ such that

$$d^m f(\xi)(x_1,\ldots,x_m) \in F$$

for all $m = 1,2,\ldots, \xi \in U$ and $x_1,\ldots,x_m \in E$.

We then have the two sheaves $\mathcal{H}$ and H. Accordingly, we shall refer to $\mathcal{H}$-holomorphy and H-holomorphy. Sometimes it is more convenient to use one of the two sheaves, but quite often it really does not matter which one we do use. For simplicity, we refer to holomorphy instead of $\mathcal{H}$-holomorphy. We shall adhere to the use of $\mathcal{H}$ whenever transcription to H is easy. When we feel that there is a reason to use H, we shall do so.

Clearly

$$\mathcal{H}(U;F) \subset H(U;F) \subset \mathcal{C}(U;F),$$

where $\mathcal{C}(U;F)$ is the vector space of all continuous mappings from U into F. We are normally interested in having $\mathcal{H}(U;F) = H(U;F)$.

§3. The Cauchy-Riemann Viewpoint

A mapping $f: U \to F$ is said to be Fréchet complex differentiable of order one on U if, for every $\xi \in U$, there is a continuous complex linear mapping $A \in \mathcal{L}(E;F)$ such that, for any continuous seminorm β on F, we can find a continuous seminorm α on E for which $x \in U$ and $\alpha(x-\xi) = 0$ imply $\beta[f(x)-f(\xi)-A(x-\xi)] = 0$, and moreover

$$\lim \frac{\beta[f(x)-f(\xi)-A(x-\xi)]}{\alpha(x-\xi)} = 0$$

provided $\alpha \neq 0$, where the limit is taken as $\alpha(x-\xi) \to 0$ and also $\alpha(x-\xi) \neq 0$. If F is separated, A is unique for each f and ξ.

It is also said that $f: U \to F$ is Fréchet real differentiable of order one on U if the preceding definition holds except that now $A \in \mathcal{L}(E_{\mathbf{R}}; F_{\mathbf{R}})$ is a continuous real linear mapping, the index R denoting passage to the real underlying spaces. By assuming that F is separated, A is unique for each f and ξ. We may write $A = A' + A''$ as the sum of a continuous complex linear mapping $A' \in \mathcal{L}(E;F)$ and a continuous complex antilinear mapping $A'' \in \mathcal{L}(E;\overline{F})$ in a unique way, where $\overline{F}$ stands for the antilinear counterpart of F. We define

$$\partial f(\xi) = A',$$

$$\overline{\partial} f(\xi) = A''.$$

We then have the mappings

$$\partial f: x \in U \mapsto \partial f(x) \in \mathcal{L}(E;F),$$

$$\overline{\partial} f: x \in U \mapsto \overline{\partial} f(x) \in \mathcal{L}(E;\overline{F}).$$

Proposition 1 (Goursat). For a mapping $f: U \to F$ to belong to $H(U;F)$ it is necessary and sufficient that f be Fréchet complex differentiable of order one on U when considered as having its values in $\hat{F}$.

Corollary (Cauchy-Riemann). If F is separated, for $f: U \to F$ to belong to $H(U;F)$ it is necessary and sufficient that f be Fréchet real differentiable of order one on U when considered as having its values in $\hat{F}$, and that the Cauchy-Riemann equation $\overline{\partial} f = 0$ be satisfied on U.

§4. Limits Of Holomorphic Mappings

A given F is said to be differentially stable if $\mathcal{H}(U;F) = H(U;F)$ holds for every E and every U; that is, by assuming that F is separated, if the following equivalent conditions are satisfied:

1) For every E, U and $f \in H(U;F)$, we have that

$$d^m f(\xi)(x_1, \ldots, x_m) \in F$$

for all $m = 1,2,\ldots, \xi \in U$ and $x_1,\ldots,x_m \in E$;

2) If U is the open disc in C of center at 0 and radius 1, for every $f \in H(U;F)$ we have that $f^{(m)}(0) \in F$ for all $m = 1,2,\cdots$.

There are two simple instances in which F is differentially stable:

1) F is sequentially complete;

2) the closed convex hull of every compact subset of F is compact.

In order that F be differentially stable it is necessary and sufficient that wF be differentially stable.

Proposition 2. **H(U;F) is always closed in $\mathcal{E}$(U;F). Once F is fixed, in order that $\mathcal{H}$(U;F) be closed in $\mathcal{E}$(U;F) for every E and every U it is necessary and sufficient that F be differentially stable.**

It is tacitly understood that $\mathcal{E}$(U;F) is endowed with the topology of uniform convergence on the compact subsets of U; however, the proposition remains true if we use only the finite dimensional compact subsets of U.

§5. Convergence Of Taylor Series

Proposition 3. **If $f \in \mathcal{H}(U;F)$, $\xi \in U$ and F is separated, the Taylor series of f at ξ, namely**

$$\sum_{m=0}^{\infty} \frac{1}{m!} d^m f(\xi)(x-\xi)^m,$$

converges uniformly to f on every compact subset of the largest ξ-balanced subset U_ξ of U; and it converges uniformly to f on some neighborhood in U of every such compact subset provided f is locally bounded.

The largest ξ-balanced subset U_ξ of U is defined as the set of all $x \in U$ such that $(1-\lambda)\xi + \lambda x \in U$ for all $\lambda \in C$, $|\lambda| \leqslant 1$; then $\xi \in U_\xi \subset U$ and U_ξ is open.

§6. Uniqueness Of Holomorphic Continuation

Proposition 4. **If $f \in \mathcal{H}(U;F)$, $\xi \in U$, F is separated and U is connected, the closed vector subspace of F generated by f(U) is equal to the closed vector subspace of F generated by**

$$d^m f(\xi)(x_1,\ldots,x_m)$$

for all $m \in N$ and $x_1,\ldots,x_m \in E$.

§7. Holomorphy Of Differentials

The limit topologies on $\mathcal{P}(^mE;F)$ and $\mathcal{L}_s(^mE;F)$ are defined as follows.

Let α and β denote arbitrary continuous seminorms on E and F, and E_α and F_β represent E and F seminormed by α and β respectively. Then $\mathcal{P}(^mE_\alpha; F_\beta)$ is seminormed accordingly. We notice that

$$\mathcal{P}(^mE;F_\beta) = \bigcup_\alpha \mathcal{P}(^mE_\alpha;F_\beta)$$

and endow $\mathcal{P}(^mE;F_\beta)$ with the corresponding inductive limit topology. We also have

$$\mathcal{P}(^mE;F) = \bigcap_\beta \mathcal{P}(^mE;F_\beta)$$

and endow $\mathcal{P}(^mE;F)$ with the corresponding projective limit topology. We thus get on $\mathcal{P}(^mE;F)$ the so-called limit topology. It contains the so-called strong topology on $\mathcal{P}(^mE;F)$ which is defined by all seminorms

$$P \in \mathcal{P}(^mE;F) \mapsto \sup\{\beta[P(x)]; x \in B\},$$

where the continuous seminorm β on F and the bounded subset B of E are arbitrary.

Likewise for the limit and strong topologies on $\mathcal{L}_s(^mE;F)$.

The natural isomorphism between $\mathcal{L}_s(^mE;F)$ and $\mathcal{P}(^mE;F)$ is a homeomorphism if these spaces are endowed with their limit, or strong, topologies.

Proposition 5. **If $f \in \mathcal{H}(U;F)$ and F is separated, then**

$$d^mf \in \mathcal{H}(U; \mathcal{L}_s(^mE;F)),$$

$$\hat{d}^mf \in \mathcal{H}(U; \mathcal{P}(^mE;F)),$$

for $m \in N$, where $\mathcal{L}_s(^mE;F)$ and $\mathcal{P}(^mE;F)$ are endowed with their limit topologies.

§8. Holomorphy Versus Finite Holomorphy

We say that f: $U \to F$ is finitely holomorphic if

$$f|(U \cap S) \in \mathcal{H}(U \cap S; F)$$

for every finite dimensional vector subspace S of E intersecting U.

We say that f: $U \to F$ is Gateaux holomorphic if the above condition holds for every one dimensional affine subspace S of E intersecting U, where holomorphy on a non-void open subset of an affine subspace S of E is defined by translating S to a vector subspace of E. If E is separated, it is equivalent to require that

$\lambda \mapsto f(a+\lambda b)$ be holomorphic from $V = \{\lambda \in C; a + \lambda b \in U\}$ into F for every $a \in U$ and $b \in E$.

There is a redundancy of terminology in the sense that finite holomorphy and Gateaux holomorphy are equivalent forms of the same concept.

Proposition 6. **A mapping is holomorphic if and only if it is finitely holomorphic and amply bounded, or equivalently continuous.**

§9. Holomorphy Versus Inductive Or Projective Limits

Proposition 7. **Let E_m ($m \in N$) be complex locally convex spaces, E a complex vector space, $\rho_m: E_m \to E$ a linear mapping and $\sigma_m: E_m \to E_{m+1}$ a compact linear mapping such that $\rho_m = \rho_{m+1} \circ \sigma_m$ ($m \in N$). Assume that**

$$E = \bigcup_{m \in N} \rho_m(E_m)$$

and endow E with the inductive limit topology. Let $U \subset E$ be open, put $U_m = \rho_m^{-1}(U)$ ($m \in N$) and assume U_0 non-void. If F is a complex locally convex space and $f: U \to F$, then $f \in \mathcal{H}(U;F)$ if and only if $f \circ \rho_m \in \mathcal{H}(U_m;F)$ for every $m \in N$.

The range of application of the above proposition is limited in the sense that the space E which appears there is of a special nature. A given E is said to be a Silva space if its topology may be defined as an inductive limit through suitable sequences (E_m), (ρ_m) and (σ_m) satisfying all the aforementioned conditions. It is known that then such sequences may be changed and chosen so that each E_m is a Banach space, if E is separated.

Proposition 8. **Let F_i ($i \in I$) be complex locally convex spaces, F a complex vector space, $\rho_i: F \to F_i$ a linear mapping for every $i \in I$, and endow F with the projective limit topology. If U is a non-void open subset of a complex locally convex space E and $f: U \to F$, then $f \in H(U;F)$ if and only if $\rho_i \circ f \in H(U;F_i)$ for every $i \in I$.**

§ 10. Some Holomorphically Significant Properties Of Locally Convex Spaces

There is a fruitful interplay between $\mathcal{L}(E;F)$ and $\mathcal{H}(U;F)$. This is due, although not exclusively, to the fact that $\mathcal{L}(E;F) \mid U \subset \mathcal{H}(U;F)$. For instance, certain studies of $\mathcal{L}(E;F)$ suggest analogous although modified considerations concerning $\mathcal{H}(U;F)$. In the following four definitions, if we omit that U is arbitrary in certain places, and write E in place of U and write $\mathcal{L}$ in place of H or $\mathcal{H}$ in other places, we get known forms of definitions that E is a Mackey space, or a bornological space, or an infrabarreled space, or a barreled space.

We say that a given E is a holomorphically Mackey space if, for U and F arbitrary, we have that $f: U \to F$ belongs to $H(U;F)$ if (and always only if) $\psi \circ f \in \mathcal{H}(U;C)$ for every $\psi \in F'$, that is $H(U;F) = H(U;wF)$.

We say that a given E is a holomorphically bornological space if, for U and F arbitrary, we have that f: U → F belongs to $\mathcal{H}(U;F)$ if (and always only if) f is finitely holomorphic, and bounded on every compact subset of U.

We say that a given E is a holomorphically infrabarreled space if, for U and F arbitrary, we have that $\mathfrak{X} \subset \mathcal{H}(U;F)$ is amply bounded if (and always only if) $\mathfrak{X}$ is bounded on every compact subset of U.

We say that a given E is a holomorphically barreled space if, for U and F arbitrary, we have that $\mathfrak{X} \subset \mathcal{H}(U;F)$ is amply bounded if (and always only if) $\mathfrak{X}$ is bounded on every finite dimensional compact subset of U.

It is then clear that a holomorphically Mackey, or holomorphically bornological, or holomorphically infrabarreled, or holomorphically barreled, space is respectively a Mackey, or bornological, or infrabarreled, or barreled, space,

Although the notion of a Mackey space is too general, that of a holomorphically Mackey space is less so; it corresponds to the very desirable fact that holomorphy be identical to weak holomorphy.

Let us introduce the following abreviations for properties of a complex locally convex space: B = Baire, S = Silva, sm = semimetrisable, hba = holomorphically barreled, hbo = holomorphically bornological, hib = holomorphically infrabarreled, hM = holomorphically Mackey.

Proposition 9. We have the following implications for the named properties of a complex locally convex space:

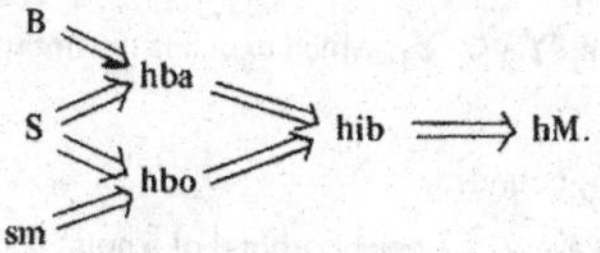

These results correspond to known ones when we replace $\mathcal{L}(E;F)$ by $\mathcal{H}(U;F)$.

§11. Bounding Subsets

A subset X of U is said to be $\mathcal{H}$-bounding in U if every $f \in \mathcal{H}(U;C)$ is bounded on X; it then follows that every $f \in \mathcal{H}(U;F)$ for any F is bounded on X. Conversely, if every $f \in \mathcal{H}(U;F)$ is bounded on X, for some separated $F \neq 0$, then X is $\mathcal{H}$-bounding in U. If X is relatively compact in U, then X is $\mathcal{H}$-bounding in U. Conversely, if from every equicontinuous sequence in E' it is possible to extract a subsequence which is convergent for $\sigma(E',E)$, then X must be precompact in E if it is $\mathcal{H}$-bounding in E. If X is $\mathcal{H}$-bounding in E, it is certainly bounded in E.

One simple way of building up plenty of entire functions is as follows.

Proposition 10. If $\varphi_m \in E'$ (m = 1,2,. . .), then

$$\sum_{m=1}^{\infty} (\varphi_m)^m$$

is the Taylor series at 0 of some $f \in \mathcal{H}(E; C)$ if and only if (φ_m) is equicontinuous and $\varphi_m \to 0$ for $\sigma(E', E)$ as $m \to \infty$. Moreover, f is bounded on every bounded subset of E if and only if $\varphi_m \to 0$ strongly as $m \to \infty$.

Corollary. If there is an equicontinuous sequence $\varphi_m \in E'$ $(m = 1, 2, \ldots)$ such that $\varphi_m \to 0$ for $\sigma(E', E)$ as $m \to \infty$, but $\varphi_m \to 0$ strongly as $m \to \infty$ breaks down, then there is a bounded subset of E which is not $\mathcal{H}$-bounding in E.

Proposition 11. The topology on $\mathcal{H}(U;F)$ of pointwise convergence and the topology on $\mathcal{H}(U;F)$ of uniform convergence on every $\mathcal{H}$-bounding subset of U induce topologies on every amply bounded subset of $\mathcal{H}(U;F)$ having the same convergent sequences.

Corollary. If E is holomorphically barreled, the topology on $\mathcal{H}(U;F)$ of uniform convergence on every finite dimensional compact subset of U and the topology on $\mathcal{H}(U;F)$ of uniform convergence on every $\mathcal{H}$-bounding subset of U have the same convergent sequences. If E is only holomorphically infrabarreled, this conclusion remains true if we replace the first topology by that of uniform convergence on every compact subset of U.

§ 12. Holomorphy Versus Perturbation Of Topology

If $\mathcal{T}_1$ and $\mathcal{T}_2$ are two locally convex topologies on a vector space, we write $\mathcal{T}_2 \prec \mathcal{T}_1$ to denote that every subset which is bounded for both $\mathcal{T}_1$ and $\mathcal{T}_2$ has a $\mathcal{T}_1$-closure which is bounded for $\mathcal{T}_2$ (and for $\mathcal{T}_1$ too free of charge). This is the case if $\mathcal{T}_2 \subset \mathcal{T}_1$, which explains the notation. We have $\mathcal{T}_2 \prec \mathcal{T}_1$ in two noteworthy cases:

1) Every $\mathcal{T}_1$-bounded subset is $\mathcal{T}_2$-bounded.

2) $\mathcal{T}_2$ is $\mathcal{T}_1$-locally closed, that is every $\mathcal{T}_2$-neighborhood of a point contains another $\mathcal{T}_2$-neighborhood of that point which is $\mathcal{T}_1$-closed.

Proposition 12. Let E and F denote two complex vector spaces, E_j and F_j denote E and F endowed with the locally convex topologies $\mathcal{T}_j(E)$ and $\mathcal{T}_j(F)$ respectively, for $j = 1, 2$. If $U \subset E$ is open for both $\mathcal{T}_1(E)$ and $\mathcal{T}_2(E)$ and non-void, let U_j denote U as a topological subspace of E_j for $j = 1, 2$. Then

$$\mathcal{H}(U_1;F_1) \cap \mathcal{AB}(U_2, F_2) \subset \mathcal{H}(U_2;F_2),$$

$$H(U_1;F_1) \cap \mathcal{AB}(U_2;F_2) \subset H(U_2;F_2),$$

provided $\mathcal{T}_2(F) \prec \mathcal{T}_1(F)$.

This general result implies several known results, such as the following one.

Corollary (Liouville). Let E be a complex locally convex space, F a complex vector space. Let F_j denote F endowed with the locally convex topology $\mathcal{T}_j(F)$ for $j = 1, 2$. Then every $f \in \mathcal{H}(E;F_1)$ such that f(E) is $\mathcal{T}_2(F)$-bounded must be a constant on E, provided $\mathcal{T}_2(F) \prec \mathcal{T}_1(F)$ and $\mathcal{T}_2(F)$ is separated.

§13. Stability Of Holomorphy By Perturbation Of Topology

Proposition 13. Let E be holomorphically bornological, F be a complex vector space, and $U \subset E$ be open and non-void. Let F_j denote F endowed with the locally convex topology $\mathcal{T}_j(F)$ for $j = 1,2$. If every $\mathcal{T}_1(F)$-bounded subset is $\mathcal{T}_2(F)$-bounded, then $\mathcal{H}(U;F_1) \subset \mathcal{H}(U;F_2)$ and $H(U;F_1) \subset H(U;F_2)$. In particular, if $\mathcal{T}_1(F)$ and $\mathcal{T}_2(F)$ have the same bounded subsets, then $\mathcal{H}(U;F_1) = \mathcal{H}(U;F_2)$ and $H(U;F_1) = H(U;F_2)$.

Proposition 14. Let E, F and G be complex locally convex spaces, E be holomorphically bornological, F be barreled, and U be a non-void open subset of E. Then $f: U \to \mathcal{L}(F;G)$ belongs to $H(U; \mathcal{L}(F;G))$, where $\mathcal{L}(F;G)$ has its limit topology, if and only if $f_y \in H(U;G)$ for every $y \in F$, where $f_y: x \in U \mapsto f(x)(y) \in G$.

§14. Holomorphy Versus Pointwise, Or Componentwise, Holomorphy

We now turn to pointwise holomorphy and function spaces of a kind which is different from that of Proposition 14.

Proposition 15. Let E, F and G be complex locally convex spaces, Y be a set, F be a vector subspace of G^Y, the inclusion mapping $F \subsetneq G^Y$ being continuous, and $U \subset E$ be open and non-void. Assume that E is holomorphically bornological, and that the original topology on F is locally closed with respect to the topology induced on F by G^Y. Then $f: U \to F$ belongs to $H(U;F)$ if and only if $f_y \in H(U;G)$ for every $y \in Y$, where $f_y: x \in U \mapsto f(x)(y) \in G$, and moreover f maps every compact subset of U into a bounded, or equivalently precompact, subset of F.

The following result specializes the preceding one to componentwise holomorphy and topological basis.

A subset $B \subset F$ is a topological basis in a topological vector space F if there is a necessarily unique mapping $\psi: b \in B \mapsto \psi_b \in F'$ such that:

1) For every $b \in B$, we have $\psi_b(b) = 1$ and $\psi_b(x) = 0$ if $x \in B$, $x \neq b$.

2) For every $x \in F$, we have that $x = \Sigma_{b \in B}\, \psi_b(x) b$ in F.

3) The original topology on F is locally closed with respect to the weak topology on F defined by all ψ_b $(b \in B)$.

Proposition 16. Assume that E is holomorphically bornological. Let B be a topological basis for F. Then $f: U \to F$ belongs to $H(U;F)$ if and only if $f_b \in \mathcal{H}(U; C)$ for every $b \in B$, where $f_b: x \in U \mapsto \psi_b[f(x)] \in C$ and $\psi: B \to F'$ is associated to B, and moreover f maps every compact subset of U into a bounded, or equivalently precompact, subset of F.

§15. Vector-Valued Versus Scalar-Valued Holomorphic Continuation

Once E is fixed, we say that weak holomorphy plus slight holomorphy imply holomorphy on E if, for every F, we have that $f \in H(V;F)$ whenever V and W are connected non-void open subsets of E with $W \subset V$,

$f \in H(V;wF)$ and $f | W \in H(W;F)$. This is the case if E is a holomorphically Mackey space.

Once F is fixed, we say that it is confined if, for every E, we have that $f^{-1}(F) = U$ whenever U is connected, $f \in \mathcal{H}(U;\hat{F})$ and $f^{-1}(F)$ has a non-void interior. To check this requirement, it suffices to take U as the open disc of center 0 and radius 1 in E = C, to assume that $f \in \mathcal{H}(U;\hat{F})$ and that 0 is interior to $f^{-1}(F)$, and conclude that $f^{-1}(F) = U$. If F is sequentially complete, it is confined. Moreover wF is confined if and only if F is confined.

Let U, V and W be connected non-void open subsets of E, with $W \subset U \cap V$. If $F \neq 0$ is separated, we say that V is a holomorphic F-valued continuation of U via W if, for every $f \in H(U;F)$ there is $g \in H(V;F)$ such that f = g on W.

Proposition 17. Assume that weak holomorphy plus slight holomorphy imply holomorphy on E, and that $F \neq 0$ is separated and confined. Then V is a holomorphic F-valued continuation of U via W if and only if V is a holomorphic C-valued continuation of U via W.

Acknowledgements

The author gratefully acknowledges partial support from Fundo Nacional de Ciência e Tecnologia (FINEP), Indústrias Klabin and Centro Brasileiro de Pesquisas Físicas, Rio de Janeiro, GB, Brasil, and National Science Foundation, Washington, D.C., USA.

Bibliography

1. N. Bourbaki, **Variétés différentielles et analytiques**, Fascicule de résultats, § 1–7, 8–15, Hermann (1967–1971).

2. L. Nachbin, **Topology on spaces of holomorphic mappings**, Springer-Verlag, Ergebnisse der Mathematik 47 (1969).

3. L. Nachbin, **Recent developments in infinite dimensional holomorphy**, Bulletin of the American Mathematical Society 79(1973).

4. Ph. Noverraz, **Pseudo-convexité, convexité polynomiale et domaines d'holomorphie en dimension infinie**, North-Holland, Notas de Matemática 48 (1973).

Added in proof:

5. H. Cartan, **Introduction à la géométrie analytique**, North-Holland, Notas de Matemática (to appear).

6. G. Coeuré, **Analytic functions and manifolds in infinite dimensional spaces**, North-Holland, Notas de Matemática (to appear).

Instituto de Matemática
Universidade Federal do Rio de Janeiro
Rio de Janeiro, Guanabara, ZC-32, Brasil

Department of Mathematics
University of Rochester
Rochester, New York 14627, USA

HOLOMORPHIC EXTENSIONS AND DOMAINS OF HOLOMORPHY FOR GENERAL FUNCTION ALGEBRAS

C. E. Rickart

§1. Introduction.

It is desirable first to give a brief outline of the setting suggested in the title by the term "General Function Algebras". These are algebras of complex-valued continuous functions defined on more-or-less arbitrary Hausdorff spaces. It is always required that the given algebra $\mathfrak{A}$ contain the constant functions and usually that the topology in the underlying Hausdorff space Σ be the $\mathfrak{A}$-topology, i.e., the coarsest under which all functions in $\mathfrak{A}$ are continuous. Under these conditions, the pair $[\Sigma, \mathfrak{A}]$ is called a **system**. Note that Σ is not even assumed to be locally compact. However, if Σ is compact, then the condition that it carry the $\mathfrak{A}$-topology is equivalent to the requirement that $\mathfrak{A}$ separate its points. Therefore, important examples of systems are the much studied Uniform Algebras on compact Hausdorff spaces. Another important example is given by $[C^n, \mathfrak{P}]$, where C^n is complex n-space and $\mathfrak{P}$ is the algebra of all polynomials on C^n. We also have the "infinite dimensional" case, $[C^\Lambda, \mathfrak{P}]$, where Λ is an arbitrary index set, C^Λ is the Cartesian product of "Λ" complex planes, and $\mathfrak{P}$ is the algebra of all ordinary polynomials each of which involves a finite number of the complex variables $\{\zeta_\lambda : \lambda \in \Lambda\}$. Two systems $[\Sigma_1, \mathfrak{A}_1]$ and $[\Sigma_2, \mathfrak{A}_2]$ are said to be **isomorphic** under a map, $\rho: \Sigma_1 \to \Sigma_2$, if ρ is a homeomorphism of Σ_1 onto Σ_2 such that $\mathfrak{A}_2 \circ \rho = \mathfrak{A}_1$. In this case, $a_2 \mapsto a_2 \circ \rho$ is an algebra isomorphism of $\mathfrak{A}_2$ with $\mathfrak{A}_1$.

Although our approach to the study of general systems has a definite function algebra character and depends heavily on certain results for Uniform Algebras, the example $[C^n, \mathfrak{P}]$ motivates the bulk of the investigation. Example $[C^\Lambda, \mathfrak{P}]$ is naturally also very important and currently provides the main overlap with the usual approach to infinite dimensional holomorphy. The point of view in the study of a system $[\Sigma, \mathfrak{A}]$ is to let $\mathfrak{A}$ generate in Σ an abstract "holomorphy theory" in analogy with the way $\mathfrak{P}$ can be regarded as generating the usual holomorphy theory in C^n. Thus we may think of $\mathfrak{A}$ as a kind of "structure algebra" on Σ. The abstract theory is based on the notion of "holomorphic function" in Σ, which is defined as follows. First let $\mathfrak{F}$ be an arbitrary family of complex-valued functions defined on arbitrary subsets of Σ and denote by loc $\mathfrak{F}$ the family of all functions that can be approximated locally uniformly by elements of $\mathfrak{F}$. If $\mathfrak{F}$ = loc $\mathfrak{F}$, then $\mathfrak{F}$ is said to be **locally closed**. The smallest locally closed family that contains a given $\mathfrak{F}$ is called the **local closure** of $\mathfrak{F}$ and is denoted by $\mathfrak{F}^{loc}$. The local closure, $\mathfrak{A}^{loc}$, of the structure algebra is, by definition, the class of all **$\mathfrak{A}$-holomorphic** functions. In the case of $[C^n, \mathfrak{P}]$, the $\mathfrak{P}$-holomorphic functions (defined on open subsets of C^n) coincide with the ordinary holomorphic functions in C^n. The $\mathfrak{A}$-holomorphic functions have many nice properties. For example, they are continuous and are closed under the usual algebraic operations, to the extent that the operations are defined. Also, arbitrary holomorphic functions of n-variables operate on the $\mathfrak{A}$-holomorphic functions, again to the extent that the functions in question are defined. The above definitions and elementary properties obviously do not depend essentially on $[\Sigma, \mathfrak{A}]$ being a system. In order to obtain some deeper and more interesting properties of $\mathfrak{A}$ -holomorphic functions, it is necessary to introduce a further restriction on $[\Sigma, \mathfrak{A}]$.

The system $[\Sigma, \mathfrak{A}]$ is said to be **natural** if every continuous homomorphism $\varphi: \mathfrak{A} \rightarrow C$ of $\mathfrak{A}$ onto C is a point evaluation; i.e., there exists a point $\sigma_\varphi \in \Sigma$ such that $\varphi: a \mapsto a(\sigma_\varphi)$ for each $a \in \mathfrak{A}$. The continuity here refers to the compact-open topology in $\mathfrak{A}$. Naturality, which turns out to be a kind of "analytic" condition on $[\Sigma, \mathfrak{A}]$, is the key to a number of analytic type properties that can be established for $[\Sigma, \mathfrak{A}]$. In the case of a uniform algebra, naturality is simply the condition that the underlying compact Hausdorff space be equal to the spectrum, or space of maximal ideals, of the algebra. Analogously, if $[\Sigma, \mathfrak{A}]$ is natural, then we call Σ the **spectrum** of $\mathfrak{A}$. The systems $[C^n, \mathfrak{P}]$ and $[C^\Lambda, \mathfrak{P}]$ are easily proved to be natural. In fact, in these cases, as for a uniform algebra on its spectrum, every homomorphism of the algebra onto C is a point evaluation. If K is a compact subset of Σ, then the set, $\hat{K} = \{\sigma : |a(\sigma)| \leq |a|_K, a \in \mathfrak{A}\}$, where $|a|_K$ is the maximum absolute value of the function a on K, is the $\mathfrak{A}$**-hull** of K. If $[\Sigma, \mathfrak{A}]$ is natural, then $\hat{K}$ **must also be compact.** An arbitrary set $\Omega \subseteq \Sigma$ is said to be $\mathfrak{A}$**-convex** if $\hat{K} \subseteq \Omega$ for every compact set $K \subseteq \Omega$. An important result for a natural system $[\Sigma, \mathfrak{A}]$ is that a set $\Omega \subseteq \Sigma$ **will be** $\mathfrak{A}$**-convex iff the system** $[\Omega, \mathfrak{A}]$ **is natural**, where the topology on Ω is that induced by Σ and the algebra is the restriction of $\mathfrak{A}$ to Ω. Another much deeper consequence of naturality is the following property:

Local Maximum Principle. **Let Ω be a compact $\mathfrak{A}$-convex subset of Σ with Silov boundary $\partial_{\mathfrak{A}}\Omega$ relative to $\mathfrak{A}$. Also let U be a relatively open subset of $\Omega \backslash \partial_{\mathfrak{A}}\Omega$. Then, if h is continuous on $\bar{U}$ and $\mathfrak{A}$–holomorphic on U, the maximum absolute value of h on U is assumed on the topological boundary of U relative to the space Ω.**

This very important principle rests ultimately on the Rossi local maximum principle for a uniform algebra (via the fact that the uniform closure of $\mathfrak{A}$ on Ω is a uniform algebra with spectrum Ω), the proof of which involves the solution of a Cousin problem in C^n [10; 2, p.62]. It is primarily through this result that the study of natural systems depends in an essential way upon a significant part of the classical SCV theory. An important result that depends on the local maximum principle for $\mathfrak{A}$-holomorphic functions is given by the following lemma [5], a version of which was first proved by Glicksberg [1].

Lemma. **Let Ω be a compact $\mathfrak{A}$–convex set in Σ and let B denote a proper $\mathfrak{A}$–boundary for Ω (i.e., B is a closed set in which each function in $\mathfrak{A}$ assumes its maximum modulus for the entire set Ω). Then there exists a point $\beta \in bd_\Omega B$ such that, for each neighborhood U of β in Ω, there exists a neighborhood $V \subseteq U$ with the property that any function that is $\mathfrak{A}$–holomorphic on U and vanishes on $U \cap B$ must vanish on V.**

If $[\Sigma_1, \mathfrak{A}_1]$ and $[\Sigma_2, \mathfrak{A}_2]$ are any two systems and $X \subseteq \Sigma_1$, then $\rho: X \rightarrow \Sigma_2$ is said to be **holomorphic** if $\mathfrak{A}_2 \circ \rho \subseteq \mathcal{O}_X$, where $\mathcal{O}_X$ denotes the family of all $\mathfrak{A}_1$-holomorphic functions defined on X. A holomorphic map is automatically continuous and one can show that $\mathcal{O}_{\rho(X)} \circ \rho \subseteq \mathcal{O}_X$. An important special case is obtained by taking $[\Sigma_2, \mathfrak{A}_2] = [C^\Lambda, \mathfrak{P}]$. Then ρ may be written in the form

$\{\rho_\lambda\}$ where $\rho_\lambda: X \to C$ for each $\lambda \in \Lambda$. In this case, ρ will be holomorphic iff each of the complex functions ρ_λ is $\mathfrak{A}_1$-holomorphic.

We note in passing that there also exists a theory of "$\mathfrak{A}$-subharmonic" functions in Σ that generalizes the theory of plurisubharmonic functions in C^n. Details concerning these and the above mentioned results, along with many other properties of natural systems, will be found in the references [5, 6, 7, 8, 9].

In this discussion, we wish to develop the notions of "holomorphic extension" and "domain of holomorphy" for the abstract situation. In §2, we introduce the notion of a $[\Sigma, \mathfrak{A}]$-domain, (Φ, p), where $[\Sigma, \mathfrak{A}]$ is a natural system and Φ is a connected Hausdorff space spread over Σ by the open, local homeomorphism p. A $[\Sigma, \mathfrak{A}]$-domain is a straightforward generalization of Riemann domain spread over C^n. This section also contains definitions of $\mathfrak{A}$-holomorphic extensions and maximal $[\Sigma, \mathfrak{A}]$-domains, or domains of $\mathfrak{A}$-holomorphy. In §3, maximal $[\Sigma, \mathfrak{A}]$-domains and maximal $\mathfrak{A}$-holomorphic extensions are constructed and related. The construction is again more-or-less straightforward involving a sheaf of germs approach. In §4 is proved the main theorem of the paper in which the equivalence of a completeness property, $\mathfrak{A}$-holomorphic convexity, and naturality for $[\Sigma, \mathfrak{A}]$-domains is established.

We wish to acknowledge at this point that our discussion of holomorphic extensions and domains of holomorphy for $[\Sigma, \mathfrak{A}]$-domains has been greatly influenced by the work of various authors who have considered these topics in the linear space setting of infinite dimensional holomorphy. In this respect, the Dissertations of M. C. Matos and M. Schottenloher were especially helpful. The list of references at the end is at best minimal. For an up-to-date bibliography on infinite dimensional holomorphy, see the Noverraz monograph [4].

§2. $\mathfrak{A}$-holomorphic extensions of $[\Sigma, \mathfrak{A}]$-domains.

We shall assume hereafter that $[\Sigma, \mathfrak{A}]$ is a natural system and that the space Σ is both connected and locally connected. A pair (Φ, p), consisting of a connected Hausdorff space Φ and a mapping $p: \Phi \to \Sigma$, is called a **$[\Sigma, \mathfrak{A}]$-domain** if p is an open local homeomorphism of Φ into Σ. Thus, p takes an open set onto an open set and each point $\varphi \in \Phi$ admits a neighborhood U_φ such that $p: U_\varphi \to p(U_\varphi)$ is a homeomorphism. Such neighborhoods will be called **p-neighborhoods**. The inverse of the homeomorphism $p|U_\varphi$ will be denoted by $p_{U_\varphi}^{-1}$ or, when it is unnecessary to exhibit the particular p-neighborhood involved, simply by p_φ^{-1}. Note that the space Φ is also locally connected.

Just as in the case of Riemann domains spread over C^n, the "holomorphic structure" of Σ lifts to Φ via the homeomorphisms p_φ^{-1}. Thus, if h is a function defined on a set $X \subseteq \Phi$, then we define h to be a **$\mathfrak{A}$-holomorphic** if for each point $\varphi \in \Phi$ and p-neighborhood U_φ, the function $h \circ p_{U_\varphi}^{-1}$ is $\mathfrak{A}$-holomorphic on the set $p(X \cap U_\varphi)$ in Σ. Now consider the pair $[\Phi, \mathfrak{A} \circ p]$. Although this is not in general a system, we can still define, just as in the case of a system, the notion of $\mathfrak{A} \circ p$-holomorphic functions in Φ. It is not very difficult to prove that a function h in Φ will be $\mathfrak{A}$-holomorphic iff it is $\mathfrak{A} \circ p$-holomorphic. We note in passing that objects of the type $[\Phi, \mathfrak{A} \circ p]$ may be considered independently of $[\Sigma, \mathfrak{A}]$ as a class of objects which we call "local (or locally natural) systems". Discussion of the more inclusive notion will be taken up at another time.

One of the problems concerning $[\Sigma, \mathfrak{A}]$-domains is whether or not for a given (Φ, p) there are enough $\mathfrak{A}$-holomorphic functions to determine the topology in Φ, or even to separate points. In special cases, e.g. when there is a concept of derivative and Taylor expansions, so a holomorphic function is determined in a neighborhood of a point by its value and the values of all of its derivatives at the point, is it easy to prove that the algebra $\mathcal{O}_\Phi$ of all holomorphic functions defined on Φ does separate the points of Φ. In the general case, we are forced on occasion simply to impose the condition that $[\Phi, \mathcal{O}_\Phi]$ be a system.

Let (Φ_0, p_0), (Φ, p) be $[\Sigma, \mathfrak{A}]$-domains and consider a mapping $\rho : \Phi_0 \to \Phi$. If ρ is continuous and $p_0 = p \circ \rho$, we call ρ a **morphism of $[\Sigma, \mathfrak{A}]$-domains** and write $\rho : (\Phi_0, p_0) \to (\Phi, p)$. It is easy to verify that ρ must be an open local homeomorphism and that $\mathcal{O}_\Phi \circ \rho \subseteq \mathcal{O}_{\Phi_0}$, so ρ is a holomorphic map. If the morphism ρ satisfies the additional condition, $\mathcal{O}_\Phi \circ \rho = \mathcal{O}_{\Phi_0}$ (i.e. every element of $\mathcal{O}_{\Phi_0}$ "extends" through ρ to an element of $\mathcal{O}_\Phi$), then we write $\rho : (\Phi_0, p_0) \Rightarrow (\Phi, p)$ and call this an **$\mathfrak{A}$-holomorphic extension** of (Φ_0, p_0). Two $\mathfrak{A}$-holomorphic extensions, $\rho_1 : (\Phi_0, p_0) \Rightarrow (\Phi_1, p_1)$ and $\rho_2 : (\Phi_0, p_0) \Rightarrow (\Phi_2, p_2)$, of (Φ_0, p_0) are said to be **isomorphic** if there exists $\rho : (\Phi_1, p_1) \Rightarrow (\Phi_2, p_2)$, where ρ is a homeomorphism of Φ_1 onto Φ_2, such that $\rho_2 = \rho \circ \rho_1$. An $\mathfrak{A}$-holomorphic extension $\rho : (\Phi_0, p_0) \Rightarrow (\Phi, p)$ is said to maximal if, for every extension $\rho' : (\Phi_0, p_0) \Rightarrow (\Phi', p')$, there exists $\rho'' : (\Phi', p') \Rightarrow (\Phi, p)$ such that $\rho = \rho'' \circ \rho'$. Thus, the following diagram commutes:

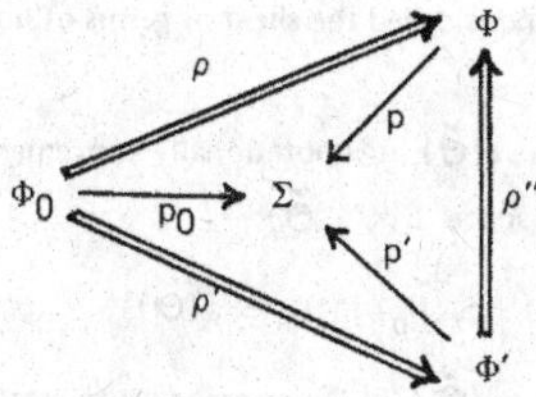

Finally, a $[\Sigma, \mathfrak{A}]$-domain (Φ_0, p_0) is said to be **maximal** if $\rho : (\Phi_0, p_0) \Rightarrow (\Phi, p)$ implies that ρ maps Φ_0 homeomorphically onto Φ. A maximal $[\Sigma, \mathfrak{A}]$-domain is also called a **domain of $\mathfrak{A}$-holomorphy**.

One can show that if $\rho : (\Phi_0, p_0) \Rightarrow (\Phi, p)$ is a maximal $\mathfrak{A}$-holomorphic extension, then (Φ, p) is a maximal $[\Sigma, \mathfrak{A}]$-domain. As a corollary we have that any two maximal $\mathfrak{A}$-holomorphic extensions of a given $[\Sigma, \mathfrak{A}]$-domain are isomorphic. Observe that, if $\mathcal{O}_{\Phi_0}$ separates the points of Φ_0 and $\rho : (\Phi_0, p_0) \Rightarrow (\Phi, p)$ then ρ is one-to-one so is a homeomorphism of Φ_0 onto an open subset of Φ. In the next section, we prove the existence of maximal extensions and domains of holomorphy. However, for this we must impose a further condition on the base system $[\Sigma, \mathfrak{A}]$.

§3. Existence of maximal $[\Sigma, \mathfrak{A}]$-domains and extensions.

We shall use here a "sheaf of germs" approach to the construction of maximal $[\Sigma, \mathfrak{A}]$-domains and maximal $\mathfrak{A}$-holomorphic extensions. However, in order for the sheaf involved to be a Hausdorff space, it is necessary to require that $[\Sigma, \mathfrak{A}]$ satisfy a **uniqueness principle** for $\mathfrak{A}$-holomorphic functions. To be precise, we shall assume through this section that any $\mathfrak{A}$-holomorphic function with an open connected domain of definition in Σ must be identically zero if it vanishes on an open subset of its domain. It is not very difficult to prove that if $[\Sigma, \mathfrak{A}]$ satisfies the uniqueness principle, then so will any $[\Sigma, \mathfrak{A}]$-domain, (Φ_0, p_0). Furthermore, if $\rho : (\Phi_0, p_0) \Rightarrow (\Phi, p)$, then the mapping $h \mapsto h \circ \rho$ is an algebra isomorphism of $\mathcal{O}_\Phi$ onto $\mathcal{O}_{\Phi_0}$.

Next let Λ be an arbitrary index set (to be associated eventually with $\mathcal{O}_\Phi$) and consider the product space C^Λ. It will be convenient to denote an element $\{\zeta_\lambda : \lambda \in \Lambda\}$ of C^Λ by $\check{\zeta}$. Similarly, a function $\sigma \mapsto \{f_\lambda(\sigma)\}$ on a subset of Σ with values in C^Λ will be denoted by $\check{f}$. As was observed in § 1, such a function will be holomorphic iff each of the complex-valued functions f_λ is $\mathfrak{A}$-holomorphic. We denote by ${}_\Lambda\check{\mathcal{O}}$ the family of all holomorphic functions with values in C^Λ and defined on open subsets of Σ. If $\check{f}$ is an element of ${}_\Lambda\check{\mathcal{O}}$ defined on a neighborhood of a point $\sigma \in \Sigma$, then $(\check{f})_\sigma$ will denote the germ of functions at σ determined by $\check{f}$. The "stalk" of all such germs at σ is denoted by ${}_\Lambda(\check{\mathcal{O}})_\sigma$, and the union of all the stalks, for $\sigma \in \Sigma$, is denoted by ${}_\Lambda(\check{\mathcal{O}})$. The set ${}_\Lambda(\check{\mathcal{O}})$ is given the usual sheaf topology in which a basic neighborhood of a point $(\check{f})_\sigma \in {}_\Lambda(\check{\mathcal{O}})$ consists of the set of all germs $(\check{g})_{\sigma'}$, where $\check{g} \in (\check{f})_\sigma$ and σ' ranges over a neighborhood V_σ in Σ on which $\check{g}$ is holomorphic. This basic neighborhood is denoted by $N_{(\check{f})_\sigma}(\check{g}, V_\sigma)$. The space ${}_\Lambda(\check{\mathcal{O}})$ is Hausdorff and is called the **sheaf of germs of holomorphic functions in Σ with values in C^Λ**.

Instead of working directly with the space ${}_\Lambda(\check{\mathcal{O}})$, it is notationally convenient to use the following homeomorphic image of ${}_\Lambda(\check{\mathcal{O}})$ in the product space $\Sigma \times {}_\Lambda(\check{\mathcal{O}})$:

$$\Sigma \#_\Lambda(\check{\mathcal{O}}) = \{(\sigma, (\check{f})_\sigma) : (\check{f})_\sigma \in {}_\Lambda(\check{\mathcal{O}})\}.$$

This is essentially a graph of the function $\pi_0 : {}_\Lambda(\check{\mathcal{O}}) \to \Sigma$, where π_0 projects each stalk ${}_\Lambda(\check{\mathcal{O}})_\sigma$ on its base point σ. With the relative product space topology in $\Sigma \#_\Lambda(\check{\mathcal{O}})$, the map $(\check{f})_\sigma \mapsto (\sigma, (\check{f})_\sigma)$ is clearly a homeomorphism of ${}_\Lambda(\check{\mathcal{O}})$ onto $\Sigma \#_\Lambda(\check{\mathcal{O}})$. Also define

$$\pi : \Sigma \#_\Lambda(\check{\mathcal{O}}) \to \Sigma, \quad (\sigma, (\check{f})_\sigma) \mapsto \sigma .$$

Then π is an open local homeomorphism. Although the space $\Sigma \#_\Lambda(\check{\mathcal{O}})$ is not connected, we can obviously still define $\mathfrak{A}$-holomorphic functions in it via the local homeomorphism π. For each $\mu \in \Lambda$, let

$$F_\mu(\sigma, (\check{f})_\sigma) = f_\mu(\sigma), \quad (\sigma, (\check{f})_\sigma) \in \Sigma \#_\Lambda(\check{\mathcal{O}}) .$$

Then F_μ is a well-defined function on $\Sigma \#_\Lambda(\check{\mathcal{O}})$ which is $\mathfrak{A}$-holomorphic. Since the space $\Sigma \#_\Lambda(\check{\mathcal{O}})$ is not connected, we shall restrict attention to its components. The following theorem provides the existence of a variety of maximal $[\Sigma, \mathfrak{A}]$-domains.

Theorem. **Let Γ be an arbitrary component of the space $\Sigma \#_\Lambda(\check{\mathcal{O}})$. Then (Γ, π) is a maximal $[\Sigma, \mathfrak{A}]$-**

domain.

Proof. That (Γ,π) is a $[\Sigma,\mathfrak{A}]$-domain is obvious. Hence let $\rho:(\Gamma,\pi) \Rightarrow (\Phi,p)$ be an arbitrary $\mathfrak{A}$-holomorphic extension of (Γ,π). Then $\rho(\Gamma)$ is an open connected subset of Φ. Let φ_0 be a limit point of $\rho(\Gamma)$ in Φ and let U_{φ_0} be a connected p-neighborhood of φ_0. Set $\sigma_0 = p(\varphi_0)$ and $V_{\sigma_0} = p(U_{\varphi_0})$, so V_{σ_0} is a connected neighborhood of σ_0 in Σ. Since $\mathcal{O}_\Phi \circ \rho = \mathcal{O}_\Gamma$, there exists for each $\mu \in \Lambda$ a function $G_\mu \in \mathcal{O}_\Phi$ such that $G_\mu \circ \rho = F_\mu$. Denote by g_μ the projection of G_μ onto V_{σ_0} so $g_\mu \circ p = G_\mu$. Define $\check{g} = \{g_\lambda\}$. Then $\check{g} \in {}_\Lambda\mathcal{O}$ so $(\sigma_0, (\check{g})_{\sigma_0}) \in \Sigma\#_\Lambda(\check{\mathcal{O}})$. Also, if $W_0 = \{(\sigma,(\check{g})_\sigma): \sigma \in V_{\sigma_0}\}$, W_0 is a connected neighborhood of the point $(\sigma_0, (\check{g})_{\sigma_0})$ in the space $\Sigma\#_\Lambda(\check{\mathcal{O}})$. Next define $W = \rho^{-1}(U_{\varphi_0} \cap \rho(\Gamma))$. Then W is an open set in Γ. By definition of holomorphic extension, we have $\pi = p \circ \rho$. Consider any point $(\sigma, (\check{f})_\sigma) = \gamma \in W$. Since W is open, there exists a neighborhood V of σ in Σ such that $\check{f}$ is holomorphic on V and $(\sigma', (\check{f})_{\sigma'}) = \gamma' \in W$ for all $\sigma' \in V$. Hence, for $\sigma' \in V$,

$$\begin{aligned} \check{f}(\sigma') &= \check{F}(\gamma') = (\check{G}\circ\rho)(\gamma') \\ &= (\check{g}\circ p\circ\rho)(\gamma') = \check{g}(\pi(\gamma')) = \check{g}(\sigma'). \end{aligned}$$

Thus $\check{f}$ and $\check{g}$ coincide in a neighborhood of the point σ so $(\check{f})_\sigma = (\check{g})_\sigma$. It follows that $W \subseteq W_0$. Since W_0 is a connected open subset of $\Sigma\#_\Lambda(\check{\mathcal{O}})$ that intersects the component Γ, it must be contained in Γ. Therefore $U_{\varphi_0} \cap \rho(\Gamma) \subseteq \rho(W_0)$. Now, shrinking the neighborhood V_{σ_0} and adjusting U_{φ_0} and W_0 accordingly, we can ensure that W_0 be a ρ-neighborhood so $\rho: W_0 \to \rho(W_0)$ is one-to-one. Moreover, the maps $p: U_{\varphi_0} \to V_{\sigma_0} = p(U_{\varphi_0})$, $\pi: W_0 \to V_{\sigma_0} = \pi(W_0)$ are also one-to-one. Since $\pi = p\circ\rho$ and U_{φ_0}, $\rho(W_0)$ are connected open sets that intersect, we conclude that $\rho(W_0) = U_{\varphi_0}$. In particular, $\varphi_0 \in \rho(\Gamma)$. Thus, $\rho(\Gamma)$ is both open and closed in Φ so we must have $\rho(\Gamma) = \Phi$. Since ρ is one-to-one on W_0, $\rho^{-1}(\varphi_0)$ consists of a single point. But φ_0 was any limit point of $\rho(\Gamma) = \Phi$ and, since Φ is connected, each of its points is a limit point. Therefore, $\rho:\Gamma \to \Phi$ is one-to-one and hence is a homeomorphism. This completes the proof of the theorem.

The space $\Sigma\#_\Lambda(\check{\mathcal{O}})$ is obviously very large. As a matter of fact, it contains, in a since, a maximal $\mathfrak{A}$-holomorphic extension of every $[\Sigma,\mathfrak{A}]$-domain. It is thus, so-to-speak, a universal representation space for $[\Sigma,\mathfrak{A}]$-domains.

Theorem. Let (Φ,p) be an arbitrary $[\Sigma,\mathfrak{A}]$-domain and choose a total indexing, $\{h_\lambda : \lambda \in \Lambda\}$, of $\mathcal{O}_\Phi$. Then $\tau:(\Phi,p) \Rightarrow (\Gamma_\Phi,\pi)$ is a maximal $\mathfrak{A}$-holomorphic extension of (Φ,p), where $\tau: \varphi \mapsto (p(\varphi), (h\circ p_\varphi^{-1})_{p(\varphi)})$ and Γ_Φ is the component of $\Sigma\#_\Lambda(\check{\mathcal{O}})$ that contains $\tau(\Phi)$.

Proof. Note that $p = \pi\circ\tau$ and, if U_φ is a p-neighborhood of φ, then $\tau(U_\varphi)$ is a π-neighborhood of $\tau(\varphi)$ in $\Sigma\#_\Lambda(\check{\mathcal{O}})$. Thus τ is an open local homeomorphism. Therefore $\tau(\Phi)$ is connected and is consequently contained in a component Γ_Φ of $\Sigma\#_\Lambda(\check{\mathcal{O}})$. Moreover, $h_\lambda = F_\lambda\circ\rho$ for each $\lambda \in \Lambda$, so $\mathcal{O}_\Phi = \mathcal{O}_{\Gamma_\Phi}\circ\tau$ and we have an $\mathcal{O}\!\mathcal{L}$-holomorphic extension $\tau : (\Phi, p) \Rightarrow (\Gamma_\Phi, \pi)$. In order to prove that this is a maximal extension, consider any $\rho : (\Phi, p) \Rightarrow (\Phi', p')$ and recall that ρ induces an isomorphism, $h' \mapsto h' \circ \rho$, of $\mathcal{O}_{\Phi'}$ onto $\mathcal{O}_\Phi$. Therefore we can index $\mathcal{O}_{\Phi'}$ by Λ so that $\check{h' \circ \rho} = \check{h}$. With this indexing of $\mathcal{O}_{\Phi'}$, construct, as for (Φ, p), the extension $\tau' : (\Phi', p') \Rightarrow (\Gamma_{\Phi'}, \pi)$ of (Φ', p') into $\Sigma\#_\Lambda(\check{\mathcal{O}})$. It is easy to verify that $\tau = \tau' \circ \rho$ which implies that $\tau(\Phi) \subseteq \tau'(\Phi')$. Therefore $\Gamma_\Phi = \Gamma_{\Phi'}$ so $\tau : (\Phi, p) \Rightarrow (\Gamma_\Phi, \pi)$ is indeed maximal.

Corollary. **An $\mathcal{O}\!\mathcal{L}$–holomorphic extension $\rho: (\Phi, p) \Rightarrow (\Phi', p')$ is maximal iff (Φ', p') is a maximal $[\Sigma, \mathcal{O}\!\mathcal{L}]$-domain.**

It is obvious that, for an arbitrary index set M, any holomorphic map of Φ into C^M extends to a holomorphic map of Γ_Φ into C^M. It is not clear to what extent results of this kind hold for more general holomorphic maps of Φ. However, for **morphisms** of $[\Sigma, \mathcal{O}\!\mathcal{L}]$-domains, it is easy to prove the following: If $\rho : (\Phi, p) \to (\Phi', p')$, then there exists $\bar{\rho} : (\Gamma_\Phi, \pi) \to (\Gamma_{\Phi'}, \pi)$ such that $\tau' \circ \rho = \bar{\rho} \circ \tau$, where $\tau: (\Phi, p) \Rightarrow (\Gamma_\Phi, \pi)$ and $\tau' : (\Phi', p') \Rightarrow (\Gamma_{\Phi'}, \pi)$.

4. Naturality of $[\Sigma, \mathcal{O}\!\mathcal{L}]$-domains.

In finite dimensions [2], as well as in certain infinite dimensional cases [3, 11], an alternative approach to the construction of domains of holomorphy involves the (appropriately defined) spectrum of the algebra of holomorphic functions on the given domain. In view of the central role played by the spectrum in our study of natural systems, it might be expected that the spectrum approach would offer the most convenient method of constructing maximal $[\Sigma, \mathcal{O}\!\mathcal{L}]$-domains. However, there are certain difficulties in the case of general $[\Sigma, \mathcal{O}\!\mathcal{L}]$-domains that we are as yet unable to overcome. An important related question concerns the naturality of a system associated with a maximal $[\Sigma, \mathcal{O}\!\mathcal{L}]$-domain. The problem is to prove naturality, under appropriate general conditions, for a system of the form $[\Gamma, \mathcal{O}_\Gamma]$, where Γ is a component of the universal space $\Sigma\#_\Lambda(\check{\mathcal{O}})$. Although these problems remain open, we have in the next theorem significant criteria for naturality of a system associated with a $[\Sigma, \mathcal{O}\!\mathcal{L}]$-domain.

Let (Φ, p) be a $[\Sigma, \mathcal{O}\!\mathcal{L}]$-domain, where, as before, we assume that the spaces Σ, Φ are connected and that $[\Sigma, \mathcal{O}\!\mathcal{L}]$ is natural. It will not be necessary, however, to require that $[\Sigma, \mathcal{O}\!\mathcal{L}]$ satisfy the uniqueness principle. The first problem here is that $[\Phi, \mathcal{O}_\Phi]$ need not be a system. We shall bypass this problem by simply assuming that $[\Phi, \mathcal{O}_\Phi]$ is a system. As usual, we denote by $\{h_\lambda : \lambda \in \Lambda\}$ a total indexing of the elements of $\mathcal{O}_\Phi$.

If K is a compact subset of Φ, then its $\mathcal{O}_\Phi$-hull is the set

$$\hat{K} = \{\varphi \in \Phi : |h(\varphi)| \leq |h|_K, h \in \mathcal{O}_\Phi\}.$$

When $[\Phi, \mathcal{O}_\Phi]$ is natural, the $\mathcal{O}_\Phi$-hull of every compact subset of Φ is also compact. Since $\mathcal{O}_\Phi$-holomorphic functions are continuous, $\hat{K}$ is always closed but will not in general be compact. If $\hat{K}$ is compact for every compact $K \subset \Phi$, then Φ is said to be **$\mathfrak{A}$-holomorphically convex** (or **$\mathcal{O}_\Phi$-convex**). A deep result contained in the next theorem is that $\mathfrak{A}$-holomorphic convexity implies naturality. We introduce next a notion of "completeness" for the space Φ.

Consider an arbitrary "directed" set Δ. This means that Δ is partially ordered by an order relation "$<$" with the property that, for δ' and δ'' in Δ, there exists $\delta \in \Delta$ such that $\delta' < \delta$ and $\delta'' < \delta$. If $\delta \mapsto \varphi_\delta$ is an association of an element of Φ with each $\delta \in \Delta$, then $\{\varphi_\delta\}$ is called a **net** in Φ. A net $\{\varphi_\delta\}$ converges to $\varphi \in \Phi$ (written $\lim_\delta \varphi_\delta = \varphi$) if there is associated with each neighborhood U of φ a $\delta_U \in \Delta$ such that $\delta_U < \delta$ implies $\varphi_\delta \in U$. If the $\lim_\delta \varphi_\delta$ exists, then it is unique. Note that, since $[\Phi, \mathcal{O}_\Phi]$ is assumed to be a system, neighborhood of the form

$$U_\varphi(L,\epsilon) = \{\varphi' \in \Phi: |h_\lambda(\varphi') - h_\lambda(\varphi)| < \epsilon, \ \lambda \in L\}$$

where L is a finite subset of Λ and $\epsilon > 0$, constitute a basis for the topology in Φ. The net $\{\varphi_\delta\}$ is said to be Cauchy if, for an arbitrary finite set $L \subset \Lambda$ and $\epsilon > 0$, there exists $\delta(L,\epsilon) \in \Delta$ such that $\delta(L,\epsilon) < \delta, \delta'$ implies

$$|h_\lambda(\varphi_\delta) - h_\lambda(\varphi_{\delta'})| < \epsilon, \ \lambda \in L.$$

If Cauchy nets converge in Φ then Φ is said to be **complete**. For our purposes, we need a weaker notion of completeness involving only Cauchy nets that are contained in the $\mathcal{O}_\Phi$-hull of some compact set. If $\{\varphi_\delta\} \subset \hat{K}$, then $\{\varphi_\delta\}$ is said to be **dominated** be the compact set K.

Definition. The space Φ is said to be **relatively complete** if each dominated Cauchy net converges in Φ.

We are now ready to state and prove our main theorem.

Theorem. If (Φ,p) is a $[\Sigma, \mathfrak{A}]$-domain such that $[\Phi, \mathcal{O}_\Phi]$ is a system, then the following are equivalent:

(i) Φ is relatively complete.

(ii) Φ is $\mathfrak{A}$-holomorphically convex.

(iii) $[\Phi, \mathcal{O}_\Phi]$ is a natural system.

Proof. That (iii) implies (ii) is a fundamental property of every natural system. For the proof that (ii) implies (i), let $\{\varphi_\delta\}$ be a Cauchy net in Φ which is domainated by the compact set K. By hypothesis, $\hat{K}$ is compact and $\{\varphi_\delta\} \subseteq \hat{K}$. Hence, for each $\delta \in \Delta$, the set $T_\delta = \overline{\{\varphi_{\delta'}: \delta < \delta'\}}$ is a non-empty compact set in Φ. Moreover, since Δ is directed, the family $\{T_\delta\}$ obviously has the finite intersection property so has a non-empty intersection. Choose a point $\varphi \in \cap T_\delta$. Let L be a finite subset of Λ and $\epsilon > 0$. Then, since $\{\varphi_\delta\}$ is Cauchy, there exists $\delta(L,\epsilon) \in \Delta$ such that $\delta(L,\epsilon) < \delta, \delta'$ implies that

$$|h_\lambda(\varphi_\delta) - h_\lambda(\varphi_{\delta'})| < \frac{\epsilon}{2}, \; \lambda \in L.$$

Since $\varphi \in T_{\delta(L,\epsilon)}$, there exists $\delta' \in \Delta$ such that $\delta(L,\epsilon) < \delta'$ and $\varphi_{\delta'} \in U_{\varphi_0}(L, \frac{\epsilon}{2})$. Therefore, $\delta(L,\epsilon) < \delta$ implies

$$|h_\lambda(\varphi_\delta) - h_\lambda(\varphi)| \leqslant |h_\lambda(\varphi_\delta) - h_\lambda(\varphi_{\delta'})| + |h_\lambda(\varphi_{\delta'}) - h_\lambda(\varphi)|$$
$$< \frac{\epsilon}{2} + \frac{\epsilon}{2} = \epsilon, \; \lambda \in L,$$

so $\varphi_\delta \in U_{\varphi_0}(L, \epsilon)$. In other words, $\lim_\delta \varphi_\delta = \varphi$, proving that (ii) implies (i). Note that the implications (iii) $\Rightarrow$ (ii) $\Rightarrow$ (i) hold for an arbitrary system so do not depend on (Φ,p) being a $[\Sigma, \mathfrak{A}]$-domain. On the other hand, the implication (i) $\Rightarrow$ (iii) does depend upon the fact that (Φ, p) is a $[\Sigma, \mathfrak{A}]$-domain and that $[\Sigma, \mathfrak{A}]$ is natural. The proof is also much more difficult than the above.

Consider first the mapping,

$$\tau : \Phi \to \Sigma \times C^\Lambda, \; \varphi \mapsto (p(\varphi), \check{h}(\varphi)),$$

and set $\tau(\varphi) = \tilde{\varphi}$. Since $[\Phi, \mathfrak{O}_\Phi]$ is a system, τ is a homeomorphism. Note that $\tilde{\Phi}$ is the graph of the function $\check{h}$: $\varphi \mapsto \{h_\lambda(\varphi)\}$ and is contained in the open set $G \times C^\Lambda$, where $p(\Phi) = G$. Since $[\Sigma, \mathfrak{A}]$ and $[C^\Lambda, \mathfrak{P}]$ are natural, the system $[\Sigma \times C^\Lambda, \mathfrak{A} \times \mathfrak{P}]$ is also natural, where $\mathfrak{A} \times \mathfrak{P}$ (the Kronecker or tensor product of $\mathfrak{A}$ and $\mathfrak{P}$) is the algebra of all polynomials in the variables $\{\zeta_\lambda\}$ with coefficients in $\mathfrak{A}$ regarded in the obvious way as functions on $\Sigma \times C^\Lambda$. The remainder of the proof that (i) $\Rightarrow$ (iii) procedes through a sequence of steps.

(1). The homeomorphism τ induces an isomorphism between the two systems $[\Phi, \mathfrak{O}_\Phi]$ and $[\tilde{\Phi}, \mathfrak{A} \times \mathfrak{P}]$.

Let F be an arbitrary element of $\mathfrak{A} \times \mathfrak{P}$. Then, for $\varphi \in \Phi$, $F(\tilde{\varphi}) = F(p(\varphi), \check{h}(\varphi))$. On the right of this equation, we have a polynomial in a finite number of the functions h_λ with coefficients of the form $a \circ p$, where $a \in \mathfrak{A}$. Since $\mathfrak{A} \circ p \subseteq \mathfrak{O}_\Phi$, it follows that the function

$$h_F : \varphi \mapsto F(p(\varphi), \check{h}(\varphi)), \; \varphi \in \Phi,$$

belongs to $\mathfrak{O}_\Phi$. The mapping $F \mapsto h_F$ is obviously a homomorphism of $\mathfrak{A} \times \mathfrak{P}$ into $\mathfrak{O}_\Phi$. Furthermore, for each $\mu \in \Lambda$ define

$$F_\mu(\sigma, \check{\zeta}) = \zeta_\mu, \; (\sigma, \check{\zeta}) \in \Sigma \times C^\Lambda.$$

Then $F_\mu \in \mathfrak{A} \times \mathfrak{P}$ and $h_{F_\mu} = h_\mu$, so $\mathfrak{A} \times \mathfrak{P}$ maps onto $\mathfrak{O}_\Phi$. In addition, $h_F = 0$ iff $F|\tilde{\Phi} = 0$, so $F|\tilde{\Phi} \mapsto h_F$ is an isomorphism the algebra $\mathfrak{A} \times \mathfrak{P} | \tilde{\Phi}$ with $\mathfrak{O}_\Phi$. Since $h_F = F \circ \tau$, this completes the proof of (1).

(2). $\tilde{\Phi}$ is a "local" $\mathfrak{A} \times \mathfrak{P}$-analytic subvariety of the open set $G \times C^\Lambda$.

This means simply that, for each $\tilde{\varphi}_0 \in \tilde{\Phi}$, there exists a neighborhood $N_{\tilde{\varphi}_0}$ of $\tilde{\varphi}_0$ in $G \times C^{\Lambda}$ such that $N_{\tilde{\varphi}_0} \cap \tilde{\Phi}$ is the set of common zeros of $\mathfrak{A} \times \mathfrak{P}$-holomorphic functions defined on $N_{\tilde{\varphi}_0}$. (Note that $\tilde{\Phi}$ need not be relatively closed in $G \times C^{\Lambda}$ so may not be an ordinary subvariety.) For the proof of (2), first choose a p-neighborhood U_{φ_0} of the point φ_0 in Φ. Since $[\Phi, \mathcal{O}_\Phi]$ is a system, we may assume that U_{φ_0} is a basic $\mathcal{O}_\Phi$-neighborhood, say $U_{\varphi_0}(L,\epsilon)$, where L is a finite subset of Λ and $\epsilon > 0$. Set $\sigma_0 = p(\varphi_0)$ and $V_{\sigma_0} = p(U_{\varphi_0})$. Then V_{σ_0} is a neighborhood of σ_0 in Σ. Also set $\check{\zeta}^0 = \check{h}(\varphi_0)$ and consider the basic neighborhood $N_{\check{\zeta}^0}(L,\epsilon)$ of the point $\check{\zeta}^0$ in C^{Λ}. Then $V_{\sigma_0} \times N_{\check{\zeta}^0} = W_{\tilde{\varphi}_0}$ is a neighborhood of $\tilde{\varphi}_0$ in $G \times C^{\Lambda}$. Now, for each $\mu \in \Lambda$, define the function

$$H_\mu(\sigma,\check{\zeta}) = (h_\mu \circ p^{-1}_{U_{\varphi_0}})(\sigma) - \zeta_\mu \;, \quad (\sigma,\check{\zeta}) \in W_{\tilde{\varphi}_0} .$$

Then H_μ is $\mathfrak{A} \times \mathfrak{P}$-holomorphic in $W_{\tilde{\varphi}_0}$ and $H_\mu(\tilde{\varphi}) = 0$ for $\tilde{\varphi} \in W_{\tilde{\varphi}_0} \cap \tilde{\Phi}$. Moreover, if $(\sigma,\check{\zeta}) \in W_{\tilde{\varphi}_0} \setminus \tilde{\Phi}$ and $p(\varphi) = \sigma$, then $\check{\zeta} \neq \check{h}(\varphi)$. Hence there exists $\mu \in \Lambda$ such that $\zeta_\mu \neq h_\mu(\varphi)$. In particular, $H_\mu(\sigma,\check{\zeta}) \neq 0$, so (2) is proved.

(3). Let K be a compact subset of Φ and let $\hat{K}$ be its $\mathcal{O}_\Phi$-hull in Φ. Then $\tilde{\hat{K}} = \hat{\tilde{K}} \cap \tilde{\Phi}$, where "$\wedge$" denotes the $\mathfrak{A} \times \mathfrak{P}$-hull in $\Sigma \times C^{\Lambda}$.

Let $F \mapsto h_F$ be the homomorphism of $\mathfrak{A} \times \mathfrak{P}$ onto $\mathcal{O}_\Phi$ defined above. Then $F(\tilde{\varphi}) = h_F(\varphi)$, for each $\varphi \in \Phi$, and $|F|_{\tilde{K}} = |h_F|_K$ for compact $K \subset \Phi$. Therefore, $|h_F(\varphi)| \leq |h_F|_K$ iff $|F(\tilde{\varphi})| \leq |F|_{\tilde{K}}$, which implies (3).

(4). The set $\tilde{\Phi}$ will be $\mathfrak{A} \times \mathfrak{P}$-convex in $\Sigma \times C^{\Lambda}$ iff $\hat{\tilde{K}} \cap \tilde{\Phi}$ is a closed (and hence compact) set for each compact set $K \subset \Phi$.

Since τ: $\Phi \to \tilde{\Phi}$ is a homeomorphism, $\tilde{K}$ is a compact set for each compact set $K \subset \Phi$, and every compact subset of $\tilde{\Phi}$ is of this form. Thus, if $\tilde{\Phi}$ is $\mathfrak{A} \times \mathfrak{P}$-convex, then, by a fundamental property of natural systems, $\hat{\tilde{K}}$ is compact and contained in $\tilde{\Phi}$ so, in particular, $\hat{\tilde{K}} \cap \tilde{\Phi}$ is closed. For the opposite implication, we must show that $\hat{\tilde{K}} \cap \tilde{\Phi}$ closed implies $\hat{\tilde{K}} \subseteq \tilde{\Phi}$. Suppose, on the contrary, that $\hat{\tilde{K}} \cap \tilde{\Phi}$ is closed but that the set $H = \hat{\tilde{K}} \setminus \tilde{\Phi}$ is non-empty. Then H is a relatively open subset of $\hat{\tilde{K}}$ and, since $\tilde{K} \subseteq \hat{\tilde{K}} \cap \tilde{\Phi}$, the set $\hat{\tilde{K}} \cap \tilde{\Phi}$ is an $\mathfrak{A} \times \mathfrak{P}$-boundary for $\hat{\tilde{K}}$. Therefore, the lemma stated in the introduction applies. However, the existence of a point on the topological boundary of $\hat{\tilde{K}} \cap \tilde{\Phi}$ in the space $\hat{\tilde{K}}$ with the properties asserted by the lemma is obviously incompatible with the fact that $\tilde{\Phi}$ is a local subvariety of $G \times C^{\Lambda}$. Hence we conclude that $\hat{\tilde{K}}$ must be contained in $\tilde{\Phi}$, completing the proof of (4).

(5). If $(\sigma_0, \check{\zeta}^0)$ is a limit point of $\hat{\tilde{K}} \cap \tilde{\Phi}$ in $\Sigma \times C^{\Lambda}$, then there exists a dominated Cauchy net $\{\varphi_\delta\}$ in Φ such that $\lim_\delta \tilde{\varphi}_\delta = (\sigma_0, \check{\zeta}^0)$.

Denote by Δ the set of all triples, (A,L,n), where A is a finite subset of $\mathfrak{A}$, L is a finite subset of Λ, and n is a positive integer. For $\delta_1 = (A_1,L_1,n_1)$ and $\delta_2 = (A_2,L_2,n_2)$, define $\delta_1 < \delta_2$ if $\delta_1 \neq \delta_2$ and

$A_1 \subseteq A_2, L_1 \subseteq L_2, n_1 \leqslant n_2$. Then Δ is a directed set. Now, for each $\delta = (A,L,n)$, consider the neighborhood

$$W_\delta = \{\sigma : |a(\sigma) - a(\sigma_0)| < \frac{1}{n}, a \in A\} \times \{\check{\zeta} : |\zeta_\lambda - \zeta_\lambda^0| < \frac{1}{n}, \lambda \in L\}$$

of the point $(\sigma_0, \check{\zeta}^0)$ in $\Sigma \times C^\Lambda$. Since $(\sigma_0, \check{\zeta}^0)$ is a limit point of the set $\widehat{\widetilde{K}} \cap \widetilde{\Phi}$, we can choose $\varphi_\delta \in W_\delta \cap \widehat{\widetilde{K}}$. By (3), we have $\varphi_\delta \in \widehat{K}$. Hence $\{\varphi_\delta\}$ is a net in Φ dominated by the compact set K. We prove that $\{\varphi_\delta\}$ is a Cauchy net. Let L_0 be an arbitrary finite subset of Λ and $\epsilon > 0$. Choose an integer $n_0 > 2\epsilon^{-1}$, and let A_0 be an arbitrary finite subset of $\mathfrak{A}$, and set $\delta_0 = (A_0, L_0, n_0)$. Consider arbitrary $\delta = (A,L,n)$ and $\delta' = (A',L',n')$ such that $\delta_0 < \delta, \delta'$. Then, in particular, $L_0 \subseteq L \cap L'$ and $n_0 \leqslant n,n'$ so $\lambda \in L_0$ implies

$$|h_\lambda(\varphi_\delta) - h_\lambda(\varphi_{\delta'})| \leqslant |h_\lambda(\varphi_\delta) - \zeta_\lambda^0| + |\zeta_\lambda^0 - h_\lambda(\varphi_{\delta'})|$$
$$< \frac{1}{n} + \frac{1}{n'} < \frac{2}{n_0} < \epsilon .$$

Thus, $\{\varphi_\delta\}$ is a Cauchy net. Finally, let W_0 be an arbitrary basic neighborhood of $(\sigma_0, \check{\zeta}^0)$ in $\Sigma \times C^\Lambda$. Since $W_0 = V_0 \times N_0$, where V_0 is a neighborhood of σ_0 in Σ and N_0 is a neighborhood of $\check{\zeta}^0$ in C^Λ, there exists a neighborhood W_δ contained in W_0. Since $\delta_0 < \delta$ implies $W_\delta \subseteq W_{\delta_0}$, it follows that $\delta_0 < \delta$ implies $\widetilde{\varphi}_\delta \in W_0$. In other words, $\lim_\delta \widetilde{\varphi}_\delta = (\sigma_0, \check{\zeta}^0)$, proving (5).

(6). (i) implies (iii).

Since Φ is assumed to be relatively complete, the Cauchy net $\{\varphi_\delta\}$ in (5) converges to a point $\varphi_0 \in \Phi$. Also, since $\tau : \Phi \to \widetilde{\Phi}$ is continuous, it follows that $(\sigma_0, \check{\zeta}^0) = \lim_\delta \widetilde{\varphi}_\delta = \widetilde{\varphi}_0$. This proves that $\widetilde{K} \cap \widetilde{\Phi}$ is closed in $\Sigma \times C^\Lambda$ and therefore, by (4), $\widetilde{\Phi}$ is $\mathfrak{A} \times \mathfrak{P}$-convex. This implies, since $[\Sigma \times C^\Lambda, \mathfrak{A} \times \mathfrak{P}]$ is natural, that $[\widetilde{\Phi}, \mathfrak{A} \times \mathfrak{P}]$ is natural and finally, by (1), that $[\Phi, \mathfrak{O}_\Phi]$ must be natural, completing the proof of the theorem.

The fact that $\mathfrak{A}$-holomorphic convexity of Φ implies naturality of the system $[\Phi, \mathfrak{O}_\Phi]$, given by the above theorem, provides a partial solution of a problem that we have mentioned elsewhere [9]; viz. the problem of proving that "Stein-like" systems are natural.

(Research supported by NSF Grant GP-30673X.)

Bibliography

1. I. Glicksberg, Maximal algebras and a theorem of Rado, Pacific J. Math. 14(1964), 919–941.

2. R. C. Gunning and H. Rossi, **Analytic Functions of Several Complex Variables**, Prentice-Hall Series in Modern Analysis, Prentice-Hall, Englewood Cliffs, N.J., 1965.

3. M. C. Matos, Holomorphic mappings and domains of holomorphy, Dissertation, University of Rochester, 1970.

4. Ph. Noverraz, **Pseudo-Convexité, Convexité Polynomiale et Domaines d'Holomorphie en Dimension Infinite**, North-Holland Math. Studies, No. 3, 1973.

5. C. E. Rickart, Analytic phenomena in general function algebras, Pacific J. Math. 18(1966), 361–377.

6. __________, Holomorphic convexity in general function algebras, Canadian J. Math. 20(1968), 272–290.

7. __________, Analytic functions of an infinite number of complex variables, Duke Math. J. 36(1969), 581–598.

8. __________, Plurisubharmonic functions and convexity properties for general function algebras, Trans. Amer. Math. Soc. 169(1972), 1–24.

9. __________, A function algebra approach to infinite dimensional holomorphy, Proc. Colloq. Analysis, Rio de Janeiro, 1972 (Ed.: L. Nachbin), Act. Sci. et Ind., Hermann, Paris, 1973.

10. H. Rossi, The local maximum modulus principle, Ann. of Math. (2) 72(1960), 1–11.

11. M. Schottenloher, Analytische Fortsetzung in Banachräumen, Dissertation, Munich 1971, Math. Annalen 199(1972), 313–336.

Yale University
New Haven, Connecticut 06520

WEAK ANALYTIC CONTINUATION FROM COMPACT SUBSETS OF C^n

Józef Siciak

§1. **Introduction.** Let X be a compact subset of C^n. Put

$$(1) \qquad L(z,X) = \sup_{\nu \geq 1} [\sup \{|f(z)| : f \in \mathcal{F}_\nu\}]^{1/\nu}, \quad z \in C^n,$$

where $\mathcal{F}_\nu = \{f\colon f$ is a polynomial of n complex variables of degree $\leq \nu$ such that $\|f\|_X \leq 1\}$, $\|f\|_X = \sup\{|f(z)| : z \in X\}$. The function (1) will be called a **polynomial extremal function** associated with the set X.

The function (1) appears to be very useful in the theory of interpolation and approximation by polynomials of n complex variables. For its basic properties and applications see [5]. Let us recall the following properties of the extremal function L:

(P1) $L(z,X) = 1$ in $\hat{X}$, where $\hat{X}$ denotes the polynomial envelope of X. $L(z,X) > 1$ in $C^n \setminus \hat{X}$.

(P2) $L(z,X) \leq L(z,Y)$ for all $z \in C^n$, if $Y \subset X$.

(P3) If X_j $(j = 1,\ldots,n)$ is a compact subset of C, then

$$L(z, X_1 \times \ldots \times X_n) = \max\{L(z_1,X_1),\ldots,L(z_n,X_n)\}, \quad z \in C^n.$$

(P4) If n = 1, then L is locally bounded in C if and only if the logarithmic capacity, cap X, of X is positive.

(P5) If n = 1 and cap X > 0, then $L(z,X) = \exp G(z)$ for $z \in D$, where D denotes the unbounded component of $C \setminus X$ and G is the Green's function for D with pole at infinity.

A compact set $X \subset C^n$ will be called **regular**, if the extremal function L associated with X is continuous at each point of X.

A compact set $X \subset R^n$ will be called regular, if X treated as a subset of C^n is regular.

A compact set $X \subset C^n$ (R^n) will be called **strongly regular**, if it is regular and satisfies the following local identity property:

Let G be an open connected subset of C^n (R^n) such that $X \cap G \neq \phi$, $X \cap \partial G = \phi$. If f_j is holomorphic (real analytic) in G, j = 1,2, and $f_1 | X \cap G = f_2 | X \cap G$, then $f_1 = f_2$ in G.

Conjecture. If X is polynomially convex then X is strongly regular if and only if it is regular.

The conjecture is true, if n = 1.

Example 1. A compact set $X \subset C^n$ is strongly regular, if for every point $a \in X$ there exist continua $Y_j \subset C$ $(j = 1,\ldots,n)$, not reduced to a point, such that $a \in Y_1 \times \ldots \times Y_n \subset X$.

Example 2. The classical Cantor set is (as a subset of C) strongly regular.

Example 3. If $n \geq 2$, the set $X = \{x \in C^n : |z_j| = 1, j = 1, \ldots, n\} \cup \{0\}$ is regular but it is not strongly regular. However X is not polynomially convex.

It follows from the definition of L that if $A: C^n \to C^n$ is any one-to-one affine mapping, then

(P5) $$L(z, AX) = L(A^{-1}z, X), \quad z \in C^n.$$

Therefore X is regular if and only if AX is regular for every one-to-one affine mapping A. This property permits one to produce further examples of regular (or strongly regular) sets.

We shall need the following approximation lemmas.

Lemma A. Let X be a polynomially convex regular compact subset of $C^n(R^n)$.

Then there exists a sequence of approximating operators $\{A_\nu\}$ with the following properties:

(i) If F is any complex (real) locally convex sequentially complete topological vector space (l.c.s.c.t.v.s.) and $f: X \to F$ is any function defined on X with values in F, then $A_\nu f$ is a polynomial of n complex (real) variables with values in F of degree $\leq \nu$. Moreover for every $u \in F'$ $A_\nu(u \circ f) = u \circ A_\nu f$.

(ii) Let $f: X \to C$ be any complex function defined on X. Then f is continuable to a holomorphic function in a neighborhood of X if and only if

$$\theta \equiv \limsup_{\nu \to \infty} \sqrt[\nu]{\|f - A_\nu f\|_X} < 1.$$

The number θ depends on f.

This Lemma is a simple consequence of Theorems 10.2 and 11.1 in [5]. A_ν may be chosen as the Lagrange interpolating operators with nodes suitably chosen in X.

Lemma B. Let D be an open connected set in the complex plane C. Let X be a compact subset of D with the positive logarithmic capacity.

Given any open set G such that $X \subset G \subset\subset D$, there exists a sequence operators $\{A_\nu\}$ with the following properties:

(i) If f is any l.c.s.c.t.v.s. over C and if $f: X \to F$ is any function defined on X with values in F, then $A_\nu f$ is a holomorphic function in D, $A_\nu(u \circ f) = u \circ A_\nu f$ for every $u \in F'$ and $A_\nu(u \circ f)$ is a rational function holomorphic in D.

(ii) There exists θ, $0 < \theta < 1$, such that

$$\limsup_{\nu \to \infty} \sqrt[\nu]{\|f - A_\nu f\|_G} \leq \theta$$

for every complex function f holomorphic in D.

This Lemma follows from the Approximation Lemma in [6], p.160.

§2. **Weak analytic continuations.** To begin with we shall prove

Theorem 1. **Let X be a polynomially convex regular compact subset of C^n (R^n). Let F be a complex (real) Banach space. Let f: $X \to F$ be a function defined on X with values in F such that for every $u \in F'$ there exists a holomorphic (analytic) function f_u in an open neighborhood $\Omega_u \subset C^n$ ($\Omega_u \subset R^n$) of X with values in C(R) such that $f_u|X = u \circ f$.**

Then there exists a function $\tilde{f}$ holomorphic (analytic) in a neighborhood Ω of X such that $\tilde{f}|X = f$.

Proof. Put

$$W_{kr} = \{u \in F': \|u\circ(f - A_\nu f)\|_X \leq k r^\nu,\ \nu \geq 1\},$$

where k = 1,2,. . ., and r is any rational number between 0 and 1. By Lemma A, $F' = \cup_{k,r} W_{kr}$. The sets W_{kr} are closed. Therefore by the Baire property of F' there exist k and r such that $\text{int}(W_{kr}) \neq \phi$. Hence by standard reasoning

(1) $$\|u\circ(f - A_\nu f)\|_X \leq M_u r^\nu,\ \nu \geq 1,\ u \in F',$$

where M_u is a positive constant depending on $u \in F'$ but not on ν. Therefore there exists $M > 0$ such that

$$\|f - A_\nu f\|_X \leq M r^\nu,\ \nu \geq 1.$$

Thus $f = \lim A_\nu f$ on X. Since $A_\nu f$ is a polynomial of n complex variables, by the Weierstrass convergence theorem it is enough to show that the sequence $\{A_\nu f\}$ converges uniformly in a neighborhood Ω of X in C^n. To this aim we put

$$f_1 = A_1 f,\ f_\nu = A_\nu f - A_{\nu-1} f \text{ for } \nu \geq 2.$$

It follows from (1) that

$$\|u\circ f_\nu\|_X \leq M_u r^\nu,\ \nu \in 2,\ u \in F',$$

where M_u is a positive constant depending on $u \in F'$. By the definition of the extremal function L we have

$$|u\circ f_\nu(z)| \leq M_u[rL(z,X)]^\nu,\ z \in C^n,\ \nu \geq 2,\ u \in F',$$

whence it follows that there exists $M > 0$ such that

$$\|f_\nu(z)\| \leq M[rL(z,X)]^\nu,\ z \in C^n,\ \nu \geq 2.$$

The required $\tilde{f}$ and Ω are now given by

$$\tilde{f} = \sum_{\nu=1}^{\infty} f_\nu \text{ for } z \text{ in } \Omega,$$

where $\Omega = \text{int}\{z \in C^n: rL(z,X) < 1\}$.

Theorem 2. Let X be a polynomially convex strongly regular compact subset of C^n (R^n). Let D be an open connected subset of C^n (R^n) containing X in its interior. Let F be a complex (real) Banach space. Let $f: X \to F$ be a function such that for every $u \in F'$ there exists a function f_u holomorphic (analytic) in D such that $f_u|X = u \circ f$.

Then there exists a unique function $\tilde{f}$ holomorphic (analytic) in D such that $\tilde{f}|X = f$.

Proof. By Theorem 1 there exists a function $g: G \to F$ holomorphic (analytic) in an open neighborhood G of X such that $g|X = f$. Of course $u \circ g|X = u \circ f = f_u|X$ for every $u \in F'$.

We may assume that each component of G intersects X. By the identity property of X we have $u \circ g = f_u$ in G for every $u \in F'$. Now we may apply Theorems 3 and 4 of [4] and we conclude that there exists $\tilde{f}$ holomorphic (analytic) in D such that $\tilde{f}|X = f$.

Remark 1. In the real case the assumption of the polynomial convexity of X is superfluous, because in view of the Stone-Weierstrass theorem every compact subset of R^n (treated as a subset of C^n) is polynomially convex.

Remark 2. Theorems 1 and 2 remain true if F' is replaced by any determining manifold $\Phi \subset F'$ (Φ is called determining if Φ is a closed subspace of F' and if there exist two constants $C_1 > 0$ and $C_2 > 0$ such that for every $X \in F$

$$\sup\{|u(x)|: u \in \Phi, \|u\| \leqslant C_1\} > C_2 \|x\|).$$

Remark 3. Theorems 1 and 2 remain true if F is any l.c.s.c.t.v.s. such that F' is a Baire space.

Theorem 3. Let D be an open connected subset of C. Let X be a compact subset of D with cap $X > 0$. Let F denote a complex l.c.s.c.t.v.s. Let $f: X \to F$ be such that for every $u \in F$ there exists a holomorphic function f_u in D such that $f_u|X = u \circ f$.

Then there exists a holomorphic function $\tilde{f}: D \to F$ such that $\tilde{f}|X = f$.

Proof. Let $\{G_k\}$ be an increasing sequence of open connected sets in C such that $X \subset G_k \subset\subset D$ and $D = \bigcup_{k=1}^{\infty} G_k$. Given a fixed positive integer k, it follows from Lemma B that there exists a sequence $\{f_\nu\}$ of holomorphic functions in D with values in F such that $\{f_\nu\}$ is uniformly convergent on G_k and $f = \lim_{\nu \to \infty} f_\nu$ on X. The function $g_k = \lim f_\nu$ is holomorphic in G_k and $g_k|X = f$.

The required function $\tilde{f}$ may be now defined by

$$\tilde{f} = \lim g_k \text{ in } D.$$

Bibliography

1. J. Bochnak and J. Siciak, Polynomials and multilinear mappings in topological vector spaces, Studia Math. 39(1) (1971), 59–76.

2. J. Bochnak and J. Siciak, Analytic functions in topological vector spaces, ibidem 39(1) (1971), 77–112.

3. M. Hervé, Analytic and plurisubharmonic functions in finite and infinite dimensional spaces, Springer Lecture Notes 198.

4. E. Ligocka and J. Siciak, Weak analytic continuation, Bull. Acad. Polon. Sci. 20(6) (1972), 461–466.

5. J. Siciak, On some extremal functions and their applications in the theory of analytic functions of several complex variables, Trans. Amer. Math. Soc. 105(2) (1962), 322–357.

6. J. Siciak, Separately analytic functions and envelopes of holomorphy of some lower dimensional subsets of C^n, Ann. Pol. Math. 22(1969), 145–171.

Institute of Mathematics
ul. Reymonta 4
Krakow, Poland

WEAK ANALYTIC FUNCTIONS AND THE CLOSED GRAPH THEOREM

Lucien Waelbroeck

Jozef SICIAK has given conditions on a compact set X in C^n which ensure that a vector-valued function f which is weakly analytic on X is extendable to a neighbourhood of X, the extension being analytic in the usual sense. The extension of f given by SICIAK is computed in a relatively explicit manner.

We shall show here how the closed graph theorem allows us often to prove the existence of analytic extensions of vector-valued, weakly analytic functions. This does not replace SICIAK's paper, my definition of an analytic function is not identical to SICIAK's. My results are often applicable when SICIAK's definition is used, but the conditions on X are not always identical.

Also, the closed graph theorem proves that extensions exist. SICIAK's methods are more computational.

§1. Let E, F be Banach spaces, and $x \in E$. An F-valued formal power series at x is a sequence $(u_n(y - x))_{n \in N}$ of homogeneous, F-valued polynomials, each u_n being homogeneous of degree n in its argument. As usual, we shall write formally

$$u(y) = \Sigma_{\oplus} u_n(y - x).$$

Let also $\phi \in F^*$ be a continuous linear form on F. We can consider the scalar-valued scalar formal power series, at $x \in E$

$$\langle u(y), \phi \rangle = \Sigma_{\oplus} \langle u_n(y - x), \phi \rangle .$$

Lemma 1. **For the formal series u(y) to be the germ of an analytic function at w, it is necessary and sufficient that some $\epsilon > 0$ and some M can be found such that $\langle u(y), \phi \rangle$ can be extended to the ϵ-neighbourhood of x for all $\phi \in F^*$, in such a way that**

$$|\langle u(y), \phi \rangle| \leqslant M \|\phi\|$$

when $\|y - x\| \leqslant \epsilon$.

This triviality is only worth being called a lemma because we shall have to refer to it later on.

§2. Let X be compact subset of the Banach space E. Assume that an F-valued formal power series $u_x(y)$ is given for all $x \in X$. Assume also that we can find for all $\varphi \in F^*$ a function u^{φ}, holomorphic on a neighbourhood V^{φ} of X and having the power series expansion $\langle u_x, \varphi \rangle$ for all $x \in X$.

Proposition 1. **When these conditions prevail, it is possible to find a neighbourhood U of X and a holomorphic mapping of U into F, say u, such that the germ of u at x is u_x for all $x \in X$.**

To prove this, we consider the mapping $\phi \mapsto u^\phi$, $F^* \to \mathcal{O}(X)$. This mapping is continuous when we put on F^* the weak-star topology, and on $\mathcal{O}(X)$ the topology of pointwise convergence of the function and each of its derivatives. A continuous mapping into a Hausdorff space has closed graph. The mapping $u \mapsto u^\phi$ has thus a closed graph when we consider on F^* and $\mathcal{O}(X)$ the weird topologies described.

We consider next the norm topology of F^*, and the natural direct limit topology of $\mathcal{O}(X)$ (this is the direct limit of the Banach spaces of bounded continuous functions on neighbourhoods of X). These topologies are stronger than the former ones. The graph is again a closed subset of $F^* \times \mathcal{O}(X)$. But we can now apply the Grothendieck closed graph theorem ([1], chapter IV, paragraph 1.5, theorem 1, corollary 1). This shows that the linear mapping $\phi \mapsto u^\phi$ maps F^* into the bounded holomorphic functions on U for some neighbourhood U of X, and that this is a bounded mapping of Banach spaces.

A straightforward application of lemma 1 proves the announced result.

§3. Proposition 1 does not satisfy Siciak. I personally have never been able to use an analytic function that did not have a Taylor expansion. I have for that reason no misgiving in putting the Taylor expansion in the definition of the function.

For Siciak, a function is a mapping. Two functions agree on a compact set X if they take the same values at each point of X. There is a philosophical difference in our points of view, and this leads to a difference in the statements of some theorems.

I shall say that a compact set X has the identity property if f vanishes identically on U as soon as U is an open set, each of whose components meets X, and f vanishes identically on X. The topology of simple convergence on X is then a Hausdorff topology on $\mathcal{O}(X)$.

Proposition 2. Let E be a Banach space, and $X \subseteq E$ be compact and have the identity property. Let $u: X \to F$ be a mapping. Assume that it is possible to find a neighbourhood V^ϕ of X for each $\phi \in F^*$ and a function u^ϕ holomorphic on V^ϕ such that $u^\phi(x) = \langle u(x), \phi\rangle$ for all $x \in X$. It is then possible to find a neighbourhood V of X and a holomorphic $v: V \to F$ such that u is the restriction of v to X.

The mapping $\phi \mapsto u^\phi$, $F^* \to \mathcal{O}(X)$ is continuous, has closed graph when we put on F^* the weak-star topology and on $\mathcal{O}(X)$ the simple topology. The same argument as in the proof of proposition 1 yields a continuous mapping $F^* \to \mathcal{O}_\infty(U)$ (the bounded holomorphic functions on U) for some neighbourhood U of X.

It is straightforward to turn this mapping $F^* \to \mathcal{O}_\infty(U)$ around to a mapping $U \to F^{**}$ which is bounded and weak-star Gâteaux holomorphic, i.e. to an element of $\mathcal{O}_\infty(U, F^{**})$. This element is of course an extension, say v, of u to U.

We must still show that v is F-valued rather than F^{**}-valued. This uses again the identity property of F and the Hahn-Banach theorem. Just let ψ be a linear form on F^{**} which vanishes on F. Then $\psi \circ v$ vanishes on X, and on U if each of the components of U meets F.

§4. Our last result is the one most directly related to Siciak's paper. We shall need the nuclearity of $\mathcal{O}(X)$, and to be sure that $\mathcal{O}(X)$ is nuclear, we shall require that X is compact in a finite dimensional space. We shall not use the condition that X has the identity property.

Proposition 3. Let $X \subseteq C^n$ be compact. Let F be a Banach space. Let $u: X \to F$ be a mapping. Assume that we can find for every $\phi \in F^*$ a function v^ϕ, holomorphic, scalar valued, on a neighbourhood of X and such that $v^\phi(x) = \langle u(x), \phi\rangle$ for every $x \in X$. It is then possible to find a function v, holomorphic and F-valued on a neighbourhood of X, and such that $v(x) = u(x)$ for all $x \in X$.

In our proof, we shall consider the algebra $\mathcal{O}(X)$ of sections of the holomorphic sheaf over X. This is a nuclear Silva algebra. We shall also need the ideal α of elements of $\mathcal{O}(X)$ which vanish at all $x \in X$ and the quotient $H(X) = \mathcal{O}(X)/\alpha$. The elements of H(X) are the functions on X which possess a holomorphic extension to some neighbourhood of X.

Also, H(X) is a separated quotient of a nuclear Silva algebra, and is therefore a nuclear Silva algebra. It will be useful later to have precise notations. We shall let $\mathcal{O}_n(X)$ be the continuous functions on a closed neighbourhood of X of order 2^{-n} whose restriction to the interior is holomorphic, with the obvious Banach space norm. We shall also let $H_n(X) = \mathcal{O}_n(X)/(\alpha \cap \mathcal{O}_n(X))$. Then $\mathcal{O}(X) = \cup\, \mathcal{O}_n(X)$, $H(X) = \cup\, H_n(X)$. The identity mapping $\mathcal{O}_n(X) \to \mathcal{O}_{n+1}(X)$, $H_n(X) \to H_{n+1}(X)$ are nuclear.

H(X) is of course interesting in this context. The element $v^\phi \in \mathcal{O}(X)$ which has $\langle u,\phi\rangle$ as restriction to X is not completely determined by u and ϕ but its quotient image $u^\phi \in H(X)$ is determined. We may speak of the mapping $\phi \to u^\phi$, $F^* \to H(X)$. This mapping is continuous when we put on F^* the weak-star topology and on H(X) the topology of pointwise convergence. And pointwise convergence on X is Hausdorff on H(X). By now we know how to apply the closed graph theorem and show that we have a bounded linear mapping $F^* \to H_n(X)$ for some large enough n.

We remember that our mapping is continuous for the weak-star topology of F^* and the pointwise topology of $H_n(X)$. The unit-ball of F^* is mapped continuously onto some bounded set of $H_n(X)$ (when each is equipped with the relevant topology). Also, pointwise convergence coincides on bounded subsets of $H_n(X)$ with the topology determined by the norm of $H_{n+1}(X)$.

The linear mapping $\phi \to u^\phi$, $F^* \to H(X)$ thus maps the unit-ball into a bounded subset of H(X). It is continuous on the unit-ball of F^* when this unit ball is equipped with the topology induced by F. We know (when H(X) has, as here, the approximation property) that this mapping is represented by an element of $H(X) \hat{\hat{\otimes}} F$ (the injective tensor product). And since H(X) is nuclear, $H(X) \hat{\hat{\otimes}} F = H(X) \hat{\otimes} F$.

It is thus possible to find a sequence of scalars $\lambda_n > 0$, $\Sigma \lambda_n < \infty$, a bounded sequence U_n in H(X) and a bounded sequence e_n in H such that the mapping $\phi \to u^\phi$ is represented by the element $\Sigma \lambda_n U_n \otimes e_n \in H(X) \hat{\otimes} F$. The bounded sequence U_n can be lifted up to a bounded sequence $V_n \in \mathcal{O}(X)$ and $\Sigma \lambda V_n \otimes e_n$ belongs to $\mathcal{O}(X) \hat{\otimes} F = \mathcal{O}(X, F)$ and has the required restriction to X.

Bibliography

1. A. Grothendieck, **Espaces vectoriels topologiques,** Pub. Soc. Math., S. Paulo, 1964.

Université Libre de Bruxelles
Faculte Des Sciences
Avenue F.-D. Roosevelt, 50
1050 Bruxelles
Belgium

THE HOLOMORPHIC FUNCTIONAL CALCULUS
AND INFINITE DIMENSIONAL HOLOMORPHY

Lucien Waelbroeck

§1. **Introduction.** My field is not infinite dimensional holomorphy. Most of my research has been about commutative topological algebras that were not Banach. The holomorphic functional calculus is one of the subjects I have worked on, and this is connected with holomorphy.

The idea is to find a way of applying results about holomorphic functions to the theory of Banach, and topological algebras. But we have of course the feedback principle. Applications tend to enrich a theory. It seems reasonable in a meeting such as this one to speak more of the analytic functions of an infinite number of variables that can be useful in Banach algebra theory, than of the specific applications.

In a first part of this talk, I would speak of the Arens-Calderón trick. This a way of getting past an obstruction in Banach algebra theory, by introducing functions of an infinite number of variables. These holomorphic functions are on "weak topological vector spaces". That means, each function depends on a finite number of variables. It is the set of all functions that depends on infinitely many of these.

I shall in a last part of this talk discuss a holomorphic functional calculus that really involves holomorphic functions on Banach spaces. The situation is essentially different from the previous one. We have good, infinite dimensional holomorphy, but no applications. Of course, the lack of applications is partly due to the embryonic state of the subject of this meeting.

§2. **About SWAC.** The standard holomorphic functional calculus was constructed by Silov [4], Waelbroeck [5], and Arens and Calderón [1] in 1953–55. I was not aware of the existence of Silov's paper when I sent my paper for publication, nor was my paper published when Arens and Calderón sent theirs.

Silov, Arens, and Calderón construct the holomorphic functional calculus with the A. Weil integral formula. It turns out that this formula is unwieldy. What these authors show is that their linear mapping combines in the manner expected with the characters of $\mathcal{A}$. This implies that it is a homomorphism when $\mathcal{A}$ is semi-simple. It is also clear that this is the homomorphism of $\mathcal{O}(S)$ into $\mathcal{A}$ if such a homomorphism exists. But a direct proof that the map $\mathcal{O}(S) \to \mathcal{A}$ given by the Weil integral formula is a homomorphism looks difficult.

In my construction, I use other techniques, involving ideals in the ring of holomorphic functions. With these techniques, it is very easy to prove the existence of the required homomorphism.

§3. **A result about holomorphic function ideals.** Let $r_1, \ldots, r_n, R_1, \ldots, R_N$ be positive real numbers, and $D \subseteq C^{n+N}$ be the polydisc of polyradius (r, R). Let also $P_1, \ldots, P_N$ be polynomials in C^n and let

$$\Delta = \{(z_1, \ldots, z_n) \mid |z_i| \leq r_i, |P_k(z)| \leq R_k\}$$

be the associated polynomial polyhedron.

Consider now $f(z,y) \in \mathcal{O}(D)$ holomorphic on a neighbourhood of D, and $f(z,P(z))$. Clearly, $f(z,P(z)) \in \mathcal{O}(\Delta)$, is holomorphic on a neighbourhood of Δ. The mapping $f(z,y) \mapsto f(z,P(z))$ is a homomorphism $\mathcal{O}(D) \to \mathcal{O}(\Delta)$.

Proposition 1. This homomorphism is surjective. Its kernel is generated by the functions $y_k - P_k(z)$, $k = 1,\ldots,N$.

It is an easy exercise to pull the surjectivity of this mapping out of the statement of lemma 7, Gunning and Rossi [3], Chapter I, section F. The proof of the result about the kernel is an advanced exercise. One must use a little diagram chasing, an induction similar to the one used in the proof of Gunning and Rossi's lemma 7 quoted above, and the fact that $f(z_1,\ldots,z_{n+1})/(z_{n+1} - P(z_1,\ldots,z_n))$ is holomorphic on the domain of f if f vanishes when $z_{n+1} = P(z_1,\ldots,z_n)$.

I like to refer the reader to Gunning and Rossi's chapter I. As the authors say, this chapter can be used as an introductory course in several variables for graduate students. It does not contain any hard core theorem.

§4. Šilov's results. I can now sketch the results that Šilov would have obtained if he had applied the above theorem about holomorphic function ideals, rather than the Weil integral formula. $\mathfrak{A}$ is a finitely generated commutative Banach algebra with unit. This means that we can find a finite system, $a_1,\ldots,a_n$, of elements of $\mathfrak{A}$, which together with the unit, generate a dense subalgebra of $\mathfrak{A}$.

Proposition 2. There is a natural, bijective, bicontinuous correspondence between the structure space $\mathfrak{M}$ of $\mathfrak{A}$ and the joint spectrum $sp(a_1,\ldots,a_n)$ of a generating system. The joint spectrum is polynomially convex. More precisely $(s_1,\ldots,s_n) \in sp(a_1,\ldots,a_n)$ iff for every polynomial P,

$$|P(s_1,\ldots,s_n)| \leqslant \|P(a_1,\ldots,a_n)\|.$$

This is well known and straightforward. If $s \in sp\, a$ and P is a polynomial, $P(s) \in sp\, P(a)$ and $|P(s)| \leqslant \|P(a)\|$. This proves one inclusion.

Conversely, if $s \in \mathbf{C}^n$ is such that $|P(s)| \leqslant \|P(a)\|$ for every polynomial P, we observe that $P(s) = 0$ if $P(a) = 0$, the mapping $P(a) \mapsto P(s)$ is therefore well defined on its domain, on the algebra generated algebraically by the unit and $a_1,\ldots,a_n$. This mapping is also continuous since $|P(s)| \leqslant \|P(a)\|$, and the domain is dense. The mapping $P(a) \mapsto P(s)$ has thus a continuous extension to $\mathfrak{A}$. This extension is a multiplicative linear form. This gives us simultaneously the polynomial convexity of the joint spectrum and the bijection between the spectrum and the structure space.

We next consider a neighbourhood U of $sp(a_1,\ldots,a_n)$. A compactness argument yields a finite number of polynomials $P_1,\ldots,P_N$ and an ϵ such that $U \supseteq V$ where

$$V = \{(z_1, \ldots, z_n) \mid |z_i| \leq \|a_i\| + \epsilon,\ |P_k(z)| \leq \|P_k(a)\| + \epsilon\}.$$

We also let D be the polydisc of polyradius $(\|a\| + \epsilon, \|P(a)\| + \epsilon)$.

Let then $F(z,y) \in \mathcal{O}(D)$, $F(z,y) = \Sigma F_{rs} z^r y^s$. The Cauchy evaluation of F_{rs} allows us to define

$$F[a, P(a)] = \Sigma F_{rs} a^r P(a)^s.$$

The series on the right hand side converges.

The mapping $F(z,y) \to F[a,P(a)]$ is an algebra homomorphism. It vanishes on the polynomials $y_k - P_k(z)$, therefore on the ideal α generated by these polynomials. Proposition 1 ensures that $\mathcal{O}(D)/\alpha$ is isomorphic with $\mathcal{O}(V)$, we therefore have a mapping $\mathcal{O}(V) \to \mathfrak{A}$, which maps z_i on a_i, and unit on unit.

The mapping $F(z,y) \mapsto F[a,P(a)]$, $\mathcal{O}(D) \to \mathfrak{A}$ is a continuous homomorphism. Grothendieck's closed graph theorem ([2], chapter IV, section 1.5, th.1, corollary 1) shows that $\mathcal{O}(V)$ is the topological quotient $\mathcal{O}(D)/\alpha$. The mapping $\mathcal{O}(V) \to \mathfrak{A}$ that we obtain is therefore a continuous homomorphism.

We now return to our original neighbourhood and map $\mathcal{O}(U) \to \mathfrak{A}$ combining the restriction $\mathcal{O}(U) \to \mathcal{O}(V)$ with our mapping $\mathcal{O}(V) \to \mathfrak{A}$. The joint spectrum has a fundamental system of polynomially convex neighbourhoods. Runge's theorem shows that at most one homomorphism $\mathcal{O}(U) \to \mathfrak{A}$ mapping z_i on a_i and unit on unit can exist if U is polynomially convex. All the homomorphisms $\mathcal{O}(U) \to \mathfrak{A}$ that we construct agree, therefore (modulo restriction). We have proved:

Proposition 3. There is a unique, continuous homomorphism

$$\mathcal{O}(\mathrm{sp}(a_1, \ldots, a_n)) \to \mathfrak{A}$$

which maps z_i on a_i and unit on unit, if $\mathfrak{A}$ is a finitely generated commutative Banach algebra with unit, and $a_1, \ldots, a_n$ is a generating system.

§5. The Arens-Calderón trick.

Let now $\mathfrak{A}$ be a commutative Banach algebra with unit. We do not assume $\mathfrak{A}$ to be finitely generated. Let $a_1, \ldots, a_n$ be elements of $\mathfrak{A}$, and $\widetilde{\mathrm{sp}}(a_1, \ldots, a_n)$ be the polynomially convex hull of their joint spectrum. The construction of paragraph 4 gives a continuous homomorphism $\mathcal{O}(\widetilde{\mathrm{sp}}\ a) \to \mathfrak{A}$. This is not satisfactory, we want a continuous $\mathcal{O}(\mathrm{sp}\ a) \to \mathfrak{A}$.

There is no problem if $\mathfrak{A}$ happens to be finitely generated, but the elements $(a_1, \ldots, a_n)$ considered are a non-generating system. We expand $(a_1, \ldots, a_n)$ to a generating $(a_1, \ldots, a_n, b_1, \ldots, b_N)$. We embed $\mathcal{O}(\mathrm{sp}\ a)$ in $\mathcal{O}(\mathrm{sp}(a,b))$ mapping f on $f \circ \Pi_b$, where Π_b: $\mathrm{sp}(a,b) \to \mathrm{sp}\ a$ is the projection map. The holomorphic functional calculus that we have constructed maps $\mathcal{O}(\mathrm{sp}(a,b))$ into $\mathfrak{A}$. We can compose it with the embedding $\mathcal{O}(\mathrm{sp}\ a) \to \mathcal{O}(\mathrm{sp}(a,b))$ and obtain a continuous homomorphism $\mathcal{O}(\mathrm{sp}\ a) \to \mathfrak{A}$ with the usual properties (We do not have any uniqueness, of course).

The Arens-Calderón trick is an extension of this idea. An element $f \in \mathcal{O}(\text{sp } a)$ extends to a neighbourhood U of sp a. We shall find new elements $b_1, \ldots, b_N$ of $\mathfrak{A}$ in such a way that $\Pi_b \, \widetilde{\text{sp}}\,(a,b) \subseteq U$, where as previously Π_b is the projection $C^{n+N} \to C^n$, $\text{sp}(a,b) \to \text{sp } a$. Then $F = f \circ \Pi_b$ is holomorphic on a neighbourhood of $\widetilde{\text{sp}}(a, b)$, and we can define $F[a,b]$ as in paragraph 4.

The existence of the elements $b_1, \ldots, b_N$ is proved once more by compactness. We notice that each $\Pi_b \, \widetilde{\text{sp}}(a,b)$ is compact. This family of sets has the finite intersection property, e.g.

$$\Pi_{bc} \, \widetilde{\text{sp}}(a, b, c) \subseteq \Pi_b \, \widetilde{\text{sp}}(a, b) \cap \Pi_c \, \widetilde{\text{sp}}(a, c).$$

All that remains to be proved is that

$$\text{sp } a = \cap_b \, \Pi_b \, \widetilde{\text{sp}}(a, b).$$

Let $(s_1, \ldots, s_n) \notin \text{sp}(a_1, \ldots, a_n)$ and choose $b_1, \ldots, b_n$ such that $\Sigma(a_i - s_i)b_i = 1$. Then $(s,t) \notin \widetilde{\text{sp}}\,(a,b)$ for all choices of $t \in C^n$ because the polynomial $\Sigma(z_i - s_i)y_i$ is identically one on sp(a,b) and therefore on $\widetilde{\text{sp}}\,(a,b)$, and vanishes on all (s,t) under consideration. This shows that $(s_1, \ldots, s_n) \notin \Pi_b \, \widetilde{\text{sp}}(a,b)$ for this choice of b.

We have therefore proved.

Proposition 4. Let $\mathfrak{A}$ be a commutative Banach algebra with unit, and let $a_1, \ldots, a_n$ be elements of $\mathfrak{A}$. There is a continuous homomorphism $\mathcal{O}(\text{sp } a_1, \ldots, a_n) \to \mathfrak{A}$ which maps z_i on a_i and unit on unit.

This is not quite proved because we have not shown that $f \circ \Pi_b [a,b]$ which we wish to define as $f[a]$ is independent of the arbitrary choice of b, but of course $f \circ \Pi_{bc} [a, b, c] = f \circ \Pi_b [a, b]$. The reader will easily supply the missing details.

In a way, we have lost the uniqueness. But we can reclaim it. We must consider simultaneously all homomorphisms $\mathcal{O}(\text{sp } a_1, \ldots, a_n) \to \mathfrak{A}$. Also, if $i_1, \ldots, i_r$ is an increasing sequence of integers from 1 to n we let $\Pi^n_{i_1 \ldots i_r} : C^n \to C^r$ map $(z_1, \ldots, z_n)$ onto $(z_{i_1}, \ldots, z_{i_r})$.

Proposition 4.bis. There is only one system of homomorphisms $\Phi_{(a_1, \ldots, a_n)} : \mathcal{O}(\text{sp}(a_1, \ldots, a_n)) \to \mathfrak{A}$ mapping z_i on a_i and unit on unit, and such that, in all possible cases

$$\Phi_{(a_{i_1} \ldots a_{i_r})}(f) = \Phi_{(a_1 \ldots a_n)}(f \circ \Pi^n_{i_1 \ldots i_r}).$$

§6. Relation with infinite dimensional holomorphy. Arens and Calderón consider functions of n variables. But along the way, they must introduce functions of n+N variables, with N large and depending on the functions. There are rings of holomorphic functions on topological vector spaces that behave just like that. These are the rings of holomorphic functions on the weak topological vector spaces.

Let (E, &) be a dual pair, consider the topology $\sigma(E, \&)$ on E. The definition of a holomorphic sheaf on E is straightforward. A holomorphic germ at $x \in E$ will be constant near x on the cosets of a weakly closed subspace F of finite codimension (and define a holomorphic germ on E/F). A simple argument (using compactness, and the identity theorems) shows that a section of the holomorphic sheaf over a compact set S extends to a neighbourhood U of S. The algebra of the sections of $\mathcal{O}$ over S can be identified with the direct limit of the algebras $\mathcal{O}(U)$, where U ranges over the neighbourhoods of S.

We assume now that U = S + V where V is an open, absolutely convex neighbourhood of 0, and that f is bounded on U. Let F be the largest subspace contained in V, F is a weakly closed subspace of finite codimension. And f induces a function f_F on a neighbourhood of S_F, the projection of S in E/F. In other words, **$\mathcal{O}(S)$ is the algebra of sections of $\mathcal{O}$ over S. It is the direct limit of the algebras $\mathcal{O}(U)$, U a neighbourhood of S. And it is the direct limit of the finite dimensional algebras $\mathcal{O}(S_F)$.**

Arens and Calderón use some polynomial convexity. Let $(a_i)_{i \in I}$ be a, possibly infinite, generating subset of $\mathcal{A}$, and

$$sp(a_i) = \{(\hat{a}_i(m)) \mid m \in \mathcal{M}\}$$

where $\mathcal{M}$ is the structure space of $\mathcal{A}$. It is clear that $sp(a_i)$ is polynomially convex in C^I, in bijective correspondence with $\mathcal{M}$.

The inclusion $\mathcal{M} \subseteq \mathcal{A}^*$, the topological dual of $\mathcal{A}$, is often considered. It is well known that $\mathcal{M}$ is $\sigma(\mathcal{A}^*, \mathcal{A})$ – polynomially convex in $\mathcal{A}^*$, and even in the completion of $\mathcal{A}^*$, i.e. in $\mathcal{A}^+$, the algebraic dual of $\mathcal{A}$.

Now the gist of the Arens-Calderón trick is the following. Assume $S \subseteq E$ is $\sigma(E, \&)$–compact, and is polynomially convex in the weak completion of E, i.e. in $\&^+$. Then $\mathcal{O}(S)$ is even the direct limit of the spaces $\mathcal{O}(\tilde{S}_F)$, where $\tilde{S}_F$ is the polynomial hull of S_F.

I must thank Mrs. E. Ligocka, who let me observe that it is not obvious that S is polynomially convex in the completion of a locally convex space E when S is polynomially convex in E. The problem of proving that this is the case or of finding a counterexample appears difficult.

§7. **Another approach.** So, the holomorphic functional calculus arising out of the Arens-Calderón trick is infinite dimensional, but not quite. Each holomorphic function on a weak space depends only on a finite number of variables. It is the set of all the holomorphic functions that leads to the introduction of infinitely many variables.

It is possible however to construct a holomorphic functional calculus that does involve genuine holomorphic functions on Banach spaces. A detailed construction will not be given here. The reader may best be referred to [6], chapter 8, paragraphs 6, 7, 8.

The construction is carried out in three steps. We first consider a Banach algebra $\mathcal{A}$, a Banach space E, and an element $a \in \mathcal{A} \hat{\otimes} E$ (the completed projective tensor product) with $\|a\| < 1$. Thus a has an expression $a = \Sigma\, a_n \otimes e_n$ with $\Sigma\, \|a_n\| \cdot \|e_n\| < 1$. We also consider a bounded holomorphic function F on the ball of radius e(= 2.718. . .) and center at the origin. Then F has a Maclaurin expansion

$$F(z) = \Sigma_0^\infty F_n(z)$$

where F_n is homogeneous of degree n. We polarize F_n and obtain a symmetric n-linear form $\hat{F}_n$. The hypotheses lead to an estimate

$$\|\hat{F}_n\| \leqslant M$$

where M is a constant.

It is possible to define

$$F[a] = \Sigma_0^\infty \Sigma_{k_1,\ldots,k_n} \hat{F}_n(e_{k_1},\ldots,e_{k_n}) a_{k_1} \cdots a_{k_n}$$

This operation has the properties that one can expect when substituting an element in a Maclaurin series.

In the second step, we take $a \in \mathcal{A} \otimes E$, i.e. a has finite rank, and $a \in \mathcal{A} \otimes E_k$ for some subspace E_k of finite dimension in E. We define

$$\mathrm{sp}\; a = \mathrm{sp}\; \Sigma\, a_n \otimes e_n = \{ \Sigma\, \hat{a}_n(m) e_n \mid m \in \mathcal{M} \}$$

so $\mathrm{sp}\; a \subseteq E_k$. We obtain a reasonable definition of f[a], with f holomorphic and bounded on a neighbourhood U of sp a when we restrict f to $U \cap E_k$ and apply the finite dimensional holomorphic functional calculus to this restriction f_k and to $a \in \mathcal{A} \otimes E_k$.

In the third step, we combine the above two constructions. We consider a function f which is holomorphic and bounded on a neighbourhood U of sp a. We split $a = a_1 + a_2$ with $a_1 \in \mathcal{A} \otimes E$ and $\|a_2\|$ small. Then U is a neighbourhood of sp a_1, and is relatively large (when compared to the small $\|a_2\|$).

We then write $g(z,y) = f(z + y)$, where z ranges on a neighbourhood of sp a_1, while y ranges on an ϵ'-ball centered at the origin; ϵ' is not too small. The function g belongs $\mathcal{O}(V) \hat{\otimes} \mathcal{O}_b(U_1)$ where V is open and finite dimensional, and $\mathcal{O}_b(U_1)$ are the holomorphic functions of bounded type on the open ball U_1. We may put a projective tensor product here because $\mathcal{O}(V)$ is nuclear.

Now, by the holomorphic functional calculus, substitution of a_1 for z, maps $\mathcal{O}(V)$ into $\mathcal{A}$. Substitution of a_2 for y maps $\mathcal{O}_b(U_1)$ into $\mathcal{A}$. Combining these two maps, we map $\mathcal{O}(V) \hat{\otimes} \mathcal{O}_b(U_1)$ into $\mathcal{A} \hat{\otimes} \mathcal{A}$, which itself is mapped into $\mathcal{A}$ by multiplication.

As already mentioned, the proof is carried out elsewhere in detail.

§8. Final comments. We have thus constructed a holomorphic functional calculus involving good holomorphic functions on Banach spaces. But this operation does not have many applications. Of course, the applications of the usual holomorphic functional calculus are due essentially to the fact that holomorphic functions on $\mathbf{C}^n$ and domains in $\mathbf{C}^n$ are more or less understood at present. There is a hope that a better understanding of infinite dimensional holomorphy could lead to application of our results.

We must mention here the fact that our construction will yield applications of the theory of holomorphic functions on compact subsets of normed spaces. But the compact sets we will encounter will not be the most general compact sets. The Gelfand homomorphism maps $\mathcal{Q} \hat{\otimes} E$ into $C(\mathcal{M}) \hat{\otimes} E$. The range of an element of $C(\mathcal{M}) \hat{\otimes} E$ is a relatively thin compact set.

Of course, if E is a nuclear space, we can study all compact subspaces of E by this method. The present author believes, prima facie, that this suggestion is defeatist. The main property of a nuclear space is that its compact sets are very thin. Each one is much thinner than the range of an $f \in C(X) \hat{\otimes} E$ can be. Instead of studying thin sets in a topological vector space, we study spaces all of whose compact sets are thin.

There is however a direction in which this suggestion of H. Hogbe-Nlend could be constructive. It may be that properties, say of holomorphic function near compact subsets of Fréchet nuclear spaces could be obtained with the holomorphic functional calculus. And these properties could be useful in the study, e.g. of nuclear algebras.

Bibliography

1. R. Arens and A. P. Calderón, **Analytic functions of several Banach algebra elements**, Ann. of Mathematics 62(1955), 204–216.

2. A. Grothendieck, **Espaces vectoriels topologiques**, Pub. Soc. Math., S. Paulo, 1964.

3. R. C. Gunning and H. Rossi, **Analytic functions of several complex variables**, Prentice Hall, 1965.

4. G. E. Silov, **On the decomposition of a commutative normed ring into a direct sum of ideals**, Math. Sbornik. 32, 2(1953), 353–364, also A.M.S. Translation, series 2.

5. L. Waelbroeck, **Le calcul symbolique dans les algèbres commutatives**, Journal de Math. P. et App. 33(1954), 147–186.

6. L. Waelbroeck, **Topological vector spaces and algebras**, Lecture Notes in Mathematics, Springer Verlag, v.230, 1971.

Université Libre de Bruxelles
Faculte Des Sciences
Avenue F.-D. Roosevelt, 50
1050 Bruxelles
Belgium

THE SPECTRUM AS ENVELOPE OF HOLOMORPHY OF A DOMAIN OVER AN ARBITRARY PRODUCT OF COMPLEX LINES

Volker Aurich

§0. Introduction.

For a domain X over $\mathbf{C}^n$ the following conditions are equivalent ([2], p.283):

(1) X is a domain of holomorphy i.e. X coincides with its envelope of holomorphy.

(2) X is holomorphically convex i.e. for any compact subset K of X, the holomorphically convex hull $\widehat{\mathcal{H}}(K,X)$ is compact.

(3) X is Stein.

(4) X is the domain of existence of a holomorphic function.

(5) X is convex with respect to the plurisubharmonic functions.

(6) $-\log d_X$ is plurisubharmonic.

(7) For any sequence $(x_n)_{n \in \mathbf{N}}$ in X with $d_X(x_n) \to 0$, there exists a holomorphic function f on X such that $f(x_n) \to \infty$.

(8) X coincides with Spec $\mathcal{H}(X)$, the space of all nonzero complex algebra homomorphisms on the algebra $\mathcal{H}(X)$ of all holomorphic functions on X.

The equivalence of (1) – (7) holds for a domain X over the product $\mathbf{C}^\Lambda$ of card(Λ) copies of $\mathbf{C}$ if the boundary distance d_X is defined in a suitable manner generalizing the usual boundary distance in finite dimensions ([1]). Besides other equivalent conditions which are analogous to conditions in the finite dimensional case there exists one which is very useful to reduce infinite dimensional problems to finite dimensions:

(9) There exists a finite subset ϕ of Λ and a Stein domain X_ϕ over $\mathbf{C}^\phi$ such that X is isomorphic (as domain over $\mathbf{C}^\Lambda$) to $X_\phi \times \mathbf{C}^{\Lambda-\phi}$.

The proof of the equivalences is based on [3] and [5]. In [3] HIRSCHOWITZ showed the equivalence of (1), (2), (4) and (9) for a domain X in $\mathbf{C}^\mathbf{N}$. In [5] MATOS generalized these results for domains over $\mathbf{C}^\mathbf{N}$; he replaced, however, (2) and (9) by more complicated technical conditions.

Up to now it is not known whether (8) is equivalent to the other conditions in case Λ is infinite. In what follows we shall prove a modification: X is a domain of holomorphy if and only if X coincides with the space of all nonzero $\mathcal{b}$-continuous algebra homomorphisms $\mathcal{H}(X) \to \mathbf{C}$ where $\mathcal{b}$ denotes the bornological topology associated with the compact-open topology on $\mathcal{H}(X)$. As we shall show in section 1 $\mathcal{b}$ has some further nice properties. Moreover the envelope of holomorphy $\mathcal{E}(X)$ is homeomorphic to Spec($\mathcal{H}(X),\mathcal{b}$) endowed with the weak topology, and this involves that $\mathcal{E}$ is a functor in the category of domains and holomorphic maps.

§1. The Space of the Holomorphic Functions on a Domain over C^Λ.

We show that the space $\mathcal{H}(X)$ of the holomorphic functions on a domain X over C^Λ is the union of subspaces $\mathcal{H}^\phi(X)$, $\phi \subset \Lambda$ finite, and each $\mathcal{H}^\phi(X)$ is isomorphic to the space $\mathcal{H}(^\phi X)$ of the holomorphic functions on a finite dimensional domain $^\phi X$ over C^ϕ. $^\phi X$ is obtained as solution of a universal problem. $X \mapsto {}^\phi X$ defines a functor $\mathcal{J}^\phi$, and there is a natural equivalence between the functors $\mathcal{H}^\phi$ and $\mathcal{H} \circ \mathcal{J}^\phi$. If $\mathcal{H}^\phi(X)$ and $\mathcal{H}(^\phi X)$ are endowed with the compact-open topology c, they are homeomorphic and $\mathcal{H}_c \circ \mathcal{J}^\phi$ and $\mathcal{H}_c^\phi$ are equivalent. The locally convex inductive topology $\mathcal{L}$ on $\mathcal{H}(X)$ with respect to the subspaces $\mathcal{H}_c^\phi(X)$ is the bornological topology associated with the compact-open topology on $\mathcal{H}(X)$. $\mathcal{L}$ is Montel and every extension pair is normal with respect to $\mathcal{L}$.

Notations: Throughout all sections Λ_0 is a fixed nonempty set. Λ, Λ', Λ_1, Λ_2 will always be subsets of Λ_0. C^Λ will denote the set of all mappings $\Lambda \to C$ endowed with the topology of pointwise convergence. If $\theta \subset \Lambda$, C^θ will be considered as subspace of C^Λ, and π_θ^Λ or π_θ will denote the projection of C^Λ onto C^θ. $F(\Lambda) := \{\phi \subset \Lambda : \phi$ is finite$\}$; $\phi \in F(\Lambda)$. (Where $:=$ indicates a defining relation).

(1.1) Definition: $p: X \to C^\Lambda$ is called a <u>domain</u> (over C^Λ) iff X is a connected nonempty Hausdorff space and p is locally a homeomorphism.

Let $\Lambda_1 \subset \Lambda_2$. A <u>morphism</u> from a domain $p_1: X_1 \to C^{\Lambda_1}$ to a domain $p_2: X_2 \to C^{\Lambda_2}$ is a continuous mapping $\omega: X_1 \to X_2$ such that the following diagram commutes:

$$\begin{array}{ccc} X_1 & \xrightarrow{\ \omega\ } & X_2 \\ \downarrow p_1 & & \downarrow p_2 \\ C^{\Lambda_1} & \xrightarrow[\ \pi_{\Lambda_2}^{\Lambda_1}\ ]{} & C^{\Lambda_2} \end{array}$$

The domains over C^θ with $\theta \subset \Lambda$ and the morphisms between them form a category $\mathcal{G}(\Lambda_0)$.

(1.2) Lemma: Every morphism is locally a projection, hence open. If Λ_2 is finite and ω is surjective, then ω is semiproper i.e. for every compact set $K_2 \subset X_2$ there exists a compact set $K_1 \subset X_1$ with $\omega(K_1) = K_2$.
The proof is found in [1].

A domain $p: X \to C^\Lambda$ will always be considered as an analytic manifold with the complex structure induced by p. Then every morphism is a holomorphic map. The C-algebra of the holomorphic functions on X will be denoted - "par abus de langage" - by $\mathcal{H}(X)$.

Let $p: X \to C^\Lambda$ be a domain.

Definition: Let $U \subset X$ and f and g be mappings defined on U. We say that f **depends on g** iff f is constant on the connected components of the fibers of g. If h_1 and h_2 are germs in $x \in X$, we say that h_1 depends on h_2 iff h_1 has a representative which depends on a representative of h_2.

Lemma: Let f_x be the germ of a holomorphic function at $x \in X$. There exists $\phi \in F(\Lambda)$ such that f_x depends on $(\pi_\phi \circ p)_x$. Moreover, there exists a smallest set ϕ with this property. It will be denoted by dep(f_x). (cf. [1], [4], [7])

Proof: x has a neighbourhood U such that $p|U \to B \times \mathbf{C}^{\Lambda-\phi}$, with $B \subset \mathbf{C}^\phi$ open, is topological and f_x has a bounded representative g: $U \to \mathbf{C}$. By Liouville's theorem, $g \circ (p|U)^{-1}$ depends on $\pi_\phi|p(U)$. Obviously, the intersection of all ϕ such that f_x depends on $(\pi_\phi \circ p)_x$ has this property, too. q.e.d.

It is easy to verify the following lemma (cf. [1], [3], [4]).

Lemma: Let $f \in \mathcal{H}(X)$. Then the mapping $X \to F(\Lambda)$, $x \mapsto \text{dep}(f_x)$ is constant.

Its value will be denoted by dep(f).

(1.3) Definition of $\mathcal{H}^\theta$: Let $p: X \to \mathbf{C}^\Lambda$ be a domain. For every $\theta \subset \Lambda_0$ $\mathcal{H}^\theta(X) := \{f \in \mathcal{H}(X): \text{dep}(f) \subset \theta\}$. $\mathcal{H}^\theta(X)$ is a $\mathbf{C}$-algebra.

A morphism ω from the domain $p: X \to \mathbf{C}^\Lambda$ to the domain $q: Y \to \mathbf{C}^{\Lambda'}$ induces a homomorphism $\omega^*: \mathcal{H}^\theta(Y) \to \mathcal{H}^\theta(X), f \mapsto f \circ \omega$ for every $\theta \subset \Lambda_0$.

Hence the $\mathcal{H}^\theta$ are contravariant functors from $\mathcal{G}(\Lambda_0)$ to the category of $\mathbf{C}$-algebras.

Since $\mathcal{H}(X) = \cup\{\mathcal{H}^\phi(X): \phi \in F(\Lambda)\}$, $\mathcal{H}(X)$ can be identified with $\lim\,\text{ind}_{\phi \in F(\Lambda)} \mathcal{H}^\phi(X)$.

(1.4) Definition of ${}^\theta X$ and $\mathcal{Y}^\theta$: Let $p: X \to \mathbf{C}^\Lambda$ be a domain. Let $\theta \subset \Lambda_0$. Consider the following equivalence relation on X:

Two points x and y are identified iff $f(x) = f(y)$ for all $f \in \mathcal{H}^\theta(X)$.

Let ${}^\theta X$ be the quotient space (endowed with the quotient topology) and ${}^\theta\sigma_p: X \to {}^\theta X$ the canonical projection.

p factors through ${}^\theta\sigma_p$ to a map ${}^\theta p: {}^\theta X \to \mathbf{C}^{\theta \cap \Lambda}$. It can be shown ([1]) that ${}^\theta p: {}^\theta X \to \mathbf{C}^{\theta \cap \Lambda}$ is a domain which has the following universal property:

For any holomorphically separable domain $q: Y \to \mathbf{C}^\Delta$ with $\Delta \subset \theta \cap \Lambda$ and any morphism $\omega: X \to Y$, there exists a unique morphism $\tilde{\omega}: {}^\theta X \to Y$ such that $\omega = \tilde{\omega} \circ {}^\theta\sigma_p$. (cf. the universal property of a complex base in [9].) This universal property involves that $X \to {}^\theta X$ defines a reflector $\mathcal{Y}^\theta$ in $\mathcal{G}(\Lambda_0)$ (for a morphism φ from p to q $\mathcal{Y}^\theta(\varphi) := \widetilde{({}^\theta\sigma_q \circ \varphi)}$).

Notation: Let $p: X \to \mathbf{C}^\Lambda$ be a domain and $\theta \subset \Lambda$. If we consider $\mathcal{H}^\theta(X)$ as topological vectorspace endowed with a topology τ we shall write $\mathcal{H}^\theta_\tau(X)$. c will denote the compact-open topology. Obviously $\mathcal{H}^\theta_c$ is a contravariant functor from $\mathcal{G}(\Lambda_0)$ to the category of locally convex spaces.

(1.5) Proposition: Let $p\colon X \to C^{\Lambda}$ be a domain and $\phi \in F(\Lambda)$. Then $^{\phi}\sigma_p^*\colon \mathcal{H}(^{\phi}X) \to \mathcal{H}^{\phi}(X)$ is a topological isomorphism.

The family $(^{\phi}\sigma_q^*$: q domain in $\mathcal{G}(\Lambda_0))$ is a natural equivalence of $\mathcal{H}_c^{\phi}$ and $\mathcal{H}_c \circ \mathcal{F}^{\phi}$.

Proof: $^{\phi}\sigma_p^*$ is injective since $^{\phi}\sigma_p$ is surjective. The surjectivity of $^{\phi}\sigma_p^*$ follows from the construction of $^{\phi}X$ and $^{\phi}\sigma_p$. Clearly $^{\phi}\sigma_p^*$ is continuous, and its inverse is continuous because $^{\phi}\sigma_p$ is semiproper by (1.2) q.e.d.

(1.6) Corollary: $\mathcal{H}_c^{\phi}(X)$ is a Fréchet-Montel space.

Let $p\colon X \to C^{\Lambda}$ be a domain. We denote by $\mathfrak{b}$ the locally convex inductive topology on $\mathcal{H}(X)$ with respect to the subspaces $\mathcal{H}_c^{\phi}(X)$ with $\phi \in F(\Lambda)$. Obviously $\mathcal{H}_{\mathfrak{b}}(X)$ is the strict inductive limit of the $\mathcal{H}_c^{\phi}(X)$.

The adjoint map of a morphism of domains $\omega\colon X \to Y$ induces a continuous linear map $\mathcal{H}^{\phi}(Y) \to \mathcal{H}^{\phi}(X)$ for every $\phi \in F(\Lambda)$, hence $\omega^*\colon \mathcal{H}_{\mathfrak{b}}(Y) \to \mathcal{H}_{\mathfrak{b}}(X)$ is continuous, too. Thus $\mathcal{H}_{\mathfrak{b}}$ is a contravariant functor from $\mathcal{G}(\Lambda_0)$ to the category of locally convex spaces.

Obviously, $\mathfrak{b}$ is finer than c. It can be shown that they are equal if and only if Λ is finite.

By virtue of (1.6), $\mathfrak{b}$ is bornological and barrelled. Hence $\mathfrak{b}$ is finer than the bornological topology c_0 associated with c. In order to show that they are equal we need the following lemma.

(1.7) Lemma: Let $p\colon X \to C^{\Lambda}$ be a domain. If $\mathfrak{B}$ is a pointwise bounded subset of $\mathcal{H}(X)$, there is a $\phi \in F(\Lambda)$ such that $\mathfrak{B} \subset \mathcal{H}^{\phi}(X)$.

Proof: Suppose the assertion is not true. Then there exists a sequence $(f_n)_{n \in N}$ in $\mathfrak{B}$ such that no finite subset of Λ contains all $\operatorname{dep}(f_n)$. We shall construct a subsequence which is not bounded in a point of X.

$n_1 := 1$. If n_k is defined, there is $n_{k+1} \in N$ so that $n_i < n_{k+1}$ for all $i \in \{1, \ldots, k\}$ and $\operatorname{dep}(f_{n_{k+1}}) \subset \bigcup_{i=1}^{k} \operatorname{dep}(f_{n_i})$. $\phi_k := \operatorname{dep}(f_{n_k})$ for all $k \in N$.

Let $a \in X$. There exist a neighbourhood U of a and a 0–neighbourhood $V = B(0,r)^{\phi'} \times C^{\Lambda - \phi'}$ in C^{Λ} so that $p|U \to p(a) + V$ is a homeomorphism.

For all $k \in N$ $g_k := f_{n_k} \circ (p|U)^{-1} \in \mathcal{H}(p(a)+V)$.

There is $k_0 \in N$ so that for all $k > k_0$ $(\phi_k - \bigcup_{i=1}^{k_0} \phi_i) \cap \phi' = \emptyset$. $\xi_0 := \pi_{\phi' \cup \phi_{k_0}}(p(a))$.

Suppose ξ_j is defined for $j \in \{0, \ldots, \ell - 1\}$.

Since $g_{k_0+1} \,\Big|\, \sum_{j=0}^{\ell-1} \xi_j + C^{\phi_{k_0+\ell} - \bigcup_{i=0}^{k_0+\ell-1} \phi_i}$ is an entire function which cannot be constant because

$$\emptyset \neq \phi_{k_0+1} - \bigcup_{i=0}^{k+\ell-1} \phi_i \subset \operatorname{dep}(f_{n_{k_0+\ell}}),$$

there is $\xi_\ell \in C^{\phi_{k_0+\ell} - \bigcup_{i=0}^{k_0+\ell-1} \phi_i}$ with $\left|g_{k_0+\ell}(\sum_{j=0}^{\ell} \xi_j)\right| > 1$. Consequently $\left|g_{k_0+q}(\sum_{j=0}^{\infty} \xi_j)\right| \to \infty$ if $q \to \infty$, whence a contradiction. (Clearly $\sum_{j=0}^{\infty} \xi_j$ is well defined.) q.e.d.

(1.8) Proposition: $\mathscr{b}$ **is the bornological topology associated with the compact-open topology.**

Proof: We need only show that $\mathscr{b}$ is coarser than c_0.

First we show that

id: $\mathcal{H}_c(X) \to \mathcal{H}_{\mathscr{b}}(X)$ is sequentially continuous. Let $(f_n)_{n\in N}$ be a convergent sequence in $\mathcal{H}_c(X)$. By (1.7) there is a $\phi \in F(\Lambda)$ so that $(f_n)_{n\in N}$ converges in $\mathcal{H}_c^{\phi}(X)$. Since $\mathcal{H}_c^{\phi}(X) \hookrightarrow \mathcal{H}_{\mathscr{b}}(X)$ is continuous, $(f_n)_{n\in N}$ converges in $\mathcal{H}_{\mathscr{b}}(X)$.

Since id: $\mathcal{H}_{c_0}(X) \to \mathcal{H}_c(X)$ is continuous, $\mathcal{H}_{c_0}(X) \to \mathcal{H}_c(X) \to \mathcal{H}_{\mathscr{b}}(X)$ is sequentially continuous, hence continuous (because c_0 is bornological). This means that $\mathscr{b}$ is coarser than c_0. q.e.d.

(1.9) Proposition: $\mathcal{H}_{\mathscr{b}}(X)$ **is a Montel space.**

Proof: Since $\mathcal{H}_{\mathscr{b}}(X)$ is barrelled it suffices to show that $\mathcal{H}_{\mathscr{b}}(X)$ is semi-Montel.

Let $\mathcal{B}$ be a bounded and closed set in $\mathcal{H}_{\mathscr{b}}(X)$. Then $\mathcal{B}$ is pointwise bounded and, by (1.7), there exists a $\phi \in F(\Lambda)$ such that $\mathcal{B} \subset \mathcal{H}^{\phi}(X)$. $\mathcal{B}$ is bounded and closed in $\mathcal{H}_c^{\phi}(X)$ since $\mathscr{b}$ induces on $\mathcal{H}^{\phi}(X)$ the compact-open topology. By virtue of (1.6), $\mathcal{B}$ is compact. q.e.d.

It follows in the same way that $\mathcal{H}_c(X)$ is semi-Montel.

(1.10) Proposition: Let $p: X \to C^{\Lambda}$ **and** $q: Y \to C^{\Lambda}$ **be domains and let** $\omega: X \to Y$ **be a morphism. Then** $\omega^*: \mathcal{H}(Y) \to \mathcal{H}(X)$ **is an algebraic isomorphism if and only if** $\omega^*: \mathcal{H}_{\mathscr{b}}(Y) \to \mathcal{H}_{\mathscr{b}}(X)$ **is a topological isomorphism.**

Proof: Suppose $\omega^*: \mathcal{H}(Y) \to \mathcal{H}(X)$ is an algebraic isomorphism. Then, for each $\phi \in F(\Lambda)$, $\omega^*: \mathcal{H}^{\phi}(Y) \to \mathcal{H}^{\phi}(X)$ is an isomorphism. $\omega^*: \mathcal{H}_c^{\phi}(Y) \to \mathcal{H}_c^{\phi}(X)$ is continuous and $\mathcal{H}_c^{\phi}(Y)$ and $\mathcal{H}_c^{\phi}(X)$ are Fréchet spaces, hence, by the open mapping theorem, $\omega^*: \mathcal{H}_c^{\phi}(Y) \to \mathcal{H}_c^{\phi}(X)$ is a homeomorphism. Consequently, $\omega^*: \mathcal{H}_{\mathscr{b}}(Y) \to \mathcal{H}_{\mathscr{b}}(X)$ is a topological isomorphism. The converse implication is trivial. q.e.d.

§2. A Special Construction of the Envelope of Holomorphy.

We need the following generalization of the intersection of domains (cf. [7] 2.3, [8] p.85).

(2.1) Proposition: Let $p: X \to C^{\Lambda}$ **be a domain and let** $(p_\phi: X_\phi \to C^{\phi})_{\phi\in F(\Lambda)}$ **be a family of domains. Suppose given a family** $(\varphi_\phi)_{\phi\in F(\Lambda)}$ **of morphisms** $\varphi_\phi: X \to X_\phi$. **Then there exists a domain** $q: Y \to C^{\Lambda}$, **a morphism** $\psi: X \to Y$ **and morphisms** $\psi_\phi: Y \to X_\phi$ **such that:**

(a) $\varphi_\phi = \psi_\phi \circ \psi$ for all $\phi \in F(\Lambda)$

(b) ψ is maximal in the following sense:

If q': $Y' \to C^{\Lambda'}$ is a domain and ψ': $X \to Y'$ is a morphism such that each φ_ϕ factors through ψ', then there is a unique morphism γ: $Y' \to Y$ with $\psi = \gamma \circ \psi'$ and $\psi'_\phi = \psi_\phi \circ \gamma$ for all $\phi \in F(\Lambda)$.

Definition: (q,ψ) or simply q or Y is called the intersection of $(\varphi_\phi)_{\phi \in F(\Lambda)}$ or of $(p_\phi)_{\phi \in F(\Lambda)}$ if no confusion can arise. q: $Y \to C^\Lambda$ is unique up to isomorphisms.

Sketch of the proof:

Let $Z \subset \prod_{\phi \in F(\Lambda)} X_\phi$ be the set of all $(x_\phi: \phi \in F(\Lambda))$ with the following properties:

(1) $\pi^\phi_{\phi \cap \phi'} \circ p_\phi(x_\phi) = \pi^{\phi'}_{\phi \cap \phi'} \circ p_{\phi'}(x_{\phi'})$ for all $\phi, \phi' \in F(\Lambda)$

(2) there is an open polydisc U with center 0 in C^Λ and there are neighbourhoods U_ϕ of x_ϕ in X_ϕ such that $p_\phi | U_\phi \to p_\phi(x_\phi) + \pi_\phi(U)$ is topological for all $\phi \in F(\Lambda)$.

We define a topology on Z in the following way: $W \subset Z$ is called a neighbourhood of $(x_\phi: \phi \in F(\Lambda)) \in Z$ iff there are U and U_ϕ satisfying (2) such that $Z \cap \prod_{\phi \in F(\Lambda)} U_\phi \subset W$. This topology on Z is finer than the trace of the product topology. ψ: $X \to Z$, $x \mapsto (\varphi_\phi(x): \phi \in F(\Lambda))$ is continuous. Let Y be the connected component which contains $\psi(X)$. The mapping q: $Y \to C^\Lambda$, $(x_\phi: \phi \in F(\Lambda)) \mapsto (p_{\{j\}}(x_{\{j\}}): j \in \Lambda)$ is well-defined according to (1) and locally a homeomorphism. Hence q: $Y \to C^\Lambda$ is a domain and ψ: $X \to Y$ is a morphism which satisfies (a) if we define ψ_ϕ: $Y \to X_\phi$, $(x_{\phi'}: \phi' \in F(\Lambda)) \mapsto x_\phi$. Condition (b) is easily verified like in [8], [9].

We shall use the following terminology:

(2.2) Definition: Let p: $X \to C^\Lambda$ and q: $Y \to C^\Lambda$ be domains and ω: $X \to Y$ a morphism. (q,ω) is called a **$\mathcal{H}(X)$–extension** of p iff ω^*: $\mathcal{H}(Y) \to \mathcal{H}(X)$ is bijective. (Notice that ω^* is always injective because of the identity theorem.)

The $\mathcal{H}(X)$–extension (q,ω) (or simply Y) is called the **envelope of holomorphy** of p iff, for any $\mathcal{H}(X)$–extension $(q'$: $Y' \to C^\Lambda$, $\omega')$, there is a morphism γ: $Y' \to Y$ such that $\omega = \gamma \circ \omega'$. On these conditions, q: $Y \to C^\Lambda$ is unique up to isomorphisms of domains.

(2.3) Construction of the envelope of holomorphy of a domain p: $X \to C^\Lambda$. For every $\phi \in F(\Lambda)$, the envelope of holomorphy of p: $X \to C^\Lambda$ exists (see e.g. [2], [6], [9]). We denote it by $({}^\phi e$: $\mathcal{E}({}^\phi X) \to C^\phi$, ${}^\phi\epsilon)$ or briefly by $\mathcal{E}({}^\phi X)$.

Set ${}^\phi\beta := {}^\phi\epsilon \circ {}^\phi\sigma_p$ for all $\phi \in F(\Lambda)$.

Let (e: $\mathcal{E}(X) \to C^\Lambda, \epsilon)$ be the intersection of $({}^\phi\beta : \phi \in F(\Lambda))$.

Then the following diagram commutes:

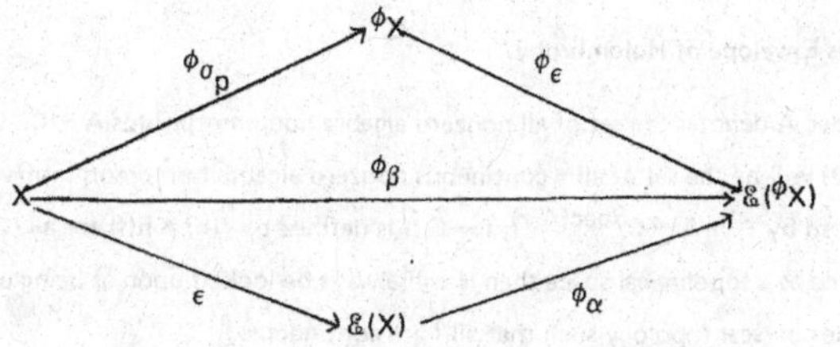

Hence the following one commutes too:

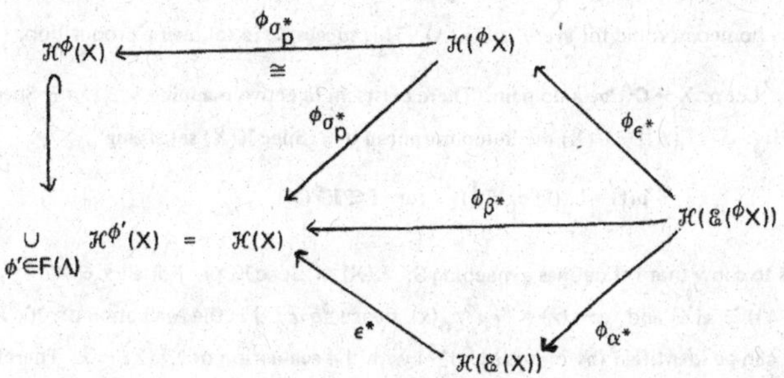

Obviously ϵ^* is surjective, hence (e, ϵ) is a $\mathcal{H}(X)$–extension of p.
Let $(q\colon Y \to C^{\Lambda},\ \delta)$ be another $\mathcal{H}(X)$–extension. Then $\delta^*\colon \mathcal{H}^{\phi}(Y) \to \mathcal{H}^{\phi}(X)$ is an isomorphism for every $\phi \in F(\Lambda)$. Hence $({}^{\phi}p, \mathcal{J}^{\phi}(\delta))$ is a $\mathcal{H}({}^{\phi}X)$–extension for every $\phi \in F(\Lambda)$. Therefore there are morphisms $\phi_{\gamma}\colon {}^{\phi}Y \to \mathcal{E}({}^{\phi}X)$ such that the following diagram commutes:

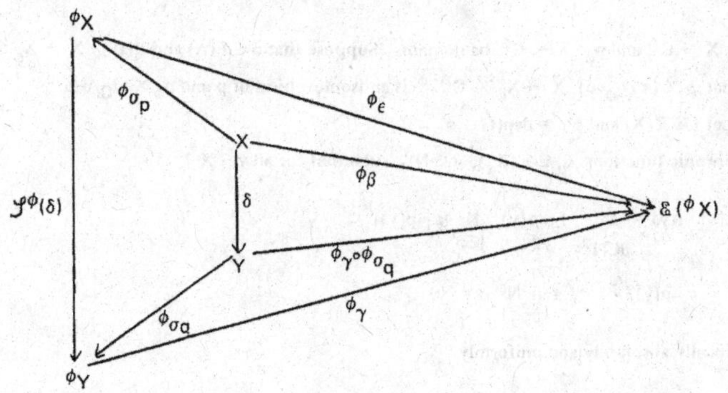

Because of the intersection property, there exists a morphism $\gamma\colon Y \to \mathcal{E}(X)$ such that $\epsilon = \gamma \circ \delta$. q.e.d.

A more detailed consideration yields that the association $[p: X \to C^{\Lambda}] \mapsto [\&(p) := e: \&(X) \to C^{\Lambda}]$ is functorial and defines a reflector $\&$ in $\mathcal{G}(\Lambda_0)$.

§3. The Spectrum as Envelope of Holomorphy.

If A is a C-algebra, Spec A denotes the set of all nonzero algebra homomorphisms $A \to C$. Let τ be a topology on A. Then Spec(A,τ) will be the set of all τ-continuous nonzero algebra homomorphisms $A \to C$. The Gelfand map is denoted by $\widehat{\ }: A \to C^{\mathrm{Spec}(A,\tau)}$, $f \to \hat{f}$ ($\hat{f}$ is defined by $\hat{f}(h) = h(f)$ for all $h \in \mathrm{Spec}(A,\tau)$). If Spec$(A,\tau)$ is considered as a topological space then it will always be looked upon as being endowed with the weak topology i.e. the coarsest topology such that all $\hat{f}$ are continuous.

For a domain $q: Y \to C^n$, $\&(Y)$, Spec $\mathcal{H}(Y)$ and Spec$(\mathcal{H}(Y), c)$ are homeomorphic (see e.g. [2]). By virtue of (1.5) this implies that, for a domain $p: X \to C^{\Lambda}$, $\&({}^{\phi}X)$, Spec $\mathcal{H}({}^{\phi}X)$, Spec $\mathcal{H}^{\phi}(X)$, Spec$(\mathcal{H}({}^{\phi}X),c)$ and Spec$(\mathcal{H}^{\phi}(X),c)$ are homeomorphic for every $\phi \in F(\Lambda)$. This suggests the following proposition.

(3.1) Proposition: Let $p: X \to C^{\Lambda}$ be a domain. There exists an injective mapping $S: \&(X) \to \mathrm{Spec}(\mathcal{H}(X), \mathcal{D})$ which assigns to $(h_{\phi}: \phi \in F(\Lambda)) \in \&(X)$ the homomorphism $h \in \mathrm{Spec}\,\mathcal{H}(X)$ satisfying

$$(*) \qquad h(f) = h_{\phi}(({}^{\phi}\sigma_p)^{-1} f) \quad \text{for} \quad f \in \mathcal{H}^{\phi}(X).$$

Proof: It suffices to show that (*) defines a mapping $S: \&(X) \to \mathrm{Spec}\,\mathcal{H}(X)$. For all $x \in X$, $({}^{\phi}\alpha\circ\epsilon\,(x): \phi \in F(\Lambda)) \in \epsilon(X)$ and ${}^{\phi}\alpha\circ\epsilon\,(x) = {}^{\phi}\epsilon \circ {}^{\phi}\sigma_p(x)$, hence ${}^{\phi}\alpha\circ\epsilon\,(x)$ is the evaluation of $\mathcal{H}({}^{\phi}X)$ in the point ${}^{\phi}\sigma_p(x)$ and can be identified (by means of ${}^{\phi}\sigma_p^{**}$) with the evaluation of $\mathcal{H}^{\phi}(X)$ in x. Therefore ${}^{\phi'}\alpha\circ\epsilon\,(x) \,|\, \mathcal{H}^{\phi''}(X) = {}^{\phi''}\alpha\circ\epsilon\,(x)$ (**) for all $\phi', \phi'' \in F(\Lambda)$ with $\phi'' \subset \phi'$. Since, for all $f \in \mathcal{H}({}^{\phi''}X)$, $\hat{f}: \&({}^{\phi''}X) \to C$, $h_{\phi''} \mapsto h_{\phi''}(f)$ is holomorphic ([2], p.49), $\hat{f}\circ{}^{\phi'}\alpha - \hat{f}\circ{}^{\phi''}\alpha$ is a holomorphic function on $\&(X)$. By virtue of (**) and the identity theorem, $\hat{f}\circ{}^{\phi'}\alpha - \hat{f}\circ{}^{\phi''}\alpha = 0$. Consequently $h_{\phi'}|\mathcal{H}^{\phi''}(X) = h_{\phi''}$ for all $(h_{\phi}: \phi \in F(\Lambda)) \in \&(X)$ and all $\phi', \phi'' \in F(\Lambda)$ with $\phi'' \subset \phi'$. Thus, (*) defines a map $S: \&(X) \to \mathrm{Spec}\,\mathcal{H}(X)$ which is obviously injective. q.e.d.

(3.2) Lemma: Let $p: X \to C^{\Lambda}$ and $p_{\phi}: X \to C^{\phi}$ be domains. Suppose that $\phi \in F(\Lambda)$ and that $\rho: X \to X_{\phi}$ is a morphism such that $\rho \times (\pi_{\Lambda-\phi}\circ p): X \to X_{\phi} \times C^{\Lambda-\phi}$ is an isomorphism of p and $p_{\phi} \times \mathrm{id}_{C^{\Lambda-\phi}}$: $X_{\phi} \times C^{\Lambda-\phi} \to C^{\Lambda}$. Let $f \in \mathcal{H}(X)$ and $\phi' := \mathrm{dep}(f) - \phi$.

Then there are holomorphic functions $f_{\mu} \in \mathcal{H}(X_{\phi})$, $\mu \in N^{\phi'}$, such that for all $y \in X$

$$f(y) = \sum_{\mu \in N^{\phi'}} f_{\mu}\circ\rho(y) \prod_{j \in \phi'} (\pi_j \circ p(y))^{\mu_j}.$$

(For $\phi' = \emptyset$, set $\prod_{j \in \phi'} (\pi_j \circ p(y))^{\mu_j} := 1$ and $N^{\phi'} := \{0\}$.)

The series converges locally absolutely and uniformly.

Proof: Set $q := p_\phi \times id_{C^{\Lambda-\phi}}$, $Y := X_\phi \times C^{\Lambda-\phi}$ and $g := f\circ(\rho \times (\pi_{\Lambda-\phi}\circ p))^{-1} \in \mathcal{H}(Y)$. Let $x \in Y$ and U be a neighbourhood of x such that $q|U \to q(U)$ is topological. For $\mu \in N^{(\Lambda)}$ we define $D^\mu g(x) := D^\mu g\circ(q|U)^{-1}(qx)$. Clearly, $D^\mu g: y \mapsto D^\mu g(y)$ is a well-defined holomorphic function on Y. Since dep(g) is finite, every point $z \in Y$ has a neighbourhood where g can be expanded into a power series

$$g(y) = \sum_{\mu\in N^{dep(g)}} D^\mu g(y) \prod_{j\in dep(g)} (\pi_j(qy - qz))^{\mu_j}$$

and where this series converges uniformly and absolutely. Hence for all $(y,z) \in X_\phi \times C^{\Lambda-\phi}$

$$g(y,z) = \sum_{\mu\in N^{\phi'}} D^\mu g(y,0) \prod_{j\in\phi'} (\pi_j(z))^{\mu_j}$$

and evidently this series converges locally uniformly and absolutely, too. Setting $f_\mu(x) = D^\mu g(x,0)$ for $x \in X_\phi$ and $\mu \in N^{\phi'}$, we obtain the desired formula. q.e.d.

Definition: Let $p: X \to C^\Lambda$ be a domain. p is called a **domain of holomorphy** iff the canonical morphism ϵ from p to $\mathcal{S}(p)$ (see (2.3)) is an isomorphism of domains (it is equivalent to say that ϵ is bijective).
$E: X \to Spec(\mathcal{S}(X),\mathfrak{b})$ denotes the map which assigns to x the evaluation homomorphism $E_x: \mathcal{H}(X) \to C$, $f \mapsto f(x)$.

(3.3) Theorem: Let $p: X \to C^\Lambda$ be a domain. p is a domain of holomorphy if and only if $E: X \to Spec(\mathcal{H}(X),\mathfrak{b})$ is bijective.

Proof: $\Rightarrow$: By virtue of condition 9 in §0 we can suppose that there is a $\phi \in F(\Lambda)$, a Stein domain $p_\phi: X_\phi \to C^\phi$ and a morphism $\rho: X \to X_\phi$ such that $\rho \times (\pi_{\Lambda-\phi}\circ p)$ is an isomorphism from p to $p_\phi \times id_{C^{\Lambda-\phi}}$.
For $h \in Spec(\mathcal{H}(X),\mathfrak{b})$ define $\mathbf{h} \in Spec\, \mathcal{H}(X_\phi)$ by $\mathbf{h}(f) := h(f\circ\rho)$. Since p_ϕ is Stein, there exists $a \in X$ so that, for all $f \in \mathcal{H}(X_\phi)$, $\mathbf{h}(f) = f\circ\rho(a)$ ([2]).
$b := (\rho \times \pi_{\Lambda-\phi}\circ p)^{-1}(a, (h(\pi_j\circ p))_{j\in\Lambda-\phi})$.

We show that $h = E_b$. Let $f \in \mathcal{H}(X)$. Using (3.2), we obtain

$$h(f) = \sum_{\mu\in N^{\phi'}} h(f_\mu) \prod_{j\in\phi'} (h(\pi_j\circ p))^{\mu_j}$$

$$= \sum_{\mu\in N^{\phi'}} f_\mu(a) \prod_{j\in\phi'} (h(\pi_j\circ p))^{\mu_j} = f(b).$$

$\Leftarrow$:

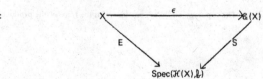

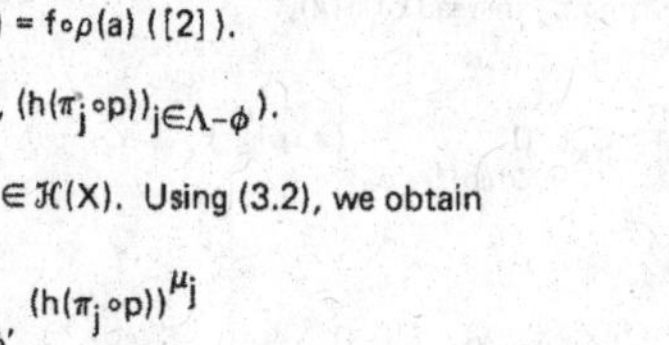

The above diagram commutes. Hence S and $\epsilon = S^{-1} \circ E$ are bijective. q.e.d.

Corollary: If p: $X \to C^{\Lambda}$ is a domain of holomorphy, E: $X \to \mathrm{Spec}(\mathcal{H}(X),c)$ is bijective, hence $\mathrm{Spec}(\mathcal{H}(X),c) = \mathrm{Spec}(\mathcal{H}(X),\mathfrak{b})$.

Proof: Follows from the proof of (3.3).

(3.4) Lemma: Let p: $X \to C^{\Lambda}$ be a domain. There exists $\phi_0 \in F(\Lambda)$ such that, for any $\phi_1 \in F(\Lambda)$ with $\phi_0 \subset \phi_1$, every $(x_\phi\colon \phi \in F(\Lambda)) \in \mathcal{E}(X)$ is determined by the components x_ϕ with $\phi \in \{\phi_1\} \cup F(\Lambda - \phi_1)$.

Proof: Since $\mathcal{E}(X)$ is a domain of holomorphy, we know by virtue of (3.2) and condition 9 in section 0 that there exists $\phi_0 \in F(\Lambda)$ such that every $f \in \mathcal{H}(X)$ can be written

$$f(x) = \sum_{\mu \in N^{\mathrm{dep}(f)-\phi_0}} f_\mu \circ p(x) \prod_{j \in \mathrm{dep}(f)-\phi_0} (\pi_j \circ p(x))^{\mu_j} \quad \text{for all } x \in \mathcal{E}(X).$$

Let $\phi_1 \in F(\Lambda)$, $\phi_0 \subset \phi_1$. For all $x \in \mathcal{E}(X)$ and $\nu \in N^{\mathrm{dep}(f)-\phi_1}$.

$$F_\nu(x) := \sum_{\substack{\mu \in N^{\mathrm{dep}(f)-\phi_0} \\ \mu|\mathrm{dep}(f)-\phi_1 - \nu}} f_\mu \circ p(x) \prod_{j \in \phi_1 - \phi_0} (\pi_j \circ p(x))^{\mu_j}.$$

Clearly, $F_\nu \in \mathcal{H}^{\phi_1}(\mathcal{E}(X)) = \mathcal{H}^{\phi_1}(X)$ and, for all $x \in \mathcal{E}(X)$,

$$f(x) = \sum_{\nu \in N^{\mathrm{dep}(f)-\phi_1}} F_\nu(x) \prod_{k \in \mathrm{dep}(f)-\phi_1} (\pi_j \circ p(x))^{\nu_k}.$$

If $x = (x_\phi\colon \phi \in F(\Lambda)) \in \mathcal{E}(X)$ then, writing S_x instead of $S(x)$,

$$S_x(f) = \sum_{\nu \in N^{\mathrm{dep}(f)-\phi_1}} S_x(F_\nu) \cdot S_x\Big(\prod_{k \in \mathrm{dep}(f)-\phi_1} (\pi_j \circ p)^{\nu_k}\Big)$$

$$= \sum_{\nu \in N^{\mathrm{dep}(f)-\phi_1}} x_{\phi_1}(F_\nu) \cdot x_{\mathrm{dep}(f)-\phi_1}\Big(\prod_{k \in \mathrm{dep}(f)-\phi_1} (\pi_j \circ p)^{\nu_k}\Big)$$

with identifying $\mathcal{E}(^{\phi}X)$, Spec $\mathcal{H}^{\phi}(X)$ and Spec$\mathcal{H}^{\phi}(\mathcal{E}(X))$. Since S is injective, the assertion follows. q.e.d.

(3.5) Lemma: Let p: $X \to C^{\Lambda}$ be a domain. The topology on $\mathcal{E}(X)$ coincides with the trace of the product topology on $\prod_{\phi \in F(\Lambda)} \mathcal{E}(^{\phi}X)$.

Proof: Obviously, the topology of $\mathcal{E}(X)$ is finer than the product topology. We show that it is coarser, too. Let $x = (x_\phi : \phi \in F(\Lambda)) \in \mathcal{E}(X)$ and U and U_ϕ like in (2.1)(2). There is $\phi' \in F(\Lambda)$ such that $\pi_{\Lambda-\phi'}(U) = C^{\Lambda-\phi'}$. Choose ϕ_0 like in (3.4). $\phi_1 := \phi_0 \cup \phi'$. For all $\phi \in F(\Lambda-\phi_1)$, $p_\phi | U \to p_\phi(x_\phi) + C^\phi$ is topological, hence ([2], p.44) $U_\phi = \mathcal{E}(^\phi X)$. Therefore (3.4) involves that

$$[\prod_{\phi\in F(\Lambda)} U_\phi] \cap \mathcal{E}(X) = [U_{\phi_1} \times \prod_{\substack{\phi\in F(\Lambda)\\ \phi\neq\phi_1}} \mathcal{E}(^\phi X)] \cap \mathcal{E}(X).$$

q.e.d.

(3.6) Theorem: Let $p: X \to C^\Lambda$ be a domain. Then $S: \mathcal{E}(X) \to \mathrm{Spec}(\mathcal{H}(X), \mathfrak{b})$ is a homeomorphism.

Proof: The following diagram commutes:

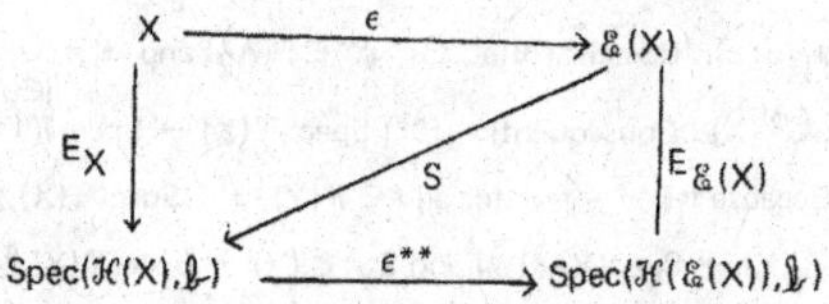

The double adjoint ϵ^{**} of ϵ is bijective by virture of (1.10), hence topological. $E_{\mathcal{E}(X)}$ is bijective by (3.3). Consequently S is bijective.

For $f \in \mathcal{H}^\phi(X)$, $\phi \in F(\Lambda)$, the extensions of f to $\mathcal{E}(^\phi X)$ and $\mathcal{E}(X)$ will be denoted by f, too. The following diagram commutes:

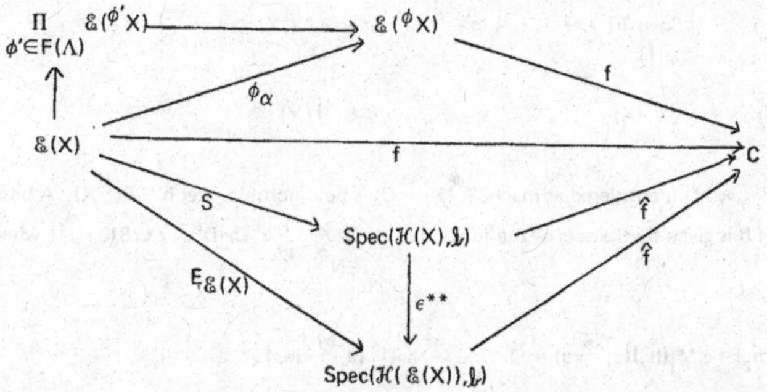

Since, for all $f \in \mathcal{H}(X)$, $\hat{f} \circ S = f$ is continuous, S is continuous. By (3.5), the topology of $\mathcal{E}(X)$ is the projective topology with respect to $(^\phi\alpha : \phi \in F(\Lambda))$. Because the $\mathcal{E}(^\phi X)$ are endowed with the projective topology with respect to $\mathcal{H}(^\phi X) \cong \mathcal{H}^\phi(X)$ and because, for all $\phi \in F(\Lambda)$ and all $f \in \mathcal{H}^\phi(X)$, $f \circ {}^\phi\alpha \circ S^{-1} = \hat{f}$ is continuous, S^{-1} is continuous.

q.e.d.

(3.7) Corollary: Let $p: X \to C^{\Lambda}$ be a domain.

p is a domain of holomorphy if and only if $E: X \to \mathrm{Spec}(\mathcal{H}(X), \mathfrak{T})$ is a homeomorphism.

Proof: Clear, because $E = S \circ \epsilon$.

(3.8) Theorem: Let $p: X \to C^{\Lambda_1}$ and $q: Y \to C^{\Lambda_2}$ be domains and let $\varphi: X \to Y$ be a holomorphic map. There exists a unique holomorphic map $\mathcal{E}(\varphi): \mathcal{E}(X) \to \mathcal{E}(Y)$ such that the following diagram commutes:

$$\begin{array}{ccc} X & \xrightarrow{\quad \varphi \quad} & Y \\ {\scriptstyle \epsilon_X}\downarrow & & \downarrow{\scriptstyle \epsilon_Y} \\ \mathcal{E}(X) & \xrightarrow[\mathcal{E}(\varphi)]{\qquad} & \mathcal{E}(Y) \end{array}$$

Proof: For $j \in \Lambda_2$ set $\phi_j := \mathrm{dep}(\pi_j \circ q \circ \varphi)$. It is clear that, for $\phi'' \in F(\Lambda_2)$ and $\phi' = \bigcup_{j \in \phi''} \phi_j$, the adjoint map φ^* maps continuously $\mathcal{H}^{\phi''}(Y)$ in $\mathcal{H}^{\phi'}(X)$. Consequently $\varphi^{**}: \mathrm{Spec}\, \mathcal{H}(X) \to \mathrm{Spec}\, \mathcal{H}(Y)$, $h \to h \circ \varphi^*$ maps $\mathrm{Spec}\,(\mathcal{H}(X), \mathfrak{T})$ in $\mathrm{Spec}(\mathcal{H}(Y), \mathfrak{T})$. Because $\hat{f} \circ \varphi^{**} = \widehat{f \circ \varphi}$ for all $f \in \mathcal{H}(Y)$, $\varphi^{**}: \mathrm{Spec}(\mathcal{H}(X), \mathfrak{T}) \to \mathrm{Spec}(\mathcal{H}(Y), \mathfrak{T})$ is continuous. Since by (3.6) $S_X: \mathcal{E}(X) \to \mathrm{Spec}(\mathcal{H}(X), \mathfrak{T})$ and $S_Y: \mathcal{E}(Y) \to \mathrm{Spec}(\mathcal{H}(Y), \mathfrak{T})$ are topological, $\mathcal{E}(\varphi) := S_Y^{-1} \circ \varphi^{**} \circ S_X$ is continuous. Moreover $\mathcal{E}(\varphi)$ makes the above diagram commutative. Thus, all we need to show is that $e_Y \circ \mathcal{E}(\varphi)$ is holomorphic or, equivalently, that $\pi_j \circ e_Y \circ \mathcal{E}(\varphi)$ is holomorphic for all $j \in \Lambda_2$. For every $j \in \Lambda_2$, φ^{**} induces a map

$$\begin{array}{ccc} \varphi^{**}{}_j : \mathrm{Spec}\, \mathcal{H}^{\phi_j}(X) & \xrightarrow{\qquad\qquad} & \mathrm{Spec}\, \mathcal{H}^{\{j\}}(Y) \\ \wr\Vert & & \wr\Vert \\ \mathrm{Spec}\, \mathcal{H}({}^{\phi_j}X) & & \mathrm{Spec}\, \mathcal{H}({}^{\{j\}}Y) \\ \wr\Vert & & \wr\Vert \\ \mathcal{E}({}^{\phi_j}X) & & \mathcal{E}({}^{\{j\}}Y) \end{array}$$

We show that ${}^{\{j\}}e_Y \circ \varphi^{**}{}_j$ (considered as map $\mathcal{E}({}^{\phi_j}X) \to C$) is holomorphic. Let $h \in \mathcal{E}({}^{\phi_j}X)$. A basis of neighbourhoods of h is given by the sets of the form $\{h_z = \sum_{\nu \in N^{\phi_j}} \frac{1}{\nu!} z^{\nu} \bullet h \circ D^{\nu} : z \in B(0,r)^{\phi_j}\}$ where $r > 0$ (cf. [2], p.50)

$${}^{\{j\}}e_Y \circ \varphi^{**}{}_j(h_z) = \varphi^{**}{}_j(h_z)(\pi_j^{\Lambda_2} \circ q) = \sum_{\nu \in N_0^{\phi_j}} \frac{1}{\nu!} h \circ D^{\nu} (\pi_j^{\Lambda_2} \circ q \circ \varphi) z^{\nu}$$

is obviously a holomorphic function of z. Hence ${}^{\{j\}}e_Y \circ \varphi^{**}{}_j$ is holomorphic.

The following diagram commutes:

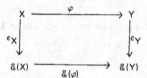

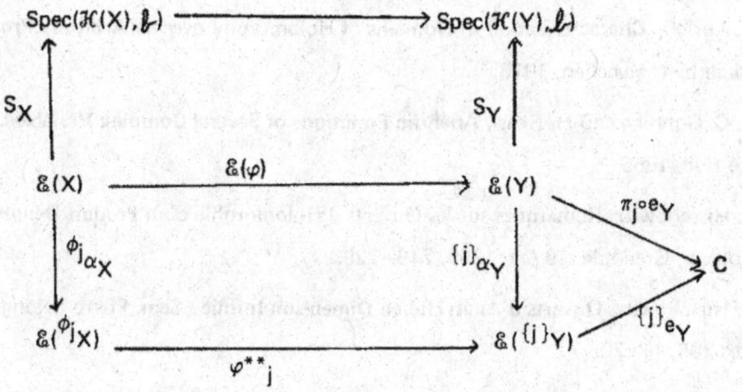

$$\pi_j \circ e_Y \circ \&(\varphi) = {}^{\{j\}}e_Y \circ {}^{\{j\}}\alpha_Y \circ \&(\varphi) = {}^{\{j\}}e_Y \circ \varphi^{**}{}_j \circ {}^{\{j\}}\alpha_X$$

is holomorphic for all $j \in \Lambda_2$. q.e.d.

Corollary: $p \leftrightarrow \&(p)$, $\varphi \leftrightarrow \&(\varphi)$ **is a reflector in the category of domains and holomorphic maps.**

Corollary: The analytic structure of &(X) depends only on the analytic structure of X (and not on the choice of p).

Bibliography

1. V. Aurich, **Characterization of Domains of Holomorphy over an Arbitrary Product of Complex Lines.** Diplomarbeit, München, 1973.

2. R. C. Gunning and H. Rossi, **Analytic Functions of Several Complex Variables.** Englewood Cliffs, N.J., Prentice-Hall, 1965.

3. A. Hirschowitz, **Remarques sur les Ouverts d'Holomorphie d'un Produit Dénombrable de Droites.** Ann. Inst. Fourier, Grenoble, 19 (1), 1969, 219–229.

4. A. Hirschowitz, **Ouverts d'Analycité en Dimension Infinie.** Sem. Pierre Lelong 1970, Springer Lecture Notes 205, 11–20.

5. M. C. Matos, **Holomorphic Mappings and Domains of Holomorphy.** Thesis, Rio de Janeiro, 1970.

6. C. E. Rickart, **Analytic Functions of an Infinite Number of Complex Variables.** Duke Mathematical Journal 36, 1969, 581–597.

7. M. Schottenloher, **Analytische Fortsetzung in Banachräumen.** Thesis, München, 1971. Math. Ann. 199(1972), 313–336.

8. K. Stein, **Einführung in die Funktionentheorie mehrerer Veränderlichen.** Vorlesungsausarbeitung, Munchen, 1962.

9. K. Stein, **Maximale holomorphe und meromorphe Abbildungen I.** Am. Journal of Math. 85(1953), 298–315.

Mathematisches Institut der Universität München
8 München 2
Theresienstrasse 39
Munich, Germany

ON BOUNDED SETS OF HOLOMORPHIC MAPPINGS

Jorge Alberto Barroso, Mário C. Matos and Leopoldo Nachbin

§1. Introduction

Let E and F be complex locally convex spaces, $U \subset E$ be open and non-void, and $\mathcal{H}(U;F)$ be the vector space of all holomorphic mappings from U into F. Motivated by uniform convergence of mappings, three natural topologies $\mathcal{T}_o$, $\mathcal{T}_\omega$ and $\mathcal{T}_\delta$ have been used on $\mathcal{H}(U;F)$; see [5], [6], [7], [8]. Let $\mathcal{B}_i(U;F)$ be the collection of all $\mathcal{T}_i$-bounded subsets of $\mathcal{H}(U;F)$, where $i = o, \omega, \delta$; and $\mathcal{B}_a(U;F)$ be the collection of all amply bounded subsets of $\mathcal{H}(U;F)$. Then $\mathcal{T}_o \subset \mathcal{T}_\omega \subset \mathcal{T}_\delta$ and $\mathcal{B}_o(U;F) \supset \mathcal{B}_\omega(U;F) \supset \mathcal{B}_\delta(U;F) \supset \mathcal{B}_a(U;F)$. We are interested in the study of the following properties of E, where $i,j = o, \omega, \delta, a$:

$A_{ij}(U;F)$: We have $\mathcal{B}_i(U;F) = \mathcal{B}_j(U;F)$.

$B_{ij}(U;F)$: A seminorm on $\mathcal{H}(U;F)$ is bounded on $\mathcal{B}_i(U;F)$ if and only if it is bounded on $\mathcal{B}_j(U;F)$.

$C(U;F)$: There is some locally convex topology on $\mathcal{H}(U;F)$ whose collection of bounded sets is $\mathcal{B}_a(U;F)$.

We can rephrase $B_{ij}(U;F)$ by saying that the bornological topologies associated to $\mathcal{B}_i(U;F)$ and $\mathcal{B}_j(U;F)$ are equal.

If one of the above properties is true for every F, or for every F and every U, we omit F, or F and U, from the notation and write $A_{ij}(U)$, $B_{ij}(U)$, $C(U)$, or A_{ij}, B_{ij}, C, respectively.

There are some routine facts about these properties, of which we point out the following: $A_{ii}(U;F)$ and $B_{ii}(U;F)$ are always true; $A_{ij}(U;F) = A_{ji}(U;F)$ and $B_{ij}(U;F) = B_{ji}(U;F)$; $A_{ij}(U;F) = B_{ij}(U;F)$ if $i \neq a$ and $j \neq a$; $A_{ia}(U;F) = B_{ia}(U;F) \cap C(U;F)$ if $i \neq a$.

This article is a continuation of [2]. Its main results give instances in which A_{oa} (Propositions 2, 3, 5, 6), $A_{\omega a}$ (Proposition 1) and $B_{oa}(E)$ (Proposition 4) hold. Some examples are given. We point out the relationship of E having A_{oa} to E being infrabarreled; it is phrased in such a way that E should be called holomorphically infrabarreled when E has A_{oa}. On the other hand, we also define E being holomorphically barreled, and point out some current examples having such a property.

For the notation and terminology we refer mostly to [6], [8], [9]; see also [10], [11].

§2. The Weak Case

Proposition 1. A weak complex locally convex space E has $A_{\omega a}$.

Proof. We first prove that E has $A_{\omega a}(E;F)$, where F is a complex normed space. Let I be an algebraic basis of the topological dual space E'. If $J \subset I$, we have the linear mapping

$$\pi_J : x \in E \mapsto (\varphi(x))_{\varphi \in J} \in \mathbb{C}^J.$$

If $f \in \mathcal{H}(E;F)$, there are $J \subset I$ finite and $a \in \mathcal{H}(C^J;F)$ such that $f = a \circ \pi_J$. If a is effective, that is it depends effectively on all variables indexed by J (as we may assume by decreasing J), then J and a are determined by f; we then denote them by J_f and a_f. If $\alpha \in N^{(I)}$, we define $D^\alpha f(0) \in F$ as being equal to $D^\alpha a_f(0)$ if $s(\alpha) \subset J_f$, and to 0 if $s(\alpha) \not\subset J_f$, where $N^{(I)}$ is the additive group formed by every $\alpha: I \to \mathbf{N}$ with finite support $s(\alpha)$, and $D^\alpha a_f$ is the α-th partial derivative of a_f. Let $b: N^{(I)} \to R$ be such that $b \geqslant 0$, $b(0) \neq 0$, and, for all finite $J \subset I$, the set formed by every α belonging to the support of b for which $s(\alpha) \subset J$ is finite. We set

$$p_b(f) = \sup\{ b(\alpha)\cdot \|D^\alpha f(0)\| \,;\, \alpha \in N^{(I)}\}.$$

This supremum is finite since , for every α, either $b(\alpha) = 0$, or $s(\alpha) \not\subset J_f$, or else $b(\alpha) \neq 0$ and $s(\alpha) \subset J_f$; in the first two cases $b(\alpha)\cdot\|D^\alpha f(0)\| = 0$, whereas the third case occurs only finitely many times.
$p_b: f \mapsto p_b(f)$ is a seminorm on $\mathcal{H}(E;F)$. It is ported by $\{0\}$. In fact, let $\epsilon > 0$, $J \subset I$ be finite, and $V_{\epsilon J}$ be formed by every $x \in E$ such that $|\varphi(x)| < \epsilon$ for any $\varphi \in J$. The family $(V_{\epsilon J})$ is a base of neighborhoods of 0 in E. Set

$$c_{\epsilon J} = \sup\{\alpha!\, b(\alpha)\, \epsilon^{-|\alpha|};\ \alpha \in N^{(I)}, s(\alpha) \subset J\},$$

where

$$\alpha! = \Pi_{\varphi\in I}\, \alpha(\varphi)!\,,\quad |\alpha| = \Sigma_{\varphi\in I}\, \alpha(\varphi)\,;$$

clearly $0 < c_{\epsilon J} < +\infty$. The estimate

$$p_b(f) \leqslant c_{\epsilon J}\cdot\sup\{\|f(x)\|\,;\ x \in V_{\epsilon J}\}$$

holds true. Indeed, if $J_f \not\subset J$, then f is unbounded on $V_{\epsilon J}$ by Liouville theorem. If $J_f \subset J$, we distinguish two alternatives. By assuming $s(\alpha) \subset J_f$, hence $s(\alpha) \subset J$, Cauchy inequality gives us

$$\|D^\alpha a_f(0)\| \leqslant \alpha!\, \epsilon^{-|\alpha|} \sup\{\|a_f(z)\|\,;\ z \in Q_{\epsilon J}\}\,,$$

where $Q_{\epsilon J}$ is formed by every $z = (z_\varphi)_{\varphi\in J_f} \in C^{J_f}$ such that $|z_\varphi| < \epsilon$ for any $\varphi \in J_f$; hence

$$\|D^\alpha f(0)\| \leqslant \alpha!\ \epsilon^{-|\alpha|}\sup\{\|f(x)\|;\, x \in V_{\epsilon J}\}.$$

By assuming $s(\alpha) \not\subset J_f$, then $D^\alpha f(0) = 0$. From these cases we conclude that the above estimate is true. Hence p_b is ported by $\{0\}$.

Let now $\mathfrak{X} \subset \mathcal{H}(E;F)$ be $\mathcal{T}_\omega$-bounded. Then $\mathfrak{X}$ depends on some finite $K \subset I$, that is $J_f \subset K$ for every $f \in \mathfrak{X}$. Indeed, if $\mathfrak{X}$ fails to do so, we argue inductively to find $f_m \in \mathfrak{X}$ ($m \in \mathbf{N}$) such that $J_{f_o} \neq \phi$ and $J_{f_m} \not\subset K_m \equiv J_{f_o} \cup \ldots \cup J_{f_{m-1}}$ ($m \geqslant 1$). Choose $\varphi_o \in J_{f_o}$ and $\varphi_m \in J_{f_m}$, $\varphi_m \notin K_m$ ($m \geqslant 1$); they are pairwise different. Since a_{f_m} is effective, choose $\alpha_m \in N^{(I)}$ so that $\varphi_m \in s(\alpha_m) \subset J_{f_m}$ and

$D^{\alpha_m} a_{f_m}(0) \neq 0$, that is $D^{\alpha_m} f_m(0) \neq 0$ $(m \in N)$. They are pairwise different since their supports are pairwise different; and, for all finite $J \subset I$, the set formed by every $m \in N$ for which $s(\alpha_m) \subset J$ is finite. Define b by letting

$$b(\alpha_m) = m.\ \|D^{\alpha_m} f_m(0)\|^{-1} \quad (m \in N),$$

$b(0) = 1$ and $b(\alpha) = 0$ for every remaining $\alpha \in N^{(I)}$. Then $p_b(f_m) \geq m$ $(m \in N)$, against the fact that $\mathcal{X}$ is $\mathcal{T}_\omega$-bounded. There results that $\mathcal{X}$ depends on some finite $K \subset I$. Since $\mathcal{X}$ is $\mathcal{T}_\omega$-bounded, hence $\mathcal{T}_o$-bounded, and C^K is locally compact, $\mathcal{X}$ is locally bounded. Thus $A_{\omega a}(E;F)$ holds.

If $\xi \in E$, $\epsilon > 0$, $H \subset I$ is finite and $U \subset E$ is the band formed by every $x \in E$ such that $|\varphi(x) - \varphi(\xi)| < \epsilon$ for every $\varphi \in H$, then $A_{\omega a}(U;F)$ holds too; the above proof applies except for minor notational changes having to do with the domain of definition of a_f for $f \in \mathcal{H}(U;F)$ and with replacing 0 by ξ. Since a non-void open $U \subset E$ is a union of bands U_λ $(\lambda \in \Lambda)$, we conclude that $A_{\omega a}(U;F)$ is true. Finally $A_{\omega a}(U;F)$ holds for every complex locally convex F once it does for every complex normed F. This proves $A_{\omega a}$. Q.E.D.

§3. The Cartesian Product Case

Proposition 2. A cartesian product $E = \Pi_{i \in I} E_i$ of semimetrizable complex locally convex spaces E_i $(i \in I)$ has A_{oa}.

Proof. It is known that a semimetrizable complex locally convex space has A_{oa}. We first prove that E has $A_{oa}(E;F)$, where F is a complex normed space. Let $\mathcal{X} \subset \mathcal{H}(E;F)$ be $\mathcal{T}_o$-bounded. Then $\mathcal{X}$ depends on some finite $K \subset I$, that is $\mathcal{X}$ is contained in the image of the mapping $\mathcal{H}(\Pi_{i \in K} E_i;F) \to \mathcal{H}(E;F)$ which results from composition by the projection $E \to \Pi_{i \in K} E_i$. To prove this it is enough to show that every sequence $f_m \in \mathcal{X}$ $(m \in N)$ depends on some finite subset of I. Since each $f \in \mathcal{H}(E;F)$ depends on a finite subset of I, there is a denumerable $J \subset I$ such that every f_m depends on a finite subset of J. Since $\Pi_{i \in J} E_i$ is semimetrizable, the sequence (f_m) is locally bounded. We can find a finite subset $K \subset J$ and a neighborhood V_i of 0 in E_i for every $i \in K$ such that, letting $V_i = E_i$ for the remaining $i \in I$, and $V = \Pi_{i \in I} V_i$, we have

$$\sup\{\|f_m(x)\|; m \in N, x \in V\} < +\infty .$$

Since every f_m is bounded on V, it follows from Liouville theorem and uniqueness of holomorphic continuation that every f_m depends on K; hence (f_m) depends on K. Therefore $\mathcal{X}$ depends on some finite $K \subset I$. Since $\mathcal{X}$ is $\mathcal{T}_o$-bounded and $\Pi_{i \in K} E_i$ is semimetrizable, $\mathcal{X}$ is locally bounded. Thus $A_{oa}(E;F)$ holds.

If U is the band $\Pi_{i \in I} U_i$, where $U_i \subset E_i$ is open and non-void for every $i \in I$ and $U_i = E_i$ except for finitely many i's, then $A_{oa}(U;F)$ holds too; the above proof applies with minor notational changes. Since a non-void open $U \subset E$ is a union of bands U_λ $(\lambda \in \Lambda)$, we conclude that $A_{oa}(U;F)$ is true. Finally, $A_{oa}(U;F)$ holds for every complex locally convex F once it does for every complex normed F. This proves A_{oa}. Q.E.D.

§4. The Denumerable Direct Sum Case

Lemma 1. Let E_m $(m \in N)$ and F be complex locally convex spaces, $E = \Sigma_{m \in N} E_m$ be the topological direct sum, $U \subset E$ be open and non-void, and $S_m = E_o + \cdots + E_m + 0 + \cdots$ $(m \in N)$. Then $\mathfrak{X} \subset \mathcal{H}(U;F)$ is amply bounded on U if, for every $\xi \in U$, there is a neighborhood V of ξ in U such that $\mathfrak{X}$ is bounded on $V \cap S_m$ for every $m \in N$.

Proof. It is enough to prove the lemma when F is seminormed. We may assume that $0 \in U$, and it suffices to show that $\mathfrak{X}$ is locally bounded at 0 on U. There is a neighborhood V_m of 0 in E_m $(m \in N)$ such that, letting $V = \Sigma_{m \in N} V_m$, then $V \subset U$ and $\mathfrak{X}$ is bounded on $V \cap S_m$ for every $m \in N$. Now, $\mathfrak{X}$ is bounded on $V \cap S_o$. Thus the set of mappings $t \in V_o \mapsto f(t,0,\ldots) \in F$, for all $f \in \mathfrak{X}$, is bounded on V_o, hence equicontinuous at 0 on V_o. Set $W_o = V_o$, and choose $M \in R$ so that

(1) $$\sup\{\|f(x_o,0,\ldots)\|;\ f \in \mathfrak{X},\ x_o \in W_o\} < M.$$

Assume that, for some $m \in N$, we have defined a neighborhood $W_k \subset V_k$ of 0 in E_k for every $k = 0,\ldots,m$ so that

(2) $$\sup\{\|f(x_o,\ldots,x_m,0,\ldots)\|;\ f \in \mathfrak{X},\ x_o \in W_o,\ldots,x_m \in W_m\} < M;$$

this is indeed the case for $m = 0$, by (1). Now, $\mathfrak{X}$ is bounded on $V \cap S_{m+1}$. Thus the set of mappings $t \in V_{m+1} \mapsto f(x_o,\ldots,x_m,t,0,\ldots) \in F$, for all $f \in \mathfrak{X}$, $x_o \in W_o,\ldots,x_m \in W_m$, is bounded on V_{m+1}, hence equicontinuous at 0 on V_{m+1}. By (2), choose a neighborhood $W_{m+1} \subset V_{m+1}$ of 0 in E_{m+1} so that

$$\sup\{\|f(x_o,\ldots,x_{m+1},0,\ldots)\|;\ f \in \mathfrak{X},\ x_o \in W_o,\ldots,x_{m+1} \in W_{m+1}\} < M.$$

In this way, we get a sequence $W_m \subset V_m$ of neighborhoods of 0 in E_m $(m \in N)$ such that, letting $W = \Sigma_{m \in N} W_m \subset V$, we will have $\|f(x)\| < M$ for every $f \in \mathfrak{X}$ and $x \in W$. Hence $\mathfrak{X}$ is locally bounded at 0 on U. Q.E.D.

A separable, or reflexive, complex normed space satisfies the following "boundedness hypothesis" [3]:

BH: A complex normed space either is finite dimensional, or else there is an entire complex function on it which is unbounded on some of its bounded subsets.

Actually, it is conjectured that BH holds true for every complex normed space.

Proposition 3. Let $E = \Sigma_{m \in N} E_m$ be a topological direct sum of complex normed spaces E_m $(m \in N)$. For E to have A_{oa} it is sufficient that all E_m be finite dimensional, or else that all $E_m = 0$ except for finitely many m's. Conversely, this condition is necessary for E to have $A_{oa}(E; C)$, provided all E_m satisfy the boundedness hypothesis.

Proof. Set $S_m = E_o + \cdots + E_m + 0 + \cdots \subset E$ $(m \in N)$. Let us prove sufficiency. Assume first that all E_m are finite dimensional. Let $U \subset E$ be open and non-void, and F be a complex locally convex space. Each $\xi \in U$ has a neighborhood V in U such that $V \cap S_m$ is compact for every $m \in N$. If $\mathfrak{X} \subset \mathcal{H}(U;F)$ is $\mathcal{T}_o$-bounded, $\mathfrak{X}$ is bounded on $V \cap S_m$ for every $m \in N$. By Lemma 1, $\mathfrak{X}$ is amply bounded. Hence E has A_{oa}. Next, if all $E_m = 0$ except for finitely many m's, then E is normable, hence it has A_{oa}.

Let us prove necessity. We first remark that, if $f \in \mathcal{H}(E_o;C)$ is bounded on every bounded subset of E_o and $\varphi_m \in (E_m)'$ $(m=1,2,\ldots)$, we have $f^* \in \mathcal{H}(E;C)$ defined by

$$f^*(x_o,\ldots,x_m,\ldots) = \Sigma_{m=1}^{\infty} f(mx_o)\,\varphi_m(x_m) \quad \text{for } x_m \in E_m \ (m \in N).$$

In fact, let $g_k \in \mathcal{H}(E;C)$ be given by

$$g_k(x_o,\ldots,x_m,\ldots) = \Sigma_{m=1}^{k} f(mx_o)\,\varphi_m(x_m) \quad \text{for } k = 1,2,\ldots$$

Set $s(r) = \sup\{|f(x_o)|;\ \|x_o\| < r\}$ for $r \in R$, $r > 0$. If $\rho = (\rho_1,\ldots,\rho_m,\ldots)$, where $\rho_m \in R$, $\rho_m > 0$, is such that

$$\Sigma_{m=1}^{\infty} s(mr)\,\rho_m < +\infty,$$

define $V_{r\rho}$ as the open subset of all $(x_m)_{m \in N} \in E$ for which $\|x_o\| < r$, $|\varphi_m(x_m)| < \rho_m$ $(m=1,2,\ldots)$. Then $g_k \to f^*$ as $k \to \infty$ uniformly on $V_{r\rho}$. Since these $V_{r\rho}$ cover E, then f^* is entire.

Let E have $A_{oa}(E,C)$. We discard the case in which all $E_m = 0$ except for finitely many m's; without loss of generality, we may assume that all $E_m \neq 0$. Choose $\varphi_m \in (E_m)'$, $\varphi_m \neq 0$ $(m \in N)$. Let us prove that E_o is finite dimensional if it satisfies BH. Assume that E_o is infinite dimensional. There is $f \in \mathcal{H}(E_o;C)$ which is unbounded on the closed unit ball B of E_o. Let f_n be the n-th partial sum of the Taylor series of f at 0 $(n \in N)$. Then $f_n \to f$ on E for $\mathcal{T}_o$ as $n \to \infty$, but $\{f_n;\ n \in N\}$ is not bounded on B. We may consider $f_n^* \in \mathcal{H}(E;C)$ since f_n is a continuous polynomial on E_o, hence it is bounded on every bounded subset of E_o. $\{f_n^*;\ n \in N\}$ is $\mathcal{T}_o$-bounded on E, since $\{f_n;\ n \in N\}$ is $\mathcal{T}_o$-bounded on E_o and every compact subset of E is contained in some S_m. However, $\{f_n^*;\ n \in N\}$ is unbounded on every neighborhood of 0 in E. In fact, for every $\epsilon_m > 0$ $(m \in N)$, if we choose $m = 1,2,\ldots$ so that $m\epsilon_o \geq 1$, we see that

$$f_n^*(x_o,0,\ldots,0,x_m,0,\ldots) = f_n(mx_o)\varphi_m(x_m)$$

is unbounded for $x_o \in E_o$, $\|x_o\| \leq \epsilon_o$, $x_m \in E_m$, $|\varphi_m(x_m)| \leq \epsilon_m$, $n \in N$. Thus E would not have $A_{oa}(E;C)$, against the assumption. It follows that E_o is finite dimensional; and likewise for all E_m. Q.E.D.

Definition 1. A non-void open subset U of a complex locally convex space E is a "Runge subset" of E if, for every complex locally convex space F and every $f \in \mathcal{H}(U;F)$, there are $f_m \in \mathcal{P}(E;F)$ and $\lambda_m \in C$ $(m \in N)$ such that $\lambda_m \to \infty$ as $m \to \infty$ and the sequence $\lambda_m(f - f_m|U)$ $(m \in N)$ is amply bounded on U.

Lemma 2. Let U be a Runge subset of a complex locally convex space E which is a topological direct sum of its vector subspaces E_1 and E_2, and $\pi: E \to E_1$ the corresponding projection. If $U \cap E_1 = \pi(U)$, for every complex locally convex space F and every $f \in \mathcal{H}(U;F)$ such that $f|(U \cap E_1) = 0$, there are $f_m \in \mathcal{P}(E;F)$ and $\lambda_m \in C$ such that $f_m|E_1 = 0$ $(m \in N)$, $\lambda_m \to \infty$ as $m \to \infty$, and the sequence $\lambda_m(f - f_m|U)$ $(m \in N)$ is amply bounded on U.

Proof. Choose f_m $(m \in N)$ according to Definition 1. It suffices to replace f_m by $g_m = f_m - f_m \circ \pi \in \mathcal{P}(E;F)$, since $g_m|E_1 = 0$, and the sequence $(\lambda_m f_m) \circ \pi$ $(m \in N)$ is amply bounded on $\pi^{-1}(U \cap E_1)$, hence on U. Q.E.D.

Remark 1. If U is ξ-balanced, where $\xi \in U$, then U is a Runge subset of E. In fact, if $f \in \mathcal{H}(U;F)$ and $f_m \in \mathcal{P}(E;F)$ is the m-th partial sum of the Taylor series of f at ξ $(m \in N)$, for every $t \in U$ and every continuous seminorm β on F there are $c \geq 0$, $0 < \theta < 1$, and a neighborhood V of t in U such that $\beta[f(x) - f_m(x)] \leq c\theta^m$ for $x \in V$, $m \in N$. If $\lambda_m \in C$ $(m \in N)$ and $\lim.\sup |\lambda_m|^{1/m} \leq 1$ as $m \to \infty$, the sequence $\lambda_m(f - f_m|U)$ $(m \in N)$ is then amply bounded. There remains to impose the additional condition $\lambda_m \to \infty$ as $m \to \infty$. In this case, if E_1 is a vector subspace of E and $f|(U \cap E_1) = 0$, then $f_m|E_1 = 0$ $(m \in N)$, so that we can discard Lemma 2, provided $\xi \in E_1$.

A subset X of a direct sum $E = \Sigma_{i \in I} E_i$ of vector spaces E_i $(i \in I)$ is "cylindrical" if X is the inverse image of its projection in $\Sigma_{i \in J} E_i$ for some finite $J \subset I$.

Proposition 4. If $E = \Sigma_{m \in N} E_m$ is a topological direct sum of complex seminormed spaces E_m $(m \in N)$ and U is a cylindrical Runge subset of E, then E has $B_{oa}(U)$.

Proof. We may assume that U is the inverse image in E of its projection in E_o; then U is the inverse image in E of its projection in $E_o \times \ldots \times E_m$ $(m \in N)$. Let F be a complex locally convex space. Take a seminorm p on $\mathcal{H}(U;F)$ which is bounded on its amply bounded subsets. Set $S_m = E_o + \ldots + E_m + 0 + \ldots \subset E$ $(m \in N)$. Let H_m be the vector subspace of all $f \in \mathcal{H}(U;F)$ such that $f|(U \cap S_m) = 0$ $(m \in N)$.

If $f \in H_m$, $\epsilon > 0$, there is $g \in \mathcal{P}(E;F)$ such that $g|S_m = 0$ and $p(f-g|U) < \epsilon$, hence $|p(f) - p(g|U)| < \epsilon$. In fact, by Lemma 2 there are $f_n \in \mathcal{P}(E;F)$ and $\lambda_n \in C$ such that $f_n|S_m = 0$ $(n \in N)$, $\lambda_n \to \infty$ as $n \to \infty$, and the sequence $\lambda_n(f - f_n|U)$ $(n \in N)$ is amply bounded. Hence $p[\lambda_n(f - f_n|U)]$ $(n \in N)$ is bounded. Then $p(f - f_n|U) \to 0$ as $n \to \infty$. It suffices to take $g = f_n$ for n large enough.

We claim that p vanishes on some H_m. Let us argue by contradiction. Assume that, for every $m \in N$, there is $f_m \in H_m$ such that $p(f_m) \neq 0$; we may assume that $p(f_m) > m$ and that f_m is the restriction to U of some element of $\mathcal{P}(E;F)$. Let $\mathcal{F} = \{f_m ; m \in N\}$. Then $\mathcal{F}|(U \cap S_n)$ is finite since it consists of at most n+1 elements, for every $n \in N$. As an element of $\mathcal{P}(E;F)$ is bounded on every bounded subset of E, Lemma 1 implies that $\mathcal{F}$ is amply bounded on U. Thus p is bounded on $\mathcal{F}$, a contradiction.

Let m be such that p vanishes on H_m. To every $f \in \mathcal{H}(U;F)$ we associate $f_m \in \mathcal{H}(U;F)$ defined by $f_m(x_o, \ldots, x_m, x_{m+1}, \ldots) = f(x_o, \ldots, x_m, 0, \ldots)$ if $(x_o, \ldots, x_m, x_{m+1}, \ldots) \in U$. Then $f - f_m \in H_m$, $p(f - f_m) = 0$ and $p(f) = p(f_m)$. Let $\mathcal{X} \subset \mathcal{H}(U;F)$ be $\mathcal{T}_o$-bounded. Call $\mathcal{X}_m \subset \mathcal{H}(U;F)$ the image of $\mathcal{X}$ by the mapping $f \mapsto f_m$. Since $\mathcal{X}|(U \cap S_m)$ is $\mathcal{T}_o$-bounded on $U \cap S_m$, it is amply bounded there because $E_o \times \ldots \times E_m$ is seminormable. Hence $\mathfrak{X}_m$ is amply bounded on U. It follows that p is bounded on $\mathfrak{X}_m$, therefore on $\mathfrak{X}$.

Q.E.D.

Concerning Proposition 4, Dineen [4] has shown that E has $B_{o\delta}(E;C)$.

We conjecture that E in Proposition 4 has B_{oa}.

§5. The Silva Case

Let E_m ($m \in N$) be complex locally convex spaces, E a complex vector space, $\rho_m: E_m \to E$ a linear mapping and $\sigma_m: E_m \to E_{m+1}$ a compact linear mapping such that $\rho_m = \rho_{m+1} \circ \sigma_m$ ($m \in N$). Assume that $E = \bigcup_{m \in N} \rho_m(E_m)$ and endow E with the inductive limit topology. Let $U \subset E$ be open, set $U_m = \rho_m^{-1}(U)$ ($m \in N$), and assume U_o non-void.

Lemma 3. If F is a complex locally convex space, then $\mathfrak{X} \subset \mathcal{H}(U;F)$ is amply bounded on U if and only if $\mathfrak{X}_m = \mathfrak{X} \circ \rho_m \subset \mathcal{H}(U_m;F)$ is amply bounded on U_m for every $m \in N$.

Proof. Necessity being clear, let us prove sufficiency. It is enough to treat F as being seminormed. We may assume that $0 \in U$, and it suffices to show that $\mathfrak{X}$ is locally bounded at 0 on U. Since $\mathfrak{X}_o$ is locally bounded at 0 on U_o, choose a convex neighborhood V_o of 0 in U_o such that $\sigma_o(V_o)$ has a compact closure contained in U_1, hence $\rho_o(V_o) \subset U$, and

$$\sup\{\|f[\rho_o(x)]\|;\ f \in \mathfrak{X},\ x \in V_o\} < M \tag{1}$$

for some $M \in R$. Assume that, for some $m \in N$, we have defined a convex neighborhood V_m of 0 in U_m such that $\sigma_m(V_m)$ has a compact closure contained in U_{m+1}, hence $\rho_m(V_m) \subset U$, and

$$\sup\{\|f[\rho_m(x)]\|;\ f \in \mathfrak{X},\ x \in V_m\} < M; \tag{2}$$

this is indeed the case for m = 0, by (1). Since $\mathfrak{X}_{m+1}$ is locally bounded at the closure of $\sigma_m(V_m)$ on U_{m+1}, hence equicontinuous there, use (2) to choose a convex neighborhood V_{m+1} of $\sigma_m(V_m)$ in U_{m+1} such that $\sigma_{m+1}(V_{m+1})$ has a compact closure contained in U_{m+2}, hence $\rho_{m+1}(V_{m+1}) \subset U$, and

$$\sup\{\|f[\rho_{m+1}(x)]\|;\ f \in \mathfrak{X},\ x \in V_{m+1}\} < M.$$

We also have $\rho_m(V_m) \subset \rho_{m+1}(V_{m+1})$. Proceeding in this way and letting

$$V = \bigcup_{m \in N} \rho_m(V_m),$$

we get a neighborhood V of 0 in U such that $\|f(x)\| < M$ for every $f \in \mathfrak{X}$, and $x \in V$. Hence $\mathfrak{X}$ is locally bounded at 0 on U.

Q.E.D.

A given E is said to be a Silva space if its topology may be defined as an inductive limit through suitable sequences (E_m), (ρ_m) and (σ_m) satisfying all the aforementioned conditions.

Proposition 5. A complex Silva space E has A_{oa}.

Proof. We may assume that E is separated. It is known that the topology of E may be defined as an inductive limit through suitable sequences (E_m), (ρ_m) and (σ_m) satisfying all the aforementioned conditions, where moreover each E_m is a Banach space. Let F be a complex locally convex space and $\mathfrak{X} \subset \mathcal{H}(U;F)$ be $\mathcal{T}_o$-bounded, $U \subset E$ being open and non-void; we may assume that U_o is non-void. Since $f \in \mathcal{H}(U;F) \mapsto f \circ \rho_m \in \mathcal{H}(U_m;F)$ is $\mathcal{T}_o$-continuous, then $\mathfrak{X} \circ \rho_m$ is $\mathcal{T}_o$-bounded, hence amply bounded for every $m \in \mathbf{N}$. By Lemma 3, $\mathfrak{X}$ is amply bounded. Q.E.D.

Remark 2. Let E be a Silva space. If F is a seminormed space, then $\mathcal{T}_o$ on $\mathcal{H}(U;F)$ is semimetrizable, hence bornological; it then follows from the fact that E has $A_{o\delta}$ (Proposition 5), that $\mathcal{T}_o = \mathcal{T}_\omega = \mathcal{T}_\delta$ on $\mathcal{H}(U;F)$. The same conclusion then results for every locally convex F.

§6. The Baire Case

Proposition 6. A Baire complex locally convex space E has A_{oa}.

Proof. Let F be a complex seminormed space. We start with two classical remarks.

If X is a non-void Baire space, and $\mathcal{U}$ is a pointwise bounded set of continuous mappings from X into F, there is at least one point of X where $\mathcal{U}$ is locally bounded.

If $p: E \to F$ is an m-homogeneous polynomial ($m \in \mathbf{N}$) and $a,b \in E$, then $\sup\{\|p(\lambda a+b)\|;\ \lambda \in \mathbf{C}, |\lambda| \leqslant 1\} = \sup\{\|p(a+\lambda b)\|;\ \lambda \in \mathbf{C}, |\lambda| \leqslant 1\}$; in fact, by the maximum principle we may replace $|\lambda| \leqslant 1$ by $|\lambda| = 1$, and then equality is clear via $\lambda \to 1/\lambda$, by m-homogeneity. In particular $\|p(b)\| \leqslant \sup\{\|p(a+\lambda b)\|;\ \lambda \in \mathbf{C}, |\lambda| \leqslant 1\}$.

Now, let $\mathfrak{X} \subset \mathcal{H}(U;F)$ be $\mathcal{T}_o$-bounded, $U \subset E$ being open and non-void, and $\xi \in U$. Take a balanced open $V \subset E$ containing 0 such that $\xi + V \subset U$. By Cauchy integral, the set

$$\mathcal{U} = \{(m!)^{-1}\hat{d}^m f(\xi);\ f \in \mathfrak{X}, m \in \mathbf{N}\}$$

is pointwise bounded on V because $\mathfrak{X}$ is bounded on every compact subset $\{\xi+\lambda x; \lambda \in \mathbf{C}, |\lambda| \leqslant 1\}$ of U, where $x \in V$. By the first remark, since V is a Baire space, there is $a \in V$ where $\mathcal{U}$ is locally bounded; let W be a balanced neighborhood of 0 in V such that $a + W \subset V$ and $\mathcal{U}$ is bounded on $a + W$. By the second remark, $\mathcal{U}$ is bounded on W. By the Taylor series at ξ, $\mathfrak{X}$ is bounded on $\xi+\theta W$ if $0 < \theta < 1$. Hence $\mathfrak{X}$ is locally bounded. Thus $A_{oa}(U;F)$ holds for every U and every seminormed F, hence for every locally convex F. This proves A_{oa}. Q.E.D.

Proposition 6 was also found independently by Dineen [4].

§7. Some Examples

Example 1. Let E be an infinite dimensional complex vector space with the weak topology associated to a norm, and $\mathfrak{X}$ be the closed unit ball in E'. Then $\mathfrak{X}$ is $\mathcal{T}_o$-bounded, but it is not locally bounded [1]; E does not have $A_{oa}(E;C)$. Actually $\mathfrak{X}$ is not $\mathcal{T}_\omega$-bounded [2]; E does not have $A_{o\omega}(E;C)$. Q.E.D.

We shall see in Examples 2 and 3 that, if X and Y are complex locally convex spaces having A_{oa}, then $E = X\times Y$ may fail to have $A_{oa}(E;C)$. However, it is easy to see that, if X has A_{ia} for some $i = o, \omega, \delta$ and Y is finite dimensional, then E has A_{ia}.

Example 2. Let X be an infinite dimensional normed space satisfying BH (section 4), and $Y = C^{(N)}$ be an infinite denumerable direct sum of C. By sufficiency in Proposition 3, X and Y have A_{oa}. By necessity, there, $E = X\times Y$ fails to have $A_{oa}(E;C)$. By Proposition 4, E has $B_{oa}(E)$. Now, $B_{oa}(E)$ implies $B_{o\delta}(E) = A_{o\delta}(E)$; hence E has $A_{o\delta}(E)$. Since $A_{oa}(E;C) = A_{\delta a}(E;C) \cap B_{oa}(E;C)$, we see that E fails to have $A_{\delta a}(E;C)$. Therefore $\mathfrak{B}_o(E;C) = \mathfrak{B}_\omega(E;C) = \mathfrak{B}_\delta(E;C) \neq \mathfrak{B}_a(E;C)$. Since $A_{oa}(E;C) = B_{oa}(E;C) \cap C(E;C)$, we see that E fails to have C(E;C). Q.E.D.

Example 3. Let $X = C^N$ and $Y = C^{(N)}$ be an infinite denumerable cartesian product and direct sum of C, respectively. By Proposition 2 and 3, X and Y have A_{oa}. Let $f_k \in \mathcal{H}(E;C)$ $(k \in N)$ be defined by $f_k(x,y) = x_k y_k$, where $E = X\times Y$, $x = (x_m)_{m\in N} \in X$ and $y = (y_m)_{m\in N} \in Y$. The sequence (f_k) is $\mathcal{T}_o$-bounded. However it is not locally bounded at 0. Hence E does not have $A_{oa}(E;C)$. Actually E does not have $A_{o\omega}(E;C)$ because we shall prove now that (f_k) is not $\mathcal{T}_\omega$-bounded. The linear mapping

$$r : g \in \mathcal{H}(E;C) \mapsto \hat{d}^2 g(0) \in \mathcal{P}(^2E;C)$$

is continuous if $\mathcal{H}(E;C)$ is given its topology $\mathcal{T}_\omega$ and $\mathcal{P}(^2E;C)$ is given its inductive limit topology. We have the natural continuous linear mapping $\mathcal{P}(^2E;C) \to \mathcal{P}(^2X;C) \times \mathcal{P}(^2Y;C) \times \mathcal{L}(X,Y;C)$, these spaces being given their inductive limit topologies. By projection, there results the natural continuous linear mapping

$$s: \mathcal{P}(^2E;C) \to \mathcal{L}(X,Y;C).$$

If $\varphi \in \mathcal{L}(X,Y;C)$, then

$$\varphi(x,y) = \Sigma_{m,n\in N}\, c_{mn}\, x_m y_n ,$$

where $c_{mn} \in C$ $(m, n \in N)$ and there is $\mu \in N$ such that $c_{mn} = 0$ for $m \geq \mu$ and all n. The linear form

$$t: \varphi \in \mathcal{L}(X,Y;C) \mapsto \Sigma_{m\in N}\, m\, c_{mm} \in C$$

is continuous. Thus $u = t \circ s \circ r : \mathcal{H}(E;C) \to C$ is a continuous linear form. Since $u(f_k) = 2k$ $(k \in N)$, we see that (f_k) is not $\mathcal{T}_\omega$-bounded. Q.E.D.

The proof of Proposition 2 shows that a cartesian product of complex locally convex spaces has A_{oa} if every denumerable cartesian subproduct has A_{oa}. This is false if we replace "denumerable" by "finite" as the following example shows.

Example 4. If we set $E_o = C^{(N)}$, $E_m = C$ $(m = 1,2,\ldots)$, $E = \Pi_{m \in N} E_m$, it follows from Example 3 that E fails to have $A_{oa}(E;C)$, or even $A_{o\omega}(E;C)$, although every $E_o \times \ldots \times E_m$ $(m \in N)$ has A_{oa} by Proposition 3. Q.E.D.

The following example shows that Proposition 4 breaks down if the E_m $(m \in N)$ are only metrizable.

Example 5. If we set $E_o = C^N$, $E_m = C$ $(m = 1,2,\ldots)$, $E = \Sigma_{m \in N} E_m$, it follows from Example 3 that E fails to have $A_{o\omega}(E;C) = B_{o\omega}(E;C)$; hence E fails to have $B_{oa}(E;C)$. Q.E.D.

The following example shows that we cannot drop compactness in Proposition 5; see Example 2.

Example 6. If E_o is a complex infinite dimensional normed space satisfying BH (section 4), $E_m = C$, $(m = 1,2,\ldots)$, $E = \Sigma_{m \in N} E_m$, necessity in Proposition 3 shows that E fails to have $A_{oa}(E;C)$. Q.E.D.

The following example shows that we cannot drop denumerability in Proposition 5, as well as in sufficiency in Proposition 3.

Example 7. If I is a set whose power is at least equal to that of the continuum, the direct sum $E = C^{(I)}$ of C indexed by I fails to have $A_{oa}(E;C)$. In fact, there is a homogeneous polynomial $p: E \to C$ of degree two which is not continuous. Let $c_{ij} \in C$ $(i,j \in I)$ be such that $p(x) = \Sigma_{i,j \in I} c_{ij} x_i x_j$ for every $x = (x_i)_{i \in I} \in E$. If $J \subset I$ is finite, define $p_J \in \mathcal{P}(^2E;C)$ by $p_J(x) = \Sigma_{i,j \in J} c_{ij} x_i x_j$. The family (p_J) is $\mathcal{T}_o$-bounded. It is not locally bounded at 0 since $p_J \to p$ pointwisely. Thus E fails to have $A_{oa}(E;C)$. Q.E.D.

§8. Holomorphically Infrabarreled, Or Barreled, Spaces

Classically, E is defined to be infrabarreled if, for every F and every $\mathfrak{X} \subset \mathcal{L}(E;F)$, if $\mathfrak{X}$ is bounded on any compact subset of E, then $\mathfrak{X}$ must be amply bounded. This form of definition suggests up to say that E is holomorphically infrabarreled if, for every F and every $\mathfrak{X} \subset \mathcal{H}(U;F)$, if $\mathfrak{X}$ is bounded on any compact subset of U, then $\mathfrak{X}$ must be amply bounded; in other words, E is called holomorphically infrabarreled if E has A_{oa}. It is then clear that a holomorphically infrabarreled space is infrabarreled.

Classically, E is defined to be barreled if, for every F and every $\mathfrak{X} \subset \mathcal{L}(E;F)$, if $\mathfrak{X}$ is bounded on any finite dimensional compact subset of E, then $\mathfrak{X}$ must be amply bounded. This form of the definition suggests us to say that E is holomorphically barreled if, for every F and every $\mathfrak{X} \subset \mathcal{H}(U;F)$, if $\mathfrak{X}$ is bounded on any finite dimensional compact subset of U, then $\mathfrak{X}$ must be amply bounded. It is then clear that a holomorphically barreled space is barreled. A superficial inspection of the proofs of Propositions 5, 6 and 2 shows us that the following complex locally convex spaces are actually not only holomorphically infrabarreled, as stated, but even holomorphically barreled: a Silva space, a Baire space, and a cartesian product of Fréchet spaces.

Acknowledgements

The authors were partially supported by Fundo Nacional de Ciência e Tecnologia (FINEP), Indústrias Klabin and Centro Brasileiro de Pesquisas Físicas, Rio de Janeiro, GB, Brasil, and National Science Foundation, Washington, D.C., U.S.A. Thanks are also due to H. Hogbe Nlend and Ph. Noverraz for their valuable criticism.

Bibliography

1. J. A. Barroso, Topologias nos espaços de aplicações holomorfas entre espacos localmente convexos, Anais da Academia Brasileira de Ciências 43(1971), 527–546.

2. J. A. Barroso and L. Nachbin, Sur certaines propriétés bornologiques des espaces d'applications holomorphes, Troisième Colloque sur l'Analyse Fonctionnelle, Liège 1970, Centre Belge de Recherches Mathématiques, Vander, Belgium (1971), 47–55.

3. S. Dineen, Unbounded holomorphic functions on a Banach space, Journal of the London Mathematical Society 4(1972), 461–465.

4. S. Dineen, Holomorphic functions on locally convex topological vector spaces, Annales de l'Institut Fourier, to appear in two parts.

5. L. Nachbin, On the topology of the space of all holomorphic functions on a given open set, Indagationes Mathematicae 29(1967), 366–368.

6. L. Nachbin, Topology on spaces of holomorphic mappings, Springer-Verlag, Germany (1969).

7. L. Nachbin, Sur les espaces vectorielles topologiques d'applications continues, Comptes Rendus de l'Académie des Sciences de Paris 271(1970), 596–598.

8. L. Nachbin, Concerning spaces of holomorphic mappings, Rutgers University, USA (1970).

9. L. Nachbin, Recent developments in infinite dimensional holomorphy, Bulletin of the American Mathematical Society 79(1973), to appear.

10. Ph. Noverraz, Pseudo-convexité, convexité polynomiale et domaines d'holomorphie en dimension infinie, North-Holland, Netherlands(1973).

11. G. Coeuré, Analytic functions and manifolds in infinite dimensional spaces, North-Holland, Netherlands, to appear.

Department of Mathematics
University of Rochester
Rochester, New York 14627, USA

Instituto de Matemática
Universidade Federal do Rio de Janeiro
Caixa Postal 1835, ZC-00
20000 Rio de Janeiro, GB, Brasil

MALGRANGE THEOREM FOR ENTIRE FUNCTIONS ON NUCLEAR SPACES

Philip J. Boland

Introduction.

If E is a locally convex topological vector space, $\mathcal{H}(E)$ will denote the space of scalar valued holomorphic functions on E. If τ is a locally convex topology on $\mathcal{H}(E)$, then a convolution operator $\mathcal{O}$ on $(\mathcal{H}(E),\tau)$ is a continuous, linear mapping $\mathcal{O}: \mathcal{H}(E) \to \mathcal{H}(E)$ which commutes with translations. The concept of a convolution operator is a generalization of that of a differential operator. A convolution operator $\mathcal{O}$ satisfies M_1 (Malgrange Theorem #1) if the kernel of $\mathcal{O}$ is the closed linear span of the exponential polynomials contained in the kernel. $\mathcal{O}$ satisfies M_2 if $\mathcal{O}$ is a surjective mapping. A subject of much study in the theory of infinite dimensional holomorphy has been for what spaces E and what topologies τ on $\mathcal{H}(E)$ does a given convolution operator $\mathcal{O}$ satisfy M_1 and M_2.

In this paper the case where E is a quasi-complete nuclear and dual nuclear space is studied. Examples of such spaces are $\mathcal{D}$, $\mathcal{S}$, $\mathcal{E}$, $\mathcal{D}'$, $\mathcal{S}'$, $\mathcal{E}'$, $\mathcal{H}(C)$, $\sum_N C$, any Fréchet Nuclear space, and the dual of any Fréchet Nuclear space. The topology τ considered on $\mathcal{H}(E)$ is the topology of uniform convergence on bounded subsets of E, or equivalently the compact open topology. It is shown that under these circumstances, $\mathcal{O}(\mathcal{H}(E)) \simeq \mathcal{H}'(E) \simeq \mathrm{Exp}\, E'$. $\mathcal{O}(\mathcal{H}(E))$ is the space of convolution operators on $(\mathcal{H}(E),\tau)$, $\mathcal{H}'(E)$ is the dual of $\mathcal{H}(E)$, and Exp E' is the space of holomorphic functions of exponential type on E'. Using the techniques of Malgrange and Gupta it follows that any convolution operator $\mathcal{O}$ on $(\mathcal{H}(E),\tau)$ satisfies M_1. Furthermore, in the case where the strong dual E' is metrizable (e.g. E is the dual of a Fréchet Nuclear space) it is shown that any non-zero convolution operator $\mathcal{O}$ is surjective.

§0. Preliminary Notation

A locally convex topological vector space E is nuclear if for any convex balanced neighborhood U of 0 in E, there exists another convex balanced neighborhood V of 0 such that and the mapping $E(V) \to E(U)$ is nuclear. (E(V), E(U) are respectively E endowed with the topologies on E generated by the Minkowski functionals of V and U). $E(V) \to E(U)$ is nuclear means that there exist $(a_n) \subseteq E'$ and $(y_n) \subseteq E$ such that for any $x \in E$, $\lim_n p_U(x - \sum_{i=1}^{n} \langle x,a_i\rangle y_i) = 0$ and $\sum_N p_{V^\circ}(a_n)p_U(y_n) < +\infty$ (p_{V°, p_V and p_U are the Minkowski functionals of V°, V and U respectively). E is dual nuclear if its strong dual E' is a nuclear space.

§1. Polynomials and Multilinear Forms

For a given locally convex topological vector space E, $\mathcal{L}(^kE)$ will denote the space of continuous k-linear forms on E. Unless otherwise specified $\mathcal{L}(^kE)$ will be endowed with the topology of uniform convergence on bounded sets. $\mathcal{L}_f(^kE)$ will denote the subspace of $\mathcal{L}(^kE)$ of all continuous k-linear forms of finite type - that is the set of $A \in \mathcal{L}(^kE)$ such that for some $(a_n^i) \subseteq E'$, A has the representation $A(x_1,\ldots,x_k) = \sum_{n=1}^{m} \langle x_1,a_n^1\rangle \cdots \langle x_k,a_n^k\rangle$. $P(^kE)$ is the space of continuous k-homogeneous polynomials on E. Given

$p \in P(^kE)$ there is a symmetric $A_p \in \mathcal{L}(^kE)$ such that $p(x) = A_p(x,\ldots,x)$. p is of finite type ($p \in P_f(^kE)$) if $A_p \in \mathcal{L}_f(^kE)$.

Definition 1.1. $A \in \mathcal{L}(^kE)$ is nuclear if it can be represented in the form $A(x_1,\ldots,x_k) = \sum_N \langle x_1,a_n^1\rangle\cdots\langle x_k,a_n^k\rangle$ (where $x_i \in E$, $a_n^i \in E'$) where for some convex balanced neighborhood U of 0 in E,

$$\sum_N p_{U^\circ}(a_n^1)\cdots p_{U^\circ}(a_n^k) < +\infty$$

$\mathcal{L}_N(^kE)$ will denote space of all nuclear continuous k-linear forms, while $P_N(^kE)$ will denote the corresponding space of continuous k homogeneous polynomials on E.

Proposition 1.1. For a nuclear locally convex space E, $\mathcal{L}_N(^kE) = \mathcal{L}(^kE)$ for each $k = 1,2,\ldots$.

Proof: Let $A \in \mathcal{L}(^kE)$ and W a convex balanced neighborhood of 0 in E such that $|A(x_1,\ldots,x_k)| \leqslant p_W(x_1)\cdots p_W(x_k)$ for all $x_i \in E$. Since E is nuclear there exists a convex balanced neighborhood V of 0 in E such that $E(V) \to E(W)$ is nuclear, and $V \subset W$. Hence there exist $(\varphi_n) \subseteq E'$ and $(y_n) \subseteq E$ such that $\lim_n p_W(x - \sum_{i=1}^n \langle x,\varphi_i\rangle y_i) = 0$ for each $x \in E$ and $\sum_N p_{V^\circ}(\varphi_n)p_W(y_n) < +\infty$. Now

$$\begin{aligned} A(x_1,\ldots,x_k) &= A(\sum_N\langle x_1,\varphi_n\rangle y_n,\ldots,\sum_N\langle x_{k-1},\varphi_n\rangle y_n,x_k) \\ &= \sum_{n_1,\ldots,n_{k-1}} \langle x_1,\varphi_{n_1}\rangle\cdots\langle x_{k-1},\varphi_{n_{k-1}}\rangle A(y_{n_1},\ldots,y_{n_{k-1}},x_k) \\ &= \sum_{n_1,\ldots,n_{k-1}} \langle x_1,\varphi_{n_1}\rangle\cdots\langle x_{k-1},\varphi_{n_{k-1}}\rangle\langle x_k,\psi_{n_1,\ldots,n_{k-1}}\rangle \end{aligned}$$

where $\langle z,\psi_{n_1,\ldots,n_{k-1}}\rangle = A(y_{n_1},\ldots,y_{n_{k-1}},z)$. Also

$$\begin{aligned} &\sum_{n_1,\ldots,n_{k-1}} p_{V^\circ}(\varphi_{n_1})\cdots p_{V^\circ}(\varphi_{n_{k-1}})p_{V^\circ}(\psi_{n_1,\ldots,n_{k-1}}) \\ &\leqslant \sum_{n_1,\ldots,n_{k-1}} p_{V^\circ}(\varphi_{n_1})\cdots p_{V^\circ}(\varphi_{n_{k-1}})p_{W^\circ}(\psi_{n_1,\ldots,n_{k-1}}) \\ &\leqslant (\sum_N p_{V^\circ}(\varphi_n)p_W(y_n))^{k-1} < +\infty. \end{aligned}$$

Therefore by redefining the above representation of A, we have that $A = \sum_N\langle\ ,a_n^1\rangle\cdots\langle\ ,a_n^k\rangle$ where $\sum_N p_{V^\circ}(a_n^1)\cdots p_{V^\circ}(a_n^k) < +\infty$, showing that $A \in \mathcal{L}_N(^kE)$.

Corollary 1.1. For a nuclear locally convex space E, $P_N(^kE) = P(^kE)$.

Corollary 1.2. Every continuous polynomial on a subspace F of a nuclear space E may be extended to a continuous polynomial on E.

Proof: It suffices to consider the homogeneous case, and in fact one need only show that if $A \in \mathcal{L}(^kF)$, then A can be extended by some $\widetilde{A} \in \mathcal{L}(^kE)$. Now F is a nuclear space and hence by Proposition 1.1, A has a representation of the form

$$A(x_1,\ldots,x_k) = \sum_N \langle x_1,a_n^1\rangle\cdots\langle x_k,a_n^k\rangle$$

where $\sum_N p_{V^\circ}(a_n^1)\cdots p_{V^\circ}(a_n^k) < +\infty$ for some convex balanced neighborhood V of 0 in F. Without loss of generality $V = W \cap F$ where W is a convex balanced neighborhood of 0 in E. For each n and i, and $x \in F$

$$\begin{aligned}|\langle x,a_n^i\rangle| &= |a_n^i(x)| \\ &\leqslant p_V(x)p_{V^\circ}(a_n^i) \\ &= p_W(x)p_{V^\circ}(a_n^i).\end{aligned}$$

By the Hahn Banach Theorem there exist $(\widetilde{a}_n^i) \subseteq E'$ such that for each n and i, $\widetilde{a}_n^i$ is an extension of a_n^i and

$$|\langle x,\widetilde{a}_n^i\rangle| \leqslant p_W(x)p_{V^\circ}(a_n^i) \quad \text{for each } x \in E.$$

Define $\widetilde{A}$ by $\widetilde{A}(x_1,\ldots,x_k) = \sum_N \langle x,\widetilde{a}_n^1\rangle\cdots\langle x_k,\widetilde{a}_n^k\rangle$. It is clear that $\widetilde{A}$ is an extension of A such that $A \in \mathcal{L}(^kE)$, since $\sup_{x_i \in W} |A(x_1,\ldots,x_k)| \leqslant \sum_N p_{V^\circ}(a_n^1)\cdots p_{V^\circ}(a_n^k) < +\infty$.

Corollary 1.3. If $\hat{E}$ is a topological completion of the nuclear space E, then any continuous polynomial on E may be extended to a continuous polynomial on $\hat{E}$.

Corollary 1.4. For a nuclear space E, $P_f(^kE)$ is dense in $P(^kE)$. ($P(^kE)$ being endowed with the topology of uniform convergence on bounded sets).

Proof: It suffices to show $\mathcal{L}_f(^kE)$ is dense in $\mathcal{L}(^kE)$. Let $A \in \mathcal{L}(^kE)$, B a bounded set in E and $\epsilon > 0$ be given. Now by Proposition 1.1, A can be represented in the form

$$A(x_1,\ldots,x_k) = \sum_N \langle x_1,a_n^1\rangle\cdots\langle x_k,a_n^k\rangle$$

where for some neighborhood V of 0 in E,

$$\sum_N p_{V^\circ}(a_n^1)\cdots p_{V^\circ}(a_n^k) < +\infty.$$

Choose $\alpha > 0$ such that $B \subseteq \frac{1}{\alpha}V = U$. Now $\sum_N p_{U^\circ}(a_n^1)\cdots p_{U^\circ}(a_n^k) < +\infty$. Choose m such that

$\sum_{n=m+1}^{\infty} p_{U^\circ}(a_n^1)\cdots p_{U^\circ}(a_n^k) < \epsilon$. Letting $A' = \sum_{n=1}^{m} \langle\ ,a_n^1\rangle\cdots\langle\ ,a_n^k\rangle$, we see that $A' \in \mathcal{L}_f({}^kE)$ and $\sup_{x_i \in B} |A(x_1,\ldots,x_k) - \hat{A}'(x_1,\ldots,x_k)| \leqslant \sum_{n=m+1}^{\infty} p_{U^\circ}(a_n^1)\cdots p_{U^\circ}(a_n^k) < \epsilon$.

Definition 1.2: For each balanced convex bounded subset B of E we may define the semi norms π_B and ϵ_B on $\mathcal{L}_f({}^kE)$ as follows:

$$\pi_B(A) = \inf\{\sum_{n=1}^{m} p_{B^\circ}(a_n^1)\cdots p_{B^\circ}(a_n^k) : A = \sum_{n=1}^{m}\langle\ ,a_n^1\rangle\cdots\langle\ ,a_n^k\rangle\}$$

$$\epsilon_B(A) = \sup\{|A(z_1,\ldots,z_k)| = |\sum_{n=1}^{m}\langle z_1,a_n^1\rangle\cdots\langle z_k,a_n^k\rangle| \text{ where } z_i \in (B^\circ)^\circ \text{ and } A = \sum_{n=1}^{m}\langle\ ,a_n^1\rangle\cdots\langle\ ,a_n^k\rangle\}.$$

$$(\ (B^\circ)^\circ = \{x : x \in (E')' \text{ and } |\langle\varphi,x\rangle| \leqslant 1 \text{ for all } \varphi \in B\ \}).$$

The topology on $\mathcal{L}_f({}^kE)$ generated by all semi-norms of the form π_B will be denoted π. Similarly that generated by all semi-norms of the form ϵ_B will be denoted by ϵ. On $P_f({}^kE)$, let

$$\pi_B(p) = \inf\{\sum_{n=1}^{m}(p_{B^\circ}(a_n))^k : p = \sum_{n=1}^{m}\langle\ ,a_n\rangle^k\}$$

and $\epsilon_B(p) = \sup\{|p(z)| : z \in (B^\circ)^\circ\}$. π and ϵ will denote the topologies on $P_f({}^kE)$ generated by respectively all semi-norms of the form π_B and ϵ_B. Note that if E is semi reflexive then ϵ_B is the sup semi-norm on B. If $p \in P_f({}^kE)$ and $A_p \in \mathcal{L}_f({}^kE)$ is the symmetric continuous k-linear form corresponding to p, then $\pi_B(A_p) \leqslant \pi_B(p) \leqslant \frac{k^k}{k!}\pi_B(A_p)$. If E is semi reflexive, then it also can be shown that $\epsilon_B(p) \leqslant \epsilon_B(A_p) \leqslant \frac{k^k}{k!}\epsilon_B(p)$.

Lemma 1.1: Let E be a dual nuclear space, and B a balanced convex bounded subset of E. Then there exists a $C_B > 0$ and a balanced convex bounded subset B_1 in E such that

$$\pi_B(A) \leqslant C_B^{k-1}\epsilon_{B_1}(A) \quad \text{for each } A \in \mathcal{L}_f({}^kE).$$

Proof: Since E' is nuclear, there exists a balanced convex bounded subset B_1 in E such that $B \subseteq B_1$ and $E'(B_1^\circ) \to E'(B^\circ)$ is nuclear. Hence there exist $(a_n) \subset (E')'$ and $(\varphi_n) \subseteq E'$ such that $\lim_m p_{B^\circ}(\varphi - \sum_{n=1}^{m}\langle\varphi,a_n\rangle\varphi_n) = 0$ for all $\varphi \in E'$ and $C_B = \sum_N p_{(B_1^\circ)^\circ}(a_n)p_{B^\circ}(\varphi_n) = \sum_N |a_n|_{B_1^\circ}|\varphi_n|_B < +\infty$. Let $A \in \mathcal{L}_f({}^kE)$ be given. By the Hahn Banach theorem there exists $\alpha \in \mathcal{L}_f'({}^kE)$ such that $\langle A, \alpha\rangle = \pi_B(A)$ and $|\langle A',\alpha\rangle| \leqslant \pi_B(A')$ for all $A' \in \mathcal{L}_f({}^kE)$. In particular if $\psi_i \in E'$ for each $i = 1,\ldots,k$, then $|\langle\psi_1\cdots\psi_k,\alpha\rangle| \leqslant |\psi_1|_B\cdots|\psi_k|_B$ and

$$\langle\psi_1\cdots\psi_k,\alpha\rangle = \langle(\sum_N\langle\psi_1,a_n\rangle\varphi_n)\cdots(\sum_N\langle\psi_{k-1},a_n\rangle\varphi_n)\psi_k,\alpha\rangle$$

$$= \sum_{n_1,\ldots,n_{k-1}}\langle\varphi_{n_1}\cdots\varphi_{n_{k-1}}(\langle\psi_1,a_{n_1}\rangle\cdots\langle\psi_{k-1}a_{n_{k-1}}\rangle\psi_k),\alpha\rangle.$$

Therefore if $A = \sum_{j=1}^{m} \langle\ \ ,\psi_1^j\rangle \cdots \langle\ \ ,\psi_k^j\rangle$, then

$$\pi_B(A) = \langle A,\alpha\rangle = \sum_{j=1}^{m} \langle \psi_1^j \ldots \psi_k^j,\alpha\rangle = \sum_{j=1}^{m} \sum_{n_1,\ldots,n_{k-1}} \varphi_{n_1}\cdots\varphi_{n_{k-1}}(\langle\psi_1^j,a_{n_1}\rangle\cdots\langle\psi_{k-1}^j,a_{n_{k-1}}\rangle\,\psi_k^j,\alpha)$$

$$= \sum_{n_1,\ldots,n_{k-1}} \langle \varphi_{n_1}\cdots\varphi_{n_{k-1}}\Big(\sum_{j=1}^{m}\langle\psi_1^j,a_{n_1}\rangle\cdots\langle\psi_{k-1}^j,a_{n_{k-1}}\rangle\psi_k^j\Big),\alpha\rangle$$

$$\leqslant \sum_{n_1,\ldots,n_{k-1}} P_{B^\circ}(\varphi_{n_1})\cdots P_{B^\circ}(\varphi_{n_{k-1}})P_{B^\circ}\Big(\sum_{j=1}^{m}\langle\psi_1^j,a_{n_1}\rangle\cdots\langle\psi_{k-1}^j,a_{n_{k-1}}\rangle\psi_k^j\Big)$$

$$\leqslant \sum_{n_1,\ldots,n_{k-1}} P_{B^\circ}(\varphi_{n_1})\ldots P_{B^\circ}(\varphi_{n_{k-1}})P_{B_1^\circ}\Big(\sum_{j=1}^{m}\langle\psi_1^j,a_{n_1}\rangle\cdots\langle\psi_{k-1}^j,a_{n_{k-1}}\rangle\psi_k^j\Big)$$

$$\leqslant \sum_{n_1,\ldots,n_{k-1}} |\varphi_{n_1}|_B\ldots|\varphi_{n_{k-1}}|_B\,|a_{n_1}|_{B_1^\circ}\ldots|a_{n_{k-1}}|_{B_1^\circ}\,\epsilon_{B_1}(A)$$

$$= \epsilon_{B_1}(A)\cdot(C_B)^{k-1}.$$

Corollary 1.5: Let E be a semi reflexive dual nuclear space and B a convex balanced bounded subset of E. Then there exist a convex balanced bounded subset B_1 of E and a constant $C_B' > 0$ such that

$$\epsilon_B(p) \leqslant \pi_B(p) \leqslant (C_B')^{k-1}\epsilon_{B_1}(p) \quad \text{for every} \quad p \in P_f({}^kE).$$

Proof: Let $A_p \in \mathcal{L}_f({}^kE)$ be the symmetric continuous k-linear form corresponding to $p \in P_f({}^kE)$. Then

$$\pi_B(p) \leqslant \frac{k^k}{k!}\pi_B(A_p) \leqslant C_B^{k-1}\frac{k^k}{k!}\epsilon_{B_1}(A_p) \leqslant C_B^{k-1}\Big(\frac{k^k}{k!}\Big)^2\epsilon_{B_1}(p)$$

where C_B and B_1 are as in Lemma 1.1. Hence choose $C_B' = \sup_k [C_B^{k-1}(\frac{k^k}{k!})^2]^{1/k-1}$ and the Corollary follows.

If E is a semi-reflexive nuclear space which is also dual nuclear, then by Corollaries 1.4 and 1.5 we may extend the definitions of the semi-norms π_B and ϵ_B to $P({}^kE)$. Hence the following Proposition:

Proposition 1.2: Let E be a semi-reflexive nuclear and dual nuclear space and B a convex balanced bounded subset of E. Then there exist a convex balanced bounded subset B_1 of E and a constant $C_B' > 0$ such that $\epsilon_B(p) \leqslant \pi_B(p) \leqslant (C_B')^{k-1}\epsilon_{B_1}(p)$ for all $p \in P({}^kE)$. In particular the π and ϵ topologies on $P({}^kE)$ coincide.

Proposition 1.3: Let E be a semi-reflexive nuclear and dual nuclear space. If $T \in P'({}^kE)$ then $\beta T \in P({}^kE')$ where $\beta T(\varphi) = T(\varphi^k)$ for $\varphi \in E'$. Furthermore β is an isomorphism between $P'({}^kE)$ and $P({}^kE')$.

Proof: Given $T \in P'({}^kE)$, there are $C > 0$ and B a convex balanced bounded subset of E such that $|T(p)| \leqslant C \sup_{z\in B} |p(z)| = C\,|p|_B$ for all $p \in P({}^kE)$. Now $|\beta T|$ is bounded on the set B° in E', and hence

$\beta T \in P(^kE')$. Since $P_f(^kE)$ is dense in $P(^kE)$ by Corollary 1.4, it follows that $\beta : P'(^kE) \to P(^kE')$ is 1-1. To finish the proof it suffices to show that β is surjective. Let $p' \in P(^kE')$ and suppose B is a convex balanced bounded subset of E such that $|p'|_{B^\circ} < r$. Define $T_{p'}: P_f(^kE) \to C$ by

$$T_{p'}(p) = \sum_{i=1}^{m} p'(\varphi_i) \text{ where } p = \sum_{i=1}^{m} \varphi_i^k .$$

$T_{p'}$ is well defined (see for example Gupta, [4]) and

$$|T_{p'}(p)| \leq |\sum_{i=1}^{m} p'(\varphi_i)| \leq r \sum_{i=1}^{m} |\varphi_i|_B^k .$$

Hence it follows that

$$|T_{p'}(p)| \leq r\,\pi_B(p) \leq r\,(C_B')^{k-1} \epsilon_{B_1}(p) \qquad \text{where}$$

C_B' and B_1 are as in Proposition 1.2. Now $T_{p'}$ may be extended to all of $P(^kE)$ and hence for all $p \in P(^kE)$, we have

$$|T_{p'}(p)| \leq r(C_B')^{k-1} \epsilon_{B_1}(p) = r(C_B')^{k-1} |p|_{B_1} .$$

Hence $T_{p'} \in P'(^kE)$. Since $\beta T_{p'} = p'$, the proposition follows.

§2. Convolution Operators on $\mathcal{H}(E)$.

In this section E will denote a quasi complete nuclear and dual nuclear locally convex space. Such a space is always semi-reflexive and therefore the results of the previous section apply. Examples of such spaces are $\mathcal{O}$, $\mathcal{D}'$, $\mathcal{S}$, $\mathcal{S}'$, $\mathcal{H}(C)$, $\sum_N C$, any Fréchet Nuclear space, and the dual of any Fréchet Nuclear space.

$\mathcal{H}(E)$ will denote the space of holomorphic functions on E endowed with the topology of uniform convergence on bounded sets. In a nuclear space which is quasi complete, any bounded subset B of E is relatively compact in E. Hence it makes sense to talk about the sup of a holomorphic function on a bounded subset B of E. If the dual E' of E is metrizable (e.g. if E = $\mathcal{E}'$, $\mathcal{S}'$ or the dual of a Fréchet-Nuclear space), then $\mathcal{H}(E)$ is a Fréchet space.

Definition 2.1. $\mathcal{Q} : \mathcal{H}(E) \to \mathcal{H}(E)$ is a convolution operator if it is linear, continuous and commutes with translations. To say that $\mathcal{Q}$ commutes with translations is to say that $\mathcal{Q}(\tau_{-z}f) = \tau_{-z}\mathcal{Q}(f)$ for arbitrary $f \in \mathcal{H}(E)$ and $z \in E$, where $\tau_{-z}f(w) = f(w+z)$ for $w \in E$. $\mathcal{O}(\mathcal{H}(E))$ will denote the space of all convolution operators on $\mathcal{H}(E)$.

The following Proposition leads to an isomorphism between $\mathcal{O}(\mathcal{H}(E))$ and $\mathcal{H}'(E)$.

Proposition 2.1. **Let $T \in \mathcal{H}'(E)$. Then $T*f(z) = T(\tau_{-z}f)$ defines a function holomorphic on E, and $f \to T*f$ is a convolution operator on $\mathcal{H}(E)$.**

Proof:

(a) $T*f$ is G-analytic. Let $x,w \in E$, $w \neq 0$ be fixed. It will be shown that $\lim_{\lambda \to 0} \frac{T*f(x+\lambda w)-T*f(x)}{\lambda}$ exists.

For each $\lambda \neq 0$, let $g_\lambda(z) = \frac{f(x+\lambda w+z)-f(x+z)}{\lambda}$ for $z \in E$, and let $g_o(z) = df(x+z).w$. Now $g_\lambda(z) \to g_o(z)$ for each $z \in E$ as $\lambda \to 0$. Suppose now B is a bounded subset of E, and hence let V be a balanced convex neighborhood of 0 such that $|f|_{x+B+V} \leqslant M$ for some constant M. Therefore there exists an $\alpha > 0$ such that $|\lambda| \leqslant \alpha$ and $z \in B$ imply that

$$\left|\frac{\hat{d}^k f(x+z).\lambda w}{k!}\right| \leqslant \frac{M}{2^k}$$

for all k = 2,3,. . . . Therefore, for $|\lambda| \leqslant \alpha$,

$$|g_\lambda - g_o|_B = \sup_{z \in B} \left|\frac{1}{\lambda} \sum_{k=2}^{\infty} \frac{\hat{d}^k f(x+z).\lambda w}{k!}\right|$$

$$\leqslant \sup_{z \in B} \left|\frac{1}{\lambda}\right| \sum_{k=2}^{\infty} \left|\frac{\lambda}{\alpha}\right|^k \left|\frac{\hat{d}^k f(x+z).\alpha w}{k!}\right|$$

$$\leqslant \frac{\lambda M}{(2\alpha)^2} \sum_{k=0}^{\infty} \left|\frac{\lambda}{2\alpha}\right|^k .$$

Hence g_λ converges to g_o in $\mathcal{H}(E)$. Therefore

$$\frac{T*f(x+\lambda w)-T*f(x)}{\lambda} = T\left(\frac{\tau_{-(x+\lambda w)}f - \tau_{-x}f}{\lambda}\right)$$

$$= T(g_\lambda)$$

$$\to T(g_o) \text{ as } \lambda \to 0.$$

$\therefore$ $T*f$ is G-analytic.

(b) $T*f \in \mathcal{H}(E)$. There exists a $C > 0$ and a bounded set B such that $|T(g)| \leqslant C|g|_B$ for all $g \in \mathcal{H}(E)$. Now for fixed $f \in \mathcal{H}(E)$ and $z \in E$ there exists a neighborhood V of 0 such that $|f|_{z+B+V} < +\infty$. Hence $|T*f|_{z+V} \leqslant C|f|_{z+B+V}$. Therefore $T*f$ is locally bounded and by part (a), $T*f \in \mathcal{H}(E)$.

(c) $f \to T*f$ is a convolution operator. It suffices to show only that the mapping is continuous. Let B be as in (b), and let B′ be another bounded set in E. Then

$$|T*f|_{B'} = \sup_{b' \in B'} |T*f(b')| \leqslant \sup_{b' \in B'} C|f|_{b'+B} = C|f|_{B+B'} .$$

$\therefore$ $f \to T*f$ is continuous.

Corollary 2.1. $\mathcal{O}(\mathcal{H}(E))$ **and** $\mathcal{H}'(E)$ **are naturally isomorphic.**

Proof: Define $\gamma : \mathcal{O}(\mathcal{H}(E)) \to \mathcal{H}'(E)$ by $\gamma(\mathcal{Q})f = \mathcal{Q}f(0)$.

Define $\tilde{\gamma}$: $\mathcal{H}'(E) \to \mathcal{O}(\mathcal{H}(E))$ by $\tilde{\gamma}T = T*$.

It follows that $\tilde{\gamma} = \gamma^{-1}$ and γ is a natural isomorphism.

Note: $\mathcal{O}(\mathcal{H}(E))$ is an algebra under composition. Hence $T_1 * T_2$ will denote $\gamma(\mathcal{Q}_1 \circ \mathcal{Q}_2)$ where $\gamma(\mathcal{Q}_i) = T_i$. It can be shown that the subspace of $\mathcal{H}(E)$ generated by all exponentials $\{e^{\varphi} : \varphi \in E'\}$ is dense in $\mathcal{H}(E)$ (see for example Gupta [4]).

Definition 2.2. Given $T \in \mathcal{H}'(E)$, the **Borel transform** $\hat{T}$ of T is the mapping $\hat{T}(\varphi) = T(e^{\varphi})$ for each $\varphi \in E'$. It is easy to see that $\hat{T} \in \mathcal{H}(E')$.

Definition 2.3. Exp E' = $\{F : F \in \mathcal{H}(E')$, F is of exponential type with respect to some semi-norm on E'$\}$ (F is of exponential type on E' if there exists a convex balanced bounded subset B of E and a constant $C > 0$ such that

$$|F(\varphi)| \leq C\, e^{P_{B^\circ}(\varphi)} = C\, e^{|\varphi|_B} \quad \text{for all } \varphi \in E').$$

Proposition 2.2. The Borel transform $\wedge$ is a 1-1 linear mapping of $\mathcal{H}'(E)$ onto Exp E'.

Proof. The subspace of $\mathcal{H}(E)$ generated by $\{e^{\varphi} : \varphi \in E'\}$ is dense in $\mathcal{H}(E)$, and hence it follows that the Borel transform is 1-1. Now suppose $T \in \mathcal{H}'(E)$. There exist $C > 0$ and B a convex balanced bounded subset of E such that $|T(g)| \leq C|g|_B$ for every $g \in \mathcal{H}(E)$. Let T_k denote the restriction of T to $P(^kE)$, and let $\beta T_k \in P(^kE')$ correspond to T_k as in Proposition 1.3. Now

$$|\hat{T}(\varphi)| = |T(e^{\varphi})| \leq C\, e^{|\varphi|_B}.$$

Hence $\hat{T} \in$ Exp E'.

The proof is completed by showing that the Borel transform maps onto Exp E'. Let $F \in$ Exp E'. Then there exist $C > 0$ and B a convex balanced bounded subset of E such that $|F(\varphi)| \leq C\, e^{|\varphi|_B}$ for all $\varphi \in E'$. Therefore there exist constants $r > 0$ and $s > 0$ such that $|\hat{d}^kF(0)|_{B^\circ} \leq rs^k$ for each k. As in the proof of Proposition 1.3, there exists a convex balanced bounded subset B_1 of E with the property that for each k, $|T_k(p)| \leq r\,s^k (C'_B)^{k-1} |p|_{B_1}$ for all $p \in P(^kE)$ (T_k being the element of $P'(^kE)$ corresponding to $\hat{d}^kF(0)$ as in Proposition 1.3, and C'_B being as in Proposition 1.2). Let $t = s \max\{C'_B, 1\}$. Now since the Taylor series of any $f \in \mathcal{H}(E)$ converges to f, define T: $\mathcal{H}(E) \to C$ by

$$T(f) = T\left(\sum_{k=0}^{\infty} \frac{\partial^k f(0)}{k!}\right) = \sum_{k=0}^{\infty} \frac{T_k}{k!}(\partial^k f(0)).$$

Note that

$$|T(f)| \leqslant \sum_{k=0}^{\infty} \left|\frac{T_k(\partial^k f(0))}{k!}\right| \leqslant r \sum_{k=0}^{\infty} t^k \left|\frac{\partial^k f(0)}{k!}\right|_{B_1}$$

$$\leqslant r \sum_{k=0}^{\infty} \frac{t^k}{(2t)^k} |f|_{2t B_1}$$

$$\leqslant 2r\,|f|_{2tB_1}.$$

Hence $T \in \mathcal{H}'(E)$. Since $\hat{T} = F$, the proof is complete.

In summary it has been established that $\mathcal{O}(\mathcal{H}(E)) \simeq \mathcal{H}'(E) \simeq \operatorname{Exp} E'$. Exp E' has the following division property: Given $F_1, F_2 \in \operatorname{Exp} E'$ such that F_1/F_2 is G-analytic on $E' \Rightarrow F_1/F_2 \in \operatorname{Exp} E'$. (see for example Gupta 4).

Theorem 2.1. **Let E be a quasi complete nuclear and dual nuclear space, and let $\mathcal{H}(E)$ be the space of holomorphic functions on E endowed with the topology of uniform convergence on bounded subsets of E. Then the kernel of any convolution operator $\mathcal{Q}$ on $\mathcal{H}(E)$ is the closed linear span of $\{pe^{\varphi} : p \in P(^kE)$ for some k, and $\varphi \in E'$, where $\mathcal{Q}(pe^{\varphi}) = 0\}$.**

Proof. See Gupta [4]. If $\mathcal{Q} = 0$, the result is trivial, so assume $\mathcal{Q} \neq 0$. There exists a $T \in \mathcal{H}'(E)$ such that $\mathcal{Q} = T*$ (Corollary 2.1). Let $S \in \mathcal{H}'(E)$ be such that $S(pe^{\varphi}) = 0$ whenever $T*pe^{\varphi} = 0$. Then it follows that $\hat{S}/\hat{T} \in \mathcal{H}(E'))$. But since Exp E' has the above mentioned division property together with Proposition 2.2 it follows that for some $U \in \mathcal{H}'(E)$, $\hat{S} = \hat{U}\hat{T}$. Therefore $S = U*T$ and if $T*f = 0$, then $S*f = (U*T)*f = U*(T*f)=0$. Hence $T*f = 0$ implies that $S(f) = S*f(0) = 0$ and an application of the Hahn Banach Theorem completes the proof.

A proof similar to that of Gupta [4] implies that in the above situation if $\mathcal{Q} \neq 0$, then ${}^t\mathcal{Q}(\mathcal{H}'(E))$ is the orthogonal of the kernel of $\mathcal{Q}$ in $\mathcal{H}(E)$ [${}^t\mathcal{Q}$ is the transpose of $\mathcal{Q}$]. Suppose that E is dual metric. Then $\mathcal{H}(E)$ is a Fréchet space and one may prove the following:

Theorem 2.2. **Let E be a quasi complete nuclear and dual nuclear space which is dual metric. Then any convolution operator on $\mathcal{H}(E)$ is surjective if $\mathcal{H}(E)$ is endowed with the topology of uniform convergence on bounded sets.**

Proof. This follows from the above remark together with the following Theorem of Dieudonne and Schwartz [2]: Let U and V be two Fréchet spaces and $h: U \to V$ be a continuous linear mapping. Then h is surjective $\Leftrightarrow {}^th$ (transpose of h) is 1-1 and ${}^th(V')$ is weakly closed in U'.

Bibliography

1. Boland, P., and Dineen, S., **Convolution Operators on G-Holomorphic Functions in Infinite Dimensions** (to appear in the Transactions of the American Mathematical Society).

2. Dieudonne, J., and Schwartz, L., **La dualite dans les espaces** ($\mathcal{F}$) **et** ($\mathcal{LF}$), Annales de l'Institut Fourier (Grenoble), tome I, 1949, pp. 61–101.

3. Dineen, S., **Holomorphic Functions on Locally Convex Topological Vector Spaces** (to appear in Annals de Institut Fourier).

4. Gupta, C.P., **Convolution Operators and Holomorphic Functions on a Banach Space**, Semanaire D'Analyse Moderne, No.2, Universite de Sherbrooke.

5. Malgrange, B., **Existence et approximation des solutions des equations aux derivees partielles et des equations des convolutions**, Annales de l'Institut Fourier Grenoble, IV, pg. 271-355, 1955-56.

6. Nachbin, Leopoldo, **Topology on Spaces of Holomorphic Mappings**, Ergebnisse der Mathematik und ihrer Grenzgebiete, Band 47, Springer Verlag, New York, 1969.

7. Pietsch, Albrecht, **Nuclear Locally Convex Spaces**, Ergebnisse der Mathematik und ihrer Grenzgebrete, Band 66, Springer Verlag, New York, 1972.

University College of Dublin
School of Mathematics
Belfield, Dublin 4
Ireland

ON SOME VARIOUS NOTIONS OF INFINITE DIMENSIONAL HOLOMORPHY

J. F. Colombeau

The aim of the present work is to show that many notions of holomorphic maps in the framework of locally convex spaces (l.c.s.) or bornological vector spaces (b.v.s.) are in fact reducible to only one definition given by J. S. e Silva in [11].

We consider only definitions satisfying the following conditions:

1) they generalize the notion of an analytic map in Banach spaces,
2) their theories are valid in the "usual" spaces,
3) the usual bilinear maps are holomorphic,

and it seems to us indispensable to demand that these three conditions should be satisfied.

§ 1. A recall of some definitions of an analytic map in complex spaces.

Let E and F be sequentially complete Hausdorff complex l.c.s., U an open set in E and f a mapping $U \to F$. We list the following definitions:

(1) f is Fantappié analytic in U (defined in [10]).

(2) f is Silva differentiable in the large sense in U:

for every convex and balanced bounded set B of E, for every convex and balanced neighbourhood V of 0 in F, the restriction of f to $U \cap E_B$ (where E_B is the normed space $\bigcup_n nB$ normed by the gauge of B) with range in the normed space F_V is holomorphic (in the usual sense of normed spaces).

(3) For every $x \in U$, there exists a sequence of bounded homogeneous polynomials ℓ_n (ℓ_n is of degree n) and a balanced neighbourhood U' of 0 in E such that, for every $h \in U'$ the series $\sum_n \ell_n(h)$ converges in F and the sum is equal to f(x+h).

(4) For every convex and balanced bounded set B of E and every $y' \in Y'$ the restriction of $y' \circ f$ at $U \cap E_B$ is holomorphic in the sense of normed spaces.

(5) f is G-analytic and "locally bounded in U" in the following sense: $\forall x \in U$ and $\forall$ B bounded in E, $\exists \epsilon > 0$ such that $f(x+\epsilon B)$ is bounded in F.

(6) f is Silva differentiable in U: $f(x+h) = f(x) + f'(x)h + r_x(h)$ where f'(x) is linear bounded from E to F and where, for every bounded set B of E, the filter basis $F_{\tau_0} = \bigcup_{0<\tau<\tau_0} \frac{r_x(\tau B)}{\tau}$ converges to zero in F in the Mackey sense ([5]).

(7) f is locally analytic between normed spaces:

for every $x \in U$, for every convex balanced bounded set B of E there exists $\epsilon > 0$ and a convex balanced bounded set B' of F such that, if $\overset{\circ}{B} = \bigcup_{0<\theta<1} \theta B$, the restriction of f to $x+\epsilon\overset{\circ}{B}$ has range in the normed space $F_{B'}$ and is analytic in $x+\epsilon\overset{\circ}{B}$ in the sense of normed spaces.

For some other definitions we denote by τE the following topology: $\Omega \subset E$ is open $\Leftrightarrow \forall x \in \Omega$, $\forall B$ bounded set in E, $\exists \epsilon > 0$ such that $x+\epsilon B \subset \Omega$. More generally this definition is valid in bornological vector spaces ([5]).

It is obvious that the injection $\tau E \to E$ is continuous and it is known ([5]) that these two topologies are identical if E is metrizable or is a Silva space.

(8) f is G analytic and continuous from U equipped by the topology induced by τE, with range in the l.c.s. F.

(9) f is G-analytic and continuous from U equipped by the topology induced by τE, with range in the topological space τF.

§2. Comparison theorems:

Theorem 1: The definitions (1), (2), (3), (4), (8), are equivalent.
The definitions (5), (6), (7) are equivalent.

Proof: (8) $\Leftrightarrow$ (9) is given in [10]; (5) $\Leftrightarrow$ (6) $\Leftrightarrow$ (7) is given in [12]. As E is sequentially complete we can suppose that we choose all the balanced convex bounded sets B of E such that the normed spaces E_B are Banach spaces; hence (8) $\Leftrightarrow$ (2) is immediate owing to the definition of the topology τE. (8) $\Rightarrow$ (3): f is G-analytic in U, so $f(a+zh) = \sum_n \frac{z^n}{n!} \delta^n f(a,h)$ and this series converges in F if $|z| < 1$; but by (2) f is indefinitely Silva differentiable in the large sense in U ([4]), so $\delta^n f(a,h) = f^{(n)}(a)\, h^{(n)}$ where $f^{(n)}(a)$ are bounded symetric n-linear maps; take $\ell_n(h) = f^{(n)}(a)\, h^n$. (3) $\Rightarrow$ (2): Recall that if E is a Banach space and F a normed space, if the ℓ_n are continuous homogeneous polynomials of degree n, and if the series $\sum_n \ell_n(h)$ converges in F for h in a neighbourhood of 0 in E, then the sum of the series is analytic in a neighbourhood of 0 in E and an application of this result with $E = E_B$ and $F = F_V$ gives the proof. (2) $\Leftrightarrow$ (4) is obvious because in Banach spaces weak analyticity implies analyticity.

The first class of equivalent definitions is called "Silva analyticity in the large sense" and the second class is called "Silva analyticity". It is clear that Silva analyticity implies Silva analyticity in the large sense. To study the connections between these notions we need the following definition.

Definition: Let E be a sequentially complete l.c.s. or more generally a complete bornological vector space ([5]); a subset K of E is said to be **strictly compact** if there exists a convex balanced bounded set B of E such that K is contained in the normed space E_B and is compact in this normed space.

The union of the Banach spaces E_K, for K belonging to the strictly compact balanced and convex sets of E, is a complete bornological vector space denoted by S(E). It is even a Schwartz bornological vector space [6].

It is obvious that the definitions (2), (3), (4), (5), (6), (7) have a meaning if E is a complete bornological vector space and U an open set for the topology τE; and that theorem 1 remains valid.

Theorem 2: The following properties are equivalent:

(1) f is Silva analytic in the large sense in U,

(2) f is Silva analytic in U if we consider that E is the bornological vector space S(E).

Proof: (1) ⇒ (2) because f is bounded on the strictly compact sets.

(2) ⇒ (1) f is continuous in U equipped by the topology $\tau S(E)$ with range in F; one shows easily that $\tau S(E) = \tau E$.

Corollary: Definition (9) is equivalent to Silva analyticity in the large sense.

Proof: (9) ⇒ (8) is obvious. (8) ⇒ (9): by theorem 2 f is Silva analytic in U if we consider that E is the bornological vector space S(E). Hence by [4], p.32, f is continuous in U equipped with the topology induced by $\tau S(E)$ with range in τF; and $\tau S(E) = \tau E$.

So we see that the definition in the large sense is exactly a particular aspect of the other definition, by using on E a suitable structure of bornological vector space. With this notion in the general framework of bornological vector spaces more general than l.c.s. one can obtain a complete theory of holomorphy. The fundamental theorems such as Hartogs, Zorn,. . . are not true in full generality but they are true in all the "usual" spaces and then their proofs are immediate consequences of the corresponding results in Banach spaces, by means of definitions (2) or (7): ([3], [4], [7], [8], [10], [11], [12]).

§3. The case of real analytic maps.

In this paragraph E is a real complete bornological vector space; let U be an open set for the topology τE, F a sequentially complete l.c.s. and f a mapping with domain in U and range in F.

Definition 1: The mapping f is said to be **analytic in the large sense** in U if, for every $x \in U$, there exists a sequence of homogeneous bounded polynomials ℓ_n of degree n and a bornivorous balanced subset P of E such that, for every $h \in P$, the series $\sum_n \ell_n(h)$ converges in the l.c.s. F and its sum is equal to f(x+h).

Definition 2: The mapping f is said to be **analytic in U** if it is "locally analytic between normed spaces" as defined in §1 definition (7)..

Theorem 2': The following properties are equivalent:

(1) f is analytic in the large sense in U.

(2) f is analytic in U if we consider that E is the bornological vector space S(E).

Proof: (2) ⇒ (3): let $x \in U$; there exists a sequence ℓ_n of bounded homogeneous polynomials of degree n, from E to F, such that, for every convex balanced strictly compact subset K of E there exists $\epsilon_K > 0$ such that the series $\sum_n \ell_n(h)$ converges in the l.c.s. F if $h \in \epsilon_K K$. To show that the union, for all the convex balanced strictly compact sets K of E, of the sets $\epsilon_K K$ is a bornivorous subset of E it suffices to show that; if

E is a Banach space, a bornivorous subset in S(E) is a neighbourhood of 0, which is easy to prove.

(1) ⇒ (2): Let B be a convex balanced bounded set of E such that E_B is a Banach space. Let $\widetilde{E}_B$ and $\widetilde{F}$ be their complexifications. By proposition 5.4, p.92 in [2] there exists an open set $\widetilde{H}$ in $\widetilde{E}_B$, containing $U \cap E_B$, such that, the restriction of f to $U \cap E_B$ can be prolongated in $\widetilde{H}$ as the sum $\widetilde{f}$ of the complexified series. If $x \in U$ and if $\widetilde{B}$ is the unit ball of $\widetilde{E}_B$, if $\overset{\circ}{\widetilde{B}}$ denotes $\bigcup_{0<\theta<1} \theta\,\widetilde{B}$, for a small $\epsilon > 0$ the restriction of $\widetilde{f}$ to $x+\epsilon\overset{\circ}{\widetilde{B}}$, with range in the l.c.s. $\widetilde{F}$, is Silva analytic in the large sense in $x+\epsilon\overset{\circ}{\widetilde{B}}$. If K is a balanced convex subset of the normed space E_B, by applying theorem 2 in the complexified spaces, we obtain the result.

These results are particular aspects of a general theory of calculus given in [4]; see [1] and [9] for surveys on calculus in l.c.s.

Bibliography

1. V. I. Averbuck and O. G. Smolyanov, **The various definitions of the derivative in linear topological spaces**, Russian Math. Surveys Vol. 23, No.4(1968), 67–113.

2. J. Bochnak and J. Siciak, **Analytic functions in topological vector spaces**, Studia Math. T. 39(1971), 77–112.

3. J. F. Colombeau, **Sur les applications G-analytiques et analytiques en dimension infinie**, Seminaire P. Lelong - Année 1971-72 - Lecture Notes in Math - Springer.

4. J. F. Colombeau, **Différentiation et Bornologie**, These - Bordeaux 1973.

5. H. Hogbé Nlend, **Théorie des Bornologies et Applications**, Lecture Notes in Math No.213, Springer.

6. H. Hogbé Nlend, **Les espaces de Fréchet Schwartz et la propriété d'approximation**, Comptes Rendus Acad. Sci. Paris A275, 1972, 1073–1075.

7. H. Hogbé Nlend, **Applications analytiques entre espaces vectoriels et algebres bornologiques**, Colloque sur les fonctions de plusieurs variables complexes, Paris, Juin 1972.

8. D. Lazet, **Applications analytiques dans les espaces bornologiques**, Séminaire Lelong - 1971-72, Lecture Notes in Math - Springer.

9. M. Z. Nashed, **Differentiability and related properties. . . Non linear functional Analysis and applications**, Academic Press, New York-London, 1971.

10. D. Pisanelli, **Applications analytiques en dimension infinie**, Bull. Sci. Math. 2nd series, 96(1972), 181–191.

11. J. S. e Silva, **Le calcul différentiel et intégral dans les espaces localement convexes réels ou complexes**, Atti. Acad. Naz. Lincei Vol.20(1956), 743–750 et Vol.21(1956), 40–46.

12. J. S. e Silva, **Conceitos de funçao differemciavel em espaços localmente convexos**, Publ. Math. Lisboa, 1957.

U E R de Mathématiques et d'Informatique
Université de Bordeaux I
Bordeaux - Talence 33405
France

APPROXIMATION AND HOMOTOPY PROPERTIES FOR HOLOMORPHIC EXTENSION IN ℓ^2

Stephen J. Greenfield*

Introduction: Consider $C^n \subset \ell^2$ via the first n coordinates. If U is an open subset of ℓ^2, put $U_n = U \cap C^n$, and $U_F = \bigcup_{n>0} U_n$. Each U_n has an envelope of holomorphy, $(\widehat{U}_n, \pi_n)$, spread over C^n. Suppose now U is extendible to an open subset, V, of ℓ^2, and V is not extendible. We try to show connections between V and the sequence $\{(\widehat{U}_n, \pi_n)\}$. In particular we want to demonstrate that V is homotopically equivalent to $\lim_{\rightarrow} \pi_n(\widehat{U}_n)$, and $V_F = \bigcup_{n>0} \pi_n(\widehat{U}_n)$. A question about Runge supports of homomorphisms of H(U) arises. The question does not seem to have been answered even in the finite-dimensional case.

§0. Here is a test case for a general conjecture.

Let $\ell^2 = \{(z_1,\ldots,z_j,\ldots), \Sigma |z_j|^2 < +\infty\}$. We consider $C^n \subset \ell^2$ in the obvious way: $(z_1,\ldots,z_n) \in C^n$ appears as $(z_1,\ldots,z_n,0,\ldots) \in \ell^2$. The projection π_n: $\ell^2 \rightarrow C^n \subset \ell^2$ just forgets all but the first n coordinates. If U is a subset of ℓ^2, then put $U_n = U \cap C^n$ and put $U_F = \bigcup_{n>0} U_n$.

A Riemann domain spread over a Banach space B is a pair (M,ϕ) where M is a complex analytic manifold modeled on B, and ϕ is an open (Frechet) complex analytic map from M to B. If M is a complex analytic manifold, we let H(M) be the (Frechet) complex analytic functions on M. A connected complex analytic manifold N is said to be a holomorphic extension of M if there is an open 1–1 holomorphic map E: M → N so that E*: H(N) → H(M) is onto. If (M,ϕ) and (N,χ) are Riemann domains, we also demand that E respect the projection maps: $\phi = \chi \circ E$. An extension (N,χ) will be called the envelope of holomorphy of (M,ϕ) if, when F: $(M,\phi) \rightarrow (R,\sigma)$ is an extension of (M,ϕ), then there is an open holomorphic map G: R → N which is the identity on M. (See [3] for information on envelopes of holomorphy in Banach space.)

In §1 we prove: if $U \subset V$ are open sets in ℓ^2, and if U is extendible to V and V is not extendible, then V is the envelope of holomorphy of U. In §2 we review and comment on some of Coueré's work [1] in obtaining an extension of U which we call E(U) as a subset of the spectrum of H(U) (with a certain topology). We try to show that E(U) = V in §3, and obtain some connection with the (finite dimensional) envelopes of holomorphy of $\{U_n\}$. We then use a theorem of Palais [6] to connect the homotopy properties of the envelopes of holomorphy of $\{U_n\}$ with V.

---

*Partially supported by NSF Grant GP-20647 and a Rutgers Research Council Fellowship.

§ 1. A result on uniqueness of extension.

Theorem: Let B be a Banach space, and $U \subset V$ be open sets in B. Suppose U extends to V, and V has no extension. Let (W,π) be a point-separating Riemann domain which is an extension of U (that means: there is an open 1−1 holomorphic map $E: U \to W$ so that $\pi \circ E = \text{id}\, U$, and E^* is onto). Then π is 1−1, and $\pi(W) \subset V$.

Proof: Consider $A = \{(Y,\chi);$ Y open and connected in B, $Y \supset U$, $\chi: Y \to W$ s.t. $\pi \circ \chi = \text{id}_Y\}$. Then A is non-empty, and is ordered under set inclusion and functional extension. Let (Y,χ) be a maximal element.

We claim $Y \subset V$, and $\chi(Y) = W$.

If $Y \not\subset V$, there is $z \in \partial V \cap Y$ and $r > 0$ so that $B_r(z) \subset Y$. Let V' be the disjoint union of V and $B_r(z)$, modulo the equivalence relation: $w \in V$ and $w' \in B_r(z)$ are equivalent iff $f(w) = f(w')$ (for all $f \in H(U)$). This is the classical method for constructing simultaneous analytic continuation - see, e.g., [4], chapter 5. The point-separation of H(W) ensures that V' will be a manifold. Since $z \in \partial V$, V' must be a proper extension of V. Hence $Y \subset V$.

Suppose $\chi(Y) \neq W$. There is $z \in \partial(\chi(Y))$ (since W is connected). Then $\pi(z) \in \bar{Y}$. Let $r > 0$ be selected so that $\pi^{-1}: B_r(\pi(z)) \to W$ is an isomorphism (we select the branch of π^{-1} mapping $\pi(z)$ to z). Choose $r' > 0$ and $y \in Y$ with $\pi(z) \in B_{r'}(\pi(z))$. Consider $Y \cup B_{r'}(y) = \tilde{Y}$ and the map $\tau: \tilde{Y} \to W$ defined by $\tau(y) = \chi(y)$ on Y, and $\tau(t) = \pi^{-1}(t)$ on $B_{r'}(y)$. If $y \in Y \cap B_{r'}(y)$, $\chi(y)$ and $\pi^{-1}(y)$ are the same points in W (follows immediately from point separation). Then $(\tilde{Y},\tau)$ is an element of A "larger" than (Y,χ), which is impossible. So $\chi(Y) = W$.

Remarks: If $\dim B < \infty$, then any point-separating Riemann domain W extending U has an injective map into E(U), the envelope of holomorphy of U. But $E(U) = V$, since E(U) is an extension of V. For $\dim B = \infty$, the result seems to follow from work of Hirshowitz [3] but a direct proof is not hard, so we give it. If "point-separating" is omitted, the result of course is false. (Easy examples even when $\dim B = 2$.)

§ 2. Some theorems of Coeuré.

(All the results in this section either appear in [1], chapter 3, or are easily derivable from the work there.)

Let (M,ϕ) be a Riemann domain spread over a separable Banach space B. If $x \in M$, we let $d(x)$ be the positive number which is the injective radius of ϕ at x; that is, $d(x)$ is the sup of the radii of balls in B centered at $\phi(x)$ on which the branch of ϕ^{-1} taking $\phi(x)$ to x is well-defined as a holomorphic map. Let r be a positive, lower semicontinuous function on M with $r \leq 1$. We shall call such functions "eligible". If r is eligible, we make the following

Definition: $H_r(M) = \{f \in H(M)$ such that f is bounded on each set $\phi^{-1}(\overline{B_t(\phi(x))})$, where $t < r(x)d(x)$, and the correct branch of ϕ^{-1} (carrying $\phi(x)$ to x) is chosen$\}$.

Proposition: Each $H_r(M)$ is a Frechet algebra, and every continuous homomorphism $h: H_r(M) \to C$ satisfies $|h(f)| \leqslant \|h\|_A$ ($= \sup\{|h(s)|, s \in A\}$) for some set A, a subset of a finite union of sets of the form $\overline{\phi^{-1}(B_t(\phi(x))}$ appearing in the above definition.

Such a set A is called a support (for $H_r(M)$) of h. There may be many different supports. Put $d(A) = \inf\{d(x), x \in A\}$. (In effect, d(A) measures how much A may be moved in M. Since $d(A) > 0$, this can be used to create a complex manifold structure on the homomorphisms of $H_r(M)$.)

Lemma: $H(M) = \cup\{H_r(M), r \text{ eligible}\}$.
H(M) is given the inductive limit topology of the $H_r(M)$. Thus a homomorphism $h: H(M) \to C$ is continuous iff its restriction to each $H_r(M)$ is continuous.

Definition: A continuous homomorphism $h: H(M) \to C$ is said to be an **interior homomorphism** ($h \in I(M)$) if, for each eligible r, there is a support A_r for $h|_{H_r(M)}$ so that $\inf\{d(A_r), r \text{ eligible}\} > 0$.

Coeuré then shows:

Theorem: I(M) can be given the structure of a Riemann domain. "Evaluation at a point" imbeds M holomorphically into I(M), and, if E(M)=those components of I(M) containing points of M, then E(M) is a point-separating extension of M.

Remark: If a continuous homomorphism is supported by a compact set, then it is interior.

If $\dim B < \infty$, then E(M) coincides with the classical envelope of holomorphy of M. (See [2], chapter 1). In this case, we will sometimes call E(M), $\widehat{M}$. Also E(M) = I(M)=all continuous homomorphisms on H(M) when $\dim B < \infty$.

Proposition: If N is an open or closed submanifold of (M, ϕ) so that $(N, \phi|_N)$ is a Riemann domain spread over some Banach subspace of B, then there is a canonical map from E(N) into E(M) extending the inclusion of N into M.

Remark: Note that this map need not be 1–1 or onto from E(N) to $E(M)|_{\phi(E(N))}$. This is true even when $\dim B < +\infty$. (Easy examples.)

An important observation of Coeuré about his topology on H(M) is the following result:

Theorem: If $E: M \to N$ is an extension of the Riemann domain M by a Riemann domain N, then $E^*: H(N) \to H(M)$ is a topological isomorphism. Thus evaluation at a point of N is a continuous homomorphism on H(M).

A digression on Runge supports:

Let A be a support of a continuous homomorphism h of $H_r(M)$. A **special neighborhood** U of A in M will be a neighborhood of A in M so that: there is some $\epsilon > 0$ so that $\phi^{-1}(\phi(x)+y) \in U$ when $x \in A$ and $\|y\| < \epsilon$, where the proper branch of ϕ^{-1} (taking $\phi(x)$ to x) is selected. Always $\epsilon \leqslant d(A)$. Not every neighborhood is a special neighborhood.

A support set A is said to be a **Runge support** A for $H_r(M)$ if, for all sufficiently small special neighborhoods U, if f is holomorphic and bounded on U and $\epsilon > 0$, there is $g \in H_r(M)$ so that

$$\|f-g\|_A = \sup_{z \in A} |f(z)-g(z)| < \epsilon.$$

This implies that any homomorphism h supported on A extends to function f holomorphic and bounded in any special neighborhood of A, and $|h(f)| \leq \|f\|_A$.

We do not know if Runge supports exists for every continuous homomorphism, even if dim $B < \infty$. We will assume they do, and use this fact in the next section to prove the main result. We hope they exist.

Remark on Runge supports: Suppose U is open in $\mathbf{C}^n$. A compact subset K of U is called **Runge** if the restriction map from H(U) to H(K) has dense image. If dim $B < \infty$, the Runge support question above reduces to: can we write $U = \bigcup_{n=1}^{\infty} K_n$, $K_n \subset \operatorname{int} K_{n+1}$, with each K_n Runge in U? (This is possible when U is pseudoconvex, and in many other cases.)

Remarks on Coeuré's work: A good topology on H(M) seems hard to find. Therefore, it would be interesting to generalize Coeuré's construction in the case when M is not a Riemann domain. In this connection we note that "pseudocompacta" and "χ-maps" might be used to replace the nice ball structure given on manifolds spread over a Banach space. See Koschorke [5] for a definition of these concepts. We hope to investigate them in connection with holomorphic functions at a later time.

§3. An approximation procedure.

In this section and the next, we let $U \subset V$ be open subsets of ℓ^2 so that U is extendible to V and V is not extendible. Then $U \subset E(U) \subset V$. Suppose $h \in V$. (We will identify evaluation at h and h. Thus $f(h) = h(f)$ for $f \in H(U)$.). Let A be a Runge support for h in $H_r(U)$. Take n large enough so that $\pi_n(A)$ is a compact subset of U_n. (We can always find such an n. For $A \subset \bigcup_{1 \leq j \leq J} \overline{B_{x_j}(r_j)} \subset \bigcup_{1 \leq j \leq J} B_{x_j}(R_j) \subset U$, with $R_j > r_j$. Since U_F is dense in U, we can first take n so large that $\pi_n(x_j) \in U_n$, for all j. Then, if $\epsilon_j = |\pi_n(x_j) - x_j|$, we can also ask that $R_j^2 - \epsilon_j^2 > r_j^2$ for all j. This is possible since $\epsilon_j \to 0$ as $n \to \infty$. Then $\pi_n(\bigcup_{1 \leq j \leq J} \overline{B_{x_j}(r_j)})$ is compact in U_n. $\pi_n(A)$ is (see Figure 1) a closed subset of that compact set, and hence compact.) If $f \in H(U_n)$, then $\tilde{f}(z) = f(\pi_n(z))$ is well defined in some special neighborhood of A. If we put $h_n(f) = h(\tilde{f}(z))$, which is well defined by the **Runge assumption**, then $h_n \in \hat{U}_n$. By a proposition of §2, there is a map $q_n : U_n \to E(U)$ extending the inclusion of U_n into U. What is $q_n(h_n)$? By testing the coordinate functions we immediately find that $h_n = \pi_n(h)$. Thus if $h \in V_F$, $h \in E(U)$ (when we take n large enough). So $V_F \subset E(U)$. But, by making an appropriate translation, we can arrange that any given point of V would lie in V_F. Thus we have shown:

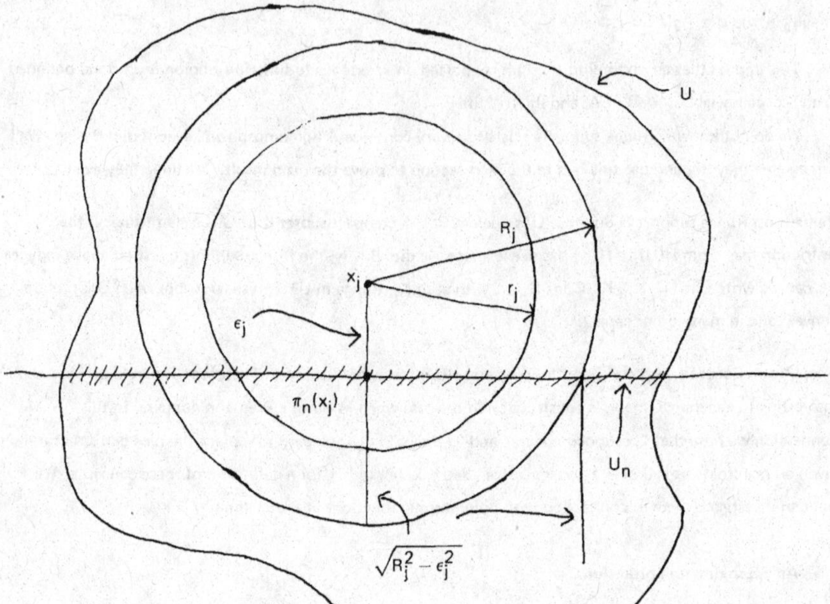

Figure 1

Theorem: Let $U \subset V$ be open subsets of ℓ^2, with U extendible to V and V not extendible. Supose the assumption on Runge supports holds. Then $E(U) = V$. Each element of $E(U)$ has a finite-dimensional compact support. If $q_n: \hat{U}_n \to E(U)$ is the canonical map of $\hat{U}_n$ to $E(U)$, then $\bigcup_{n>0} q_n(U_n) = V_F$.

We use this fact in the next section to obtain some facts on the homotopy structure of V relative to the sequence $\{q_n(\hat{U}_n)\}$.

§4. Conclusions on homotopy.

$U \subset V \subset \ell^2$ satisfy the same hypotheses as in §4. Palais [6] shows the following result:

Theorem: Let V be any open set in ℓ^2. Then the natural map from $\varinjlim V_n$ to V (the direct limit of the V_n's, using the obvious linking inclusion $V_n \subset V_{n+1}$) is a homotopy equivalence.

Actually he proved the map is a weak homotopy equivalence, but remarked that if V were dominated by a simplicial complex (as it is here) a weak homotopy equivalence is the same as a homotopy equivalence.

The following result follows easily from work of Palais.

Proposition: Suppose W_n is an open submanifold of V_n, with $W_n \subset W_{n+1}$ and $\cup W_n = \cup V_n$. Then $\varinjlim W_n$ is homotopically equivalent to $\varinjlim V_n$ (and so to V).

This situation actually occurs with $W_n = q_n(\hat{U}_n)$ (from §3).

Theorem: Let $U \subset V$ be open subsets of ℓ^2, with U extendible to V and V not extendible. Let $q_n: \hat{U}_n \to E(U) = V$ be the canonical map of $\hat{U}_n$ to E(U), extending the inclusion $U_n \subset U$. Suppose the assumption on Runge supports holds. Then $\varinjlim q_n(\hat{U}_n)$ is homotopically equivalent to V.

Remarks: Since $U_n \subset U_m$, $m > n$, there is a natural map $s_{nm}: \hat{U}_n \to \hat{U}_m$. Thus $\{U_n, s_{nm}\}$ is a directed system of topological spaces. We would naturally like to conclude that $\varinjlim(U_n, s_{nm})$ is homotopically equivalent to V. But we cannot yet do that (we know of no counterexample). One difficulty: the s_{nm}'s are not at all 1–1 maps and are probably quite complicated in general. Note also that, in general, $q_n(\hat{U}_n) \subsetneq V_n$.

Example: The simplest example for U is as a tube over an open set 0 in ℓ^2_R. Then V is a tube over the open subset ch(0) (convex hull=ch) of ℓ^2_R. Unfortunately the homotopy type of everything involved is then quite trivial. (However, even here it is not immediately clear how to verify the Runge assumption.)

We hope to construct other, more interesting examples (such as considering the complement of a variety of codim > 1 in ℓ^2 as U).

Of course we can probably state the above theorems more generally (ℓ^2 is certainly not necessary for the work of Palais; also we can consider these problems when U is a domain spread over a Banach space.) But the essential difficulties (analytic: the Runge problem – and topological: the intricacy of the linking maps s_{nm}) seem present in the case exposed here.

Bibliography

1. G. Coeuré, **Fonctions plurisousharmoniques sur les espaces vectoriels topologiques et applications à l'étude des functions analytiques**, Ann. Inst. Fourier, Grenoble, 20, 1(1969), 361–432.

2. R. C. Gunning and H. Rossi, **Analytic Functions of Several Complex Variables**, Prentice Hall, Englewood Cliffs, N.J. (1965).

3. A. Hirschowitz, **Prolongement analytique en dimension infinie**, Ann. Inst. Fourier, Grenoble, 22, 2(1972), 255–292.

4. L. Hörmander, **An Introduction to Complex Analysis in Several Complex Variables**, Van Nostrand, Princeton, N.J. (1966).

5. U. Koschorke, **Pseudo-compact subset of infinite-dimensional manifolds**, J. Diff. Geom., 5(1971), 127–134.

6. R. S. Palais, **Homotopy theory of infinite-dimensional manifolds**, Topology, 5(1966), 1–16.

Department of Mathematics
Rutgers University
New Brunswick, New Jersey 08903

ON THE WEIERSTRASS PROBLEM IN BANACH SPACES

Yves Hervier

Let E and F be Banach spaces (Ω,φ) a domain of holomorphy over E, and $(x^n)_1^\infty$ a sequence in Ω.

Proposition 1. - The following statements are equivalent:

(i) $\exists\, f \in \mathcal{O}(\Omega)$, such that $(f(x^n))_1^\infty$ is injective, and $\lim_n |f(x^n)| = +\infty$.

(ii) $\exists\, g \in \mathcal{O}(\Omega)$ taking on $(x^n)_1^\infty$ prescribed scalar values $(a^n)_1^\infty$.

(iii) $\exists\, h \in \mathcal{O}(\Omega,F)$ taking on $(x^n)_1^\infty$ prescribed vector values $(b^n)_1^\infty$.

If $\Omega \subset E$, these conditions are equivalent to the following one:

(iv) (x^n) is injective, and $\exists\, f \in \mathcal{O}(\Omega)$ such that $\lim_n |f(x^n)| = +\infty$.

One can conjecture that this proposition remains true for a general Ω.

Definition. - A sequence $(x^n)_1^\infty$ in Ω will be said to be a Weierstrass sequence if there exists an $f \in \mathcal{O}(\Omega)$ such that $\lim_n |f(x^n)| = +\infty$.

Obviously, a Weierstrass sequence in Ω has no cluster point in Ω.

Conversely, it is natural to ask the following

Question. - Is any sequence in Ω without a cluster point a Weierstrass sequence?

It is well known (by Cartan, Theorem B, for instance) that it is true if E is finite dimensional.

If E is infinite dimensional, we know only some conditions for Ω to be holomorphically convex (i.e., for all sequence $(x^n)_1^\infty$ in Ω, there exists an $f \in \mathcal{O}(\Omega)$ unbounded on (x^n)).

- if $\Omega = E$, E being separable or reflexive (cf. Dineen [2] and Hirshowitz [5]). For a counter-example in ℓ_∞ cf. [2].
- if E has a countable basis [4].

A method due to Gruman and Kiselmann [3] allows us to look at other sequences;

First, we show that:

Theorem 2. - In a L.C.S., a sequence $(x^n)_1^\infty$ such that $\lim_n p(x^n) = +\infty$ for a continuous semi-norm p is a Weierstrass sequence.

More precisely, one sees that if such a sequence is injective, it verifies the equivalent conditions (i), (ii), (iii).

We use then theorem 2 to show (in § III) two lemmas, which simplify the problem.

In § IV, we begin the study of other sequences, and in particular bounded sequences. A lemma describing a kind of Weierstrass sequence is proved, from which the following theorem is obtained.

Theorem 3. - **If E is separable, a weakly convergent sequence without a cluster point is a Weierstrass sequence.**

However, we do not know the general answer to the previous question.

Finally, in § V, we obtain analogous results for a general Ω.

I would like to express my gratitude to André Hirschowitz for his continual encouragement and help during the preparation of this paper.

§ I. Proof of the Proposition 1.

Obviously, we have (iii) ⇒ (ii) ⇒ (i) ⇒ (iv).

That (i) ⇒ (ii) is shown by the Weierstrass theorem.

(ii) ⇒ (iii): this was shown in [1] (more precisely, it is shown in [1] that the Cartan theorem B is still valid when F is infinite dimensional).

If $\Omega \subset E$, (iv) ⇒ (i):

One can first suppose that, for all n, $|f(x^n)| \geqslant \|x^n\|^2$. Now, for i and j given, the space of linear forms ℓ on E such that $\ell(x^i) - \ell(x^j) \neq f(x^i) - f(x^j)$ is an open subset of E^* which is dense in E^*. By the Baire theorem, there exists a linear form ℓ on E such that, for all distinct i and j, $f(x^i) + \ell(x^i) \neq f(x^j) + \ell(x^j)$. It is clear then that $(f+\ell)$ satisfies (i).

§ II. Sequences which tend to infinity.

1.- The case of a Banach Space.

Theorem 1: **In a Banach Space, every sequence which tends to infinity is a Weierstrass sequence.**

Proof: (I thank Martin Schottenloher for his suggestion which simplified the proof).

Let E be a Banach Space, and let $(x^n)_1^\infty$ be a sequence in E such that $\lim_n \|x^n\| = +\infty$. One can assume that the sequence values $\|x^n\|$ is increasing.

More, one can assume that the sequence $(\|x^n\|)_1^\infty$ is strictly increasing: indeed, for n and p fixed ($n \neq p$), the set $\{x \in E \mid \|x-x^n\| = \|x-x^p\|\}$ is a set of the first category. The union, for all n and p ($n \neq p$), of these sets is still of the first category. Thus its complement, being of the second category, is not empty. Then, it suffices to take a new origin in this last set.

By the Hahn-Banach theorem, one can find, for each $n \in \mathbf{N}$, a $\mathbf{R}$-linear form u_n on E such that

$$|u_n(x^n)| > \sup_{x \in B(0,r_n)} |u_n(x)|$$

where $\|x^{n-1}\| < r_n < \|x^n\|$, and $\bar{B}(0,r_n)$ is the closed ball with center 0, and radius r_n. Therefore, there exists a holomorphic function v_n on E such that:

$$|v_n(x^n)| > \sup_{x \in B(0,r_n)} |v_n(x)|.$$

Indeed, we put $v_n(x) = u_n(x) - i\, u_n(ix) + \lambda$, (where $i^2 = -1$), for a suitable λ (see for instance Schottenloher [8]). Then we construct by induction a sequence $(f_n)_1^\infty$, $f_n \in \mathcal{O}(E)$, such that

$$|f_n(x^n)| > n + 1 + \sum_{j=1}^{n-1} f_j(x^n), \quad \sup_{x \in B(0,r_n)} |f(x)| < \frac{1}{2^n},$$

for every $n > 1$.

Define a function f by

$$f(x) = \sum_1^\infty f_n(x) \text{ for all } x \in E.$$

The function f is holomorphic on E (see [8] for instance), and $\lim_n |f(x^n)| = +\infty$.

Note: It is very simple to find a linear form unbounded on $(x^n)_1^\infty$. But linear forms are not sufficient to find a function which tends to infinity on $(x^n)_1^\infty$. Consider for instance, in c_0, the sequence defined by: $x^n = n\, e^n ((e^n)_1^\infty$ being the canonical basis in $c_0)$. A linear form ℓ on c_0 has the form $\ell((u^n)_1^\infty) = \sum_1^\infty c_n u^n$, where $(c_n) \in \ell^1$. But it is impossible to find any sequence $(c_n)_1^\infty$ in ℓ^1 such that: $\lim_n nc_n = +\infty$.

2. - The case of a locally convex space (l.c.s.).

It is natural to say that a sequence in a l.c.s. "tends to infinity" if it has no bounded subsequence.

We cannot hope to obtain in a general l.c.s. the result obtained in separable Banach spaces: indeed we are going to see that there are sequences which "tend to infinity" in the abovesense, without being Weierstrass sequences.

a) Example. - Let $(y^i)_1^\infty$ be the sequence in **N**: $(y^i)_1^\infty = (1, 2, 1, 2, 3, 1, 2, 3, 4, 1, \ldots)$ (that is: $\forall i, j \in \mathbf{N}$, $j \leq i$, $y^{\frac{i(i+1)}{2} + j} = j + 1$).

Consider in $\mathbf{C}^\mathbf{N}$ the sequence $(u^n)_1^\infty$ defined by $u^n = (u_i^n)_{i=0}^\infty$, with:

$$\begin{cases} u_0^n = y^n \\ u_i^n = 0 \quad \text{for} \quad 0 < i < y^n \\ u_i^n = n \quad \text{for} \quad i \geq y^n \end{cases}$$

that is:

$$u^1 = (1, 1, 1, \ldots)$$
$$u^2 = (2, 0, 2, 2, \ldots)$$
$$u^3 = (1, 3, 3, \ldots)$$
$$u^4 = (2, 0, 4, 4, \ldots)$$
$$u^5 = (3, 0, 0, 5, 5, \ldots)$$
$$u^6 = (1, 6, 6, \ldots)$$
$$\ldots\ldots\ldots\ldots$$

and, for $\alpha = \frac{n(n+1)}{2}$:

$$u^{\alpha} = (1, \alpha, \alpha, \ldots)$$
$$u^{\alpha+1} = (2, 0, \alpha+1, \alpha+1, \ldots)$$
$$u^{\alpha+2} = (3, 0, 0, \alpha+2, \alpha+2, \ldots)$$
$$\ldots\ldots\ldots\ldots$$
$$u^{\alpha+n} = (n, \underbrace{0, \ldots, 0}_{n \text{ times}}, \alpha+n, \alpha+n, \ldots)$$

(i) Now we claim that (u^n) "tends to infinity". Indeed let $(u^{j_n})_1^\infty$ be a subsequence of (u^n), and let k be a bound for $(u_0^{j_n})_1^\infty$. Then, $\lim_n u_k^{j_n} = +\infty$ (more precisely, $u_k^{j_n} = j_n$ for all n).

(ii) (u^n) is not a Weierstrass sequence. Indeed, it is known (cf. Hirshowitz [6], Rickart [7]) that a holomorphic function on $\mathbf{C}^{\mathbf{N}}$ depends only on a finite number of coordinates. But the projection of $(x^n)_1^\infty$ on a finite number of coordinates has many cluster points.

However, we have the following theorem, which is analogous to theorem 1:

b) Theorem 2. - **If there is a semi-norm p on F such that $\lim_n p(x^n) = +\infty$, then (x^n) is a Weierstrass sequence.**

Proof. - We shall define a linear mapping $\ell : F \to C_0$ such that $\lim_n \|\ell(x^n)\| = +\infty$ and then apply theorem 1 in C_0 to $(\ell(x^n))_1^\infty$.

We can suppose that, for all i, $p(x^i) \geq 1$. Now, for each i, let ℓ_i be a linear form on F such that:

$$\begin{cases} |\ell_i(x^i)| = p(x^i) \\ |\ell_i(x)| \leq p(x), \qquad x \in F. \end{cases}$$

For $x \in F$, we set

$$\ell(x) = \left(\frac{\ell_1(x)}{\sqrt{p(x^1)}}, \frac{\ell_2(x)}{\sqrt{p(x^2)}}, \ldots, \frac{\ell_i(x)}{\sqrt{p(x^i)}}, \ldots \right)$$

and this defines a linear mapping $F \to \mathbf{C}^{\mathbf{N}}$.

Moreover, for all $x \in F$, $\frac{\|\ell_n(x)\|}{\sqrt{p(x^n)}} \leqslant \frac{p(x)}{\sqrt{p(x^n)}}$, and hence $\|\ell(x)\| \leqslant p(x)$. Therefore ℓ defines a continuous linear mapping $F \to C_0$.

Now, for all n, $\|\ell(x^n)\| \geqslant \sqrt{p(x^n)}$, and so $\lim_n \|\ell(x^n)\| = +\infty$. Therefore, by theorem 1, there exists an $f \in \mathcal{O}(C_0)$ such that $\lim_n |f \circ \ell(x^n)| = +\infty$ and so $(x^n)_1^\infty$ is a Weierstrass sequence.

Remark . – If we suppose that (x^n) is injective, we can add to ℓ_i a linear form h_i on F such that, for $j < i$, $(\ell_i + h_i)(x^j) \neq (\ell_i + h_i)(x^i)$. The new sequence $(\ell(x^n))$ is then injective. Thus $(x^n)_1^\infty$ will verify the equivalent conditions (i) to (iii).

§ III. Technical Lemmas.

Let E be a Banach space, and let (Ω,φ) be a domain spread over E. Let $x \in \Omega$; for $\epsilon > 0$ small enough, $B(x,\epsilon)$ will denote the neighborhood of x in Ω which is isomorphic by φ to $B(\varphi(x),\epsilon)$ ($\subset E$). $d(x)$ will denote the distance of x from the boundary of Ω, that is the least upper bound of the set $\{\epsilon : \epsilon > 0$ and $B(x,\epsilon)$ is defined$\}$.

Lemma 1. - Let $(y^n)_1^\infty$, $(x^n)_1^\infty$ be sequences in Ω such that, for all n, $x^n \in B(y^n, \epsilon_n)$ where

$$0 < \epsilon_n \leqslant d(y^n)$$

and

$$\lim_n \epsilon_n = 0.$$

Then, if $(y^n)_1^\infty$ is a Weierstrass sequence, so is $(x^n)_1^\infty$.

Proof. - We are going to construct a holomorphic function $\gamma : \Omega \to H_0$, where H_0 is the space of holomorphic germs at zero of functions on E. Then, we shall find a continuous semi-norm on H_0 which tends to infinity on $(\gamma(x^n))_1^\infty$.

(1) Let w be a function in $\mathcal{O}(\Omega)$ such that, for all n, $\|w(y^n)\| > 2n^3$ (there exists such a w, because $(y^n)_1^\infty$ is a Weierstrass sequence).

For $x \in \Omega$, $\gamma(x)$ will denote the germ at zero of the function:

$$B(0,d(x)) \ni z \mapsto w \circ \varphi_x^{-1}(\varphi(x) + z),$$

where $\varphi_x = \varphi|_{B(x,d(x))}$.

Then $\gamma : \Omega \to H_0$ defined in this way is holomorphic (cf. Hirschowitz [5]).

(2) Let $f \in H_0$. For any sequence $(z^n)_1^\infty$ in E such that $\lim_n \|z^n\| = 0$, and for any increasing sequence $(k_n)_1^\infty$ in $\mathbb{N}$, the sequence $(|T_{k_n}(f)(z^n)|)_1^\infty$ is bounded (where $T_n(f)$ denotes the Taylor polynomial of f of order n at zero).

Therefore,

$$P_{(z^n),(k_n)}(f) = \sum_{1}^{\infty} \frac{1}{n^2} |T_{k_n}(f)(z^n)|$$

defines a continuous semi-norm on H_0.

Now, in order to find such a semi-norm which tends to infinity on $(\gamma(x^n))$, it suffices to take

$$z^n = \varphi(y^n) - \varphi(x^n)$$

and k_n such that $|T_{k_n}(\gamma(x^n))(z^n)| > n^3$.

This is possible, because:

(i) $\gamma(x^n)(z^n) = w \circ \varphi_{x^n}^{-1}(\varphi(y^n)) = w(y^n)$,

(ii) $|w(y^n)| > 2n^3$,

and (iii) $\lim_j |T_j(\gamma(x^n))(z^n) - \gamma(x^n)(z^n)| = 0$.

Q.E.D.

Lemma 2. - Let $(x^n)_1^\infty \subset \Omega$. Suppose I, J are subsets of N such that $(x^n)_{n\in I}$ and $(x^n)_{n\in J}$ are Weierstrass sequences, then $(x^n)_{n\in I\cup J}$ is a Weierstrass sequence.

Proof. - (We shall suppose that $I \cap J = \phi$). Let $(\epsilon_n)_{n\in I}$ be such that $0 < \epsilon_n < d(x^n)/2$ and $\lim_{n\in I} \epsilon_n = 0$.

(1) There are $(y^n)_{n\in I}$ in Ω, and $f_1 \in \mathcal{O}(\Omega)$ such that:

(i) $\forall n \in I,\ y^n \in B(x^n, \epsilon_n)$,

(ii) $\forall n \in I,\ f_1(y^n) \neq 0$,

(iii) $\forall n \in J,\ f_1(x^n) = 0$.

In fact, $(x^n)_{n\in J}$ is a Weierstrass sequence. Hence there exists an $f_1 \in O(\Omega)$, $f_1 \neq 0$, such that, for all $n \in J$, $f_1(x^n) = 0$.

One takes now, for each $n \in I$, a $y^n \in B(x^n, \epsilon_n)$ such that $f_1(y^n) \neq 0$.

(2) $\exists f \in \mathcal{O}(\Omega)$ such that $\lim_{n\in I} |f(y^n)| = +\infty$ and $\lim_{n\in J} |f(x^n)| = +\infty$.

It suffices to take a function f of the form $f(x) = f_1(x)f_2(x) + g(x)$, where:

(i) $g \in \mathcal{O}(\Omega)$ is such that $\lim_{n\in J} |g(x^n)| = +\infty$ ((x^n) is a Weierstrass sequence),

(ii) $f_2 \in \mathcal{O}(\Omega)$ is such that, for all $n \in I$:

$$|f_2(y^n)| > \frac{n+|g(y^n)|}{|f_1(y^n)|}$$ (by lemma 1, $(y^n)_{n\in I}$ is a Weierstrass sequence).

(3) The sequence $(z^n)_{n\in I\cup J}$, with $z^n = \begin{cases} y^n \text{ for } n\in I \\ x^n \text{ for } n\in J \end{cases}$ is a Weierstrass sequence. Therefore, by Lemma 1, so is $(x^n)_{n\in I\cup J}$ (it is easy to see that $y^n \in B(x^n, d(x^n)/2) \Rightarrow x^n \in B(y^n, d(y^n))$).

IV. General Sequences.

Theorem 3. - **In a separable Banach space, a weakly convergent sequence with no cluster point is a Weierstrass sequence.**

Notes. – (1) We cannot hope to obtain this result for all Banach spaces, since there exists a sequence in ℓ_∞ which converges weakly, but not strongly, and on which every holomorphic function is bounded (cf. Dineen [2]).

(2) Every separable Banach space being isometric to a subspace of a Banach space with countable basis, we may assume that E has a countable basis. Let $(e_1, e_2, \ldots, e_n, \ldots)$ be this basis, E_n the space spanned by $(e_1, \ldots, e_n)$, and $\pi_n : E \to E_n$ the natural projection.

By lemma 1 we can also suppose that, for all n, $x^n \in E_n \setminus E_{n-1}$.

In order to prove the theorem 3, we first prove a more general lemma. $(x^n)_1^\infty$ will henceforth denote a sequence without a cluster point in E, where $x^n \in E_n \setminus E_{n-1}$. For ϵ, such that $0 < \epsilon < \inf \|x^n\|$, let $p_n^\epsilon = \sup\{ j < n \mid \|x^n - \pi_j(x^n)\| > \epsilon\}$.

In the next section, we shall prove the following:

Lemma 3. - **If, for some $\epsilon > 0$, $\lim_n p_n^\epsilon = +\infty$, then (x^n) is a Weierstrass sequence.**

Some examples will indicate which sequences satisfy this hypothesis. If, for some $\epsilon > 0$, the sequence $(p_n^\epsilon)_1^\infty$ is bounded by N, then, for all n, and all $j > N$,

$$\|x^n - \pi_j(x^n)\| \leq \epsilon .$$

Therefore it is easy to show that, if $(x^n)_1^\infty$ is bounded, there is an $\epsilon > 0$ for which $(p_n^\epsilon)_1^\infty$ is not bounded (which is not sufficient for lemma 3). However, the following example shows we cannot hope to improve this result:

Example. - In c_0, we consider the sequence $(y^n)_1^\infty$ where

$y^1 = (1, 0, \ldots)$

$y^2 = (1, ½, 0, \ldots)$ $\qquad y^3 = (1, 1, 0, \ldots)$

$y^4 = (1, ½, ½, \ldots)$ $\qquad y^5 = (1, 1, 1/3, 0, \ldots)$ $\qquad y^6 = (1, 1, 1, 0, \ldots)$

.... and if $\alpha_n = \dfrac{n(n+1)}{2}$:

$$1 \leqslant i \leqslant n+1 \Rightarrow y^{\alpha_n+i} = (\underbrace{1,\dots,1}_{i \text{ times}}, \underbrace{\tfrac{1}{1+i},\dots,\tfrac{1}{1+i}}_{(n-i-1) \text{ times}}, 0,\dots).$$

It is clear that this sequence is bounded, and has no cluster point. However, for every $\epsilon > 0$, it has a subsequence $(y^{j_n^\epsilon})_n$ such that $(p^\epsilon_{j_n^\epsilon})_n$ is bounded. (More precisely, if $i > \frac{1}{\epsilon}$, then $p^\epsilon_{\alpha_n+i} = i$ for all n).

However, one can find some sequences satisfying the hypothesis of lemma 3:

Lemma 4. - *If for every j the sequence $(\pi_j(x^n))_{n=1}^\infty$ is convergent, then $(x^n)_1^\infty$ is a Weierstrass sequence.*

(This will prove, in particular, theorem 3).

Proof. - For all j, let $y^j = \lim_n \pi_j(x^n)$.

(a) If, for ϵ, $p^\epsilon_n < N_\epsilon$ for all n, then: $m,p \geqslant N_\epsilon \Rightarrow \|y^p - y^m\| < 4\epsilon$. More precisely $\exists\, k = k(m,p)$, such that $x^k \in B(y^p, 2\epsilon) \cap B(y^m, 2\epsilon)$.

Indeed, we have in that case ($p^\epsilon_n < N_\epsilon$): $n, j > N_\epsilon \Rightarrow \|x^n - \pi_j(x^n)\| \leqslant \epsilon$.

Let $m, p > N_\epsilon$, and let $k = k(m,p)$ be such that:

$$\|\pi_m(x^k) - y^m\| < \epsilon$$

$$\|\pi_p(x^k) - y^p\| < \epsilon.$$

Then we have: $x^k \in B(y^m, 2\epsilon) \cap B(y^p, 2\epsilon)$, and in particular

$$\|y^m - y^p\| < 4\epsilon.$$

(b) Let us suppose now that, for every $\epsilon > 0$, $(p^\epsilon_n)_{n=1}^\infty$ does not tend to infinity. That is this sequence contains a bounded subsequence. Then, applying (a) to the corresponding subsequence of $(x^n)_1^\infty$, one proves first that $(y^k)_1^\infty$ is a Cauchy sequence, and, second, that $y = \lim_k y^k$ is a cluster point for $(x^n)_1^\infty$. This is contrary to the hypothesis.

Other applications of lemma 3.

(1) One can see, (either directly or by lemma 2), that the same result is still true for sequences $(x^n)_1^\infty$ such that the number of cluster points of the sequence $(\pi_j(x^n))_{n=1}^\infty$ is finite and does not depend on j.

(2) Moreover, it is clear that the definition of p^ϵ_n does not depend essentially on the first coordinates. More precisely, let σ_n: $E \to E$ denote the projection $\sum_1^\infty x_i e_i \to \sum_n^\infty x_i e_i$. It is easy to see that:

(i) $\lim_n p^\epsilon_n(x_n) = +\infty \Leftrightarrow \lim_n p^\epsilon_n[\sigma_p(x^n)] = +\infty$,

(ii) $(x^n)_1^\infty$ has no cluster point $\Leftrightarrow$ $(\sigma_p(x^n))_{n=1}^\infty$ has no cluster point,

(iii) $(x^n)_1^\infty$ weakly convergent $\Rightarrow$ $(\sigma_p(x^n))_{n=1}^\infty$ weakly convergent.

This proves, for instance, that if all the weak cluster points of $(x^n)_1^\infty$ are in only one E_p, then $(x^n)_1^\infty$ is a Weierstrass sequence. More generally we have:

Proposition 2. - A bounded sequence in a separable Banach Space, which has no cluster points and such that its weak cluster points lie in a finite dimensional subspace, is a Weierstrass sequence.

§V. Sequences in Domains Spread over E.

Let E be a separable Banach space with a basis, and (Ω, φ) a domain of holomorphy over E.

a) Let us recall the following facts (cf. [4]):

Let $\Omega_n = \varphi^{-1}(E_n)$, $\Omega_\infty = \bigcup_n \Omega_n$.

We shall say that a sequence $(K_n)_1^\infty$ of compact sets $K_n \subset \Omega_n$ verifies (C) if:

(C) $\quad \forall x \in \Omega, \ \exists r > 0, \ \exists N, \ n \geq N \Rightarrow B(x,r) \cap \Omega_n \subset K_n$.

The interest in this condition becomes evident from the following lemmas:

Lemma 5. - Let $(x^n)_1^\infty$ be a sequence in Ω such that $x^n \in \Omega_n \backslash \Omega_{n-1}$. If there exists a sequence $(K_n)_1^\infty$ of compact subsets of Ω such that

(i) K_n is a Runge compact in Ω_n ;

(ii) $(K_n)_1^\infty$ verify (c) ;

(iii) $x^n \notin K_n$.

Then $(x^n)_1^\infty$ is a Weierstrass sequence.

This lemma was proved in [3] for $\Omega \subset E$, and in [4] for a general Ω.

b) Examples of sequences (K_n) which verify (C).

Let $(A_n)_1^\infty$, $(\epsilon_n)_1^\infty$ be sequences in R_+, $\lim_n A_n = +\infty$, $\lim_n \epsilon_n = 0$.

1) $\Omega = E$, $\epsilon > 0$, $K_n = \overline{B}(0, A_n) \cap \{x \in E_n \mid \|x - \pi_{j_n}(x)\| \leq \epsilon\}$.

If $\lim_n j_n = +\infty$, $(K_n)_1^\infty$ verifies (C):

Proof. - Let $x \in \Omega$. Let N_1, N_2, N_3 be integers such that:

(i) $n > N_1 \Rightarrow A_n > \|x\| + \epsilon$; (ii) $j > N_2 \Rightarrow \|x - \pi_j(x)\| < \frac{\epsilon}{2}$; (iii) $n > N_3 \Rightarrow j_n > N_2$.

Then, if $r < \epsilon/4$, $n > \sup(N_1, N_3) \Rightarrow B(x,r) \cap \Omega_n \subset K_n$. Therefore, we obtain lemma 3 with $j_n = p_n^\epsilon$.

2) (Ω, φ) a domain of holomorphy over E. We denote by $\Omega^\epsilon = \{x \in \Omega \mid d(x) > \epsilon\}$, $\Omega_n^\epsilon = \Omega_n \cap \Omega^\epsilon$.

Let $\rho_n^\epsilon : \Omega_n^\epsilon \to R_+$ the geodesic distance of Ω_n^ϵ to a fixed point $z_0 \in \Omega_0$. ρ_n^ϵ is continuous on $\Omega_n^{2\epsilon}$.

Suppose that the sequence $(\epsilon_n)_1^\infty$ is decreasing. If we take $K_n = K_n((\epsilon_n), (A_n))$ = the Runge hull in Ω_n of $\{x \in \overline{\Omega_n^{\epsilon_n}} \mid \rho_n^{\epsilon_n/2}(x) \leqslant A_n\}$, the sequence $(K_n)_1^\infty$ is increasing and verifies (C) (cf. [4]).

We are going to use this kind of compact sets to prove the following theorem.

c) **Theorem 4.** - **Let (Ω, φ) be a domain of holomorphy over a Banach space E with countable basis. Every sequence in Ω, the boundary distance of which tends to zero, is a Weierstrass sequence.**

Proof. - By lemma 1, we can suppose that $x^n \in \Omega_n \setminus \Omega_{n-1}$. We shall suppose moreover that the sequence $(d(x^n))$ is decreasing. Then it suffices to apply lemma 5 to $(x^n)_1^\infty$ for the compact sets $K_n = K_n((2d(x^n)),(n))$.

Remark. - If $x^n \in \Omega_n \setminus \Omega_{n-1}$ this method allows us to obtain an $f \in \mathcal{O}(\Omega)$ such that $\lim_n |f(x^n)| = +\infty$, the sequence $|f(x^n)|$ being injective.

By a direct proof one can also obtain this result, without assuming $x^n \in \Omega_n \setminus \Omega_{n-1}$.

It is a very technical work, but it gives the following more precise result:

Theorem 5. - **Let (Ω, φ) be a domain of holomorphy over a separable Banach space with basis. Let (x^n) be an injective sequence in Ω. If $\lim_n d(x^n) = 0$, then there exists an $f \in \mathcal{O}(\Omega)$ taking on (x^n) arbitrary prescribed values (a^n).**

d) **Other sequences.** - As in E, where we cannot conclude for general bounded sequences, we do not know a general answer for sequences in Ω which remain far from the boundary. There are two problems:

(1) sequences for which $(\varphi(x^n))_1^\infty$ has no cluster point (for instance sequences without cluster point which are in a finite number of leaves). For these sequences the problem is the same as in E.

(2) sequences for which $(\varphi(x^n))_1^\infty$ has cluster points. For these sequences, we have just the following result:

Lemma 6. - **Let $(x^n)_1^\infty$ be a sequence without cluster point in Ω such that $d(x^n) > \epsilon > 0$. Then if $(\varphi(x^n))$ is convergent, (x^n) is a Weierstrass sequence.**

Proof. - We suppose that $\lim_n \varphi(x^n) = 0$. Then, for n large enough, we can suppose that $0 < \|\varphi(x^n)\| < \epsilon < d(x^n)$.

Now, if $a^n = \varphi_{x^n}^{-1}(0)$, (where $\varphi_{x^n} = \varphi|_{B(x^n)d(x^n)}$) the sequence (a^n) is obviously a Weierstrass sequence and so (x^n) by lemma 1.

Bibliography

1. L. Bungart, **Holomorphic functions with values in locally convex spaces and applications to integral formulas,** Trans. Am. Math. Soc. 111 (64).

2. S. Dineen, **Unbounded holomorphic functions on a Banach space,** J. London Math. Soc. 4, 3 (72).

3. L. Gruman and C. O. Kiselman, **Sur le problème de Lévi dans un espace de Banach à base,** Comptes Rendus de l'Académie des Sciences, t. 274, série A, p. 1296.

4. Y. Hervier, **Sur le problème de Lévi pour les espaces étalés banachiques,** C. R. Acad. Sc. t. 275, série A, p. 821.

5. A. Hirschowitz, **Bornologie des espaces de fonctions analytiques en dimension infinie,** Séminaire P. Lelong, 1970, Lecture Notes no. 205.

6. ____________, **Remarques sur les ouverts d'holomorphie d'un produit dénombrable de droites,** Ann. Inst. Fourier, t. XIX, fasc. 1 (1969).

7. C. E. Rickart, **Analytic functions of an infinite number of complex variables,** Duke Math. J. 36(1969), p. 581.

8. M. Schottenloher, **Analytische Fortsetsung in Banachräumen,** Thesis, Munchen 1971.

Departement de Mathematiques
Institut de Mathematiques et Sciences Physiques
Parc Valrose
06031, Nice, Cedex
France

A COUNTEREXAMPLE IN THE LEVI PROBLEM

Bengt Josefson

We shall give an example of an open pseudoconvex domain in $C_0(A)$ which is not a domain of holomorphy in any sense. A is an uncountable index set and $C_0(A)$ is the Banach space of complex functions on A, which are arbitrarily small off of finite sets, with the supremum norm. The same construction also gives an open pseudoconvex domain in $\ell^\infty(A)$ which is not a domain of holomorphy. In addition we will prove that every open convex set in $\ell^\infty(A)$ is a domain of existence for holomorphic functions which is not true in $C_0(A)$. [Theorem 1]

An example of an extensionpair which is not normal in the sense of Alexander for the compact-open topology or the Nachbin topology will also be given. This answers a question of Alexander, G. Coeure and M. Schottenloher negatively.

Finally an example will be given of two complete locally convex topologies on a vectorspace which are different but have the same holomorphic functions. This answers negatively a question of P. Boland and L. Nachbin.

§ 1.

Let $1 \in A$ and let $z = x + iy = \{z_\alpha\}_{\alpha\in A} = \{x_\alpha + iy_\alpha\}_{\alpha\in A}$ denote a vector in $C_0(A)$. For every $\beta \in A$ let $e_\beta = \{z_\alpha\}_{\alpha\in A}$ where $z_\beta = 1$ and $z_\alpha = 0$ if $\alpha \neq \beta$.

If $P \in \mathcal{P}(^nC_0(A), C)$, the space of n-homogeneous, continuous polynomials on $C_0(A)$, then it can be shown that

$$P(z) = \sum_{\substack{\alpha_1 \neq \alpha_2 \neq \ldots \neq \alpha_n \\ \alpha_j \in A \\ r_{\alpha_j} \in N \cup \{0\} \\ \sum_{j=1}^n r_{\alpha_j} = n}} a_{\alpha_1 \ldots \alpha_n}^{r_{\alpha_1} \ldots r_{\alpha_n}} \cdot (z_{\alpha_1})^{r_{\alpha_1}} \cdot (z_{\alpha_2})^{r_{\alpha_2}} \ldots (z_{\alpha_n})^{r_{\alpha_n}}$$

where $a_{\alpha_1 \ldots \alpha_n}^{r_{\alpha_1} \ldots r_{\alpha_n}} \in C$.

Definition 1. $\|P\| = \sup_{z\in B} |P(z)|^{1/n}$, where B is the open unit ball in $C_0(A)$.

$$\|P\|_2 = \Big(\sum_{\substack{\alpha_j \in A \\ \alpha_1 \neq \alpha_2 \neq \ldots \neq \alpha_n \\ r_{\alpha_j} \in N \cup \{0\} \\ \sum_{j=1}^n r_{\alpha_j} = n}} \left| a_{\alpha_1 \ldots \alpha_n}^{r_{\alpha_1} \ldots r_{\alpha_n}} \right|^2 \Big)^{1/2}.$$

Definition 2. Let $\mathcal{U}$ be an open set in a locally convex vectorspace. Let $\mathcal{H}(\mathcal{U})$ be the set of G-analytic, locally bounded functions in $\mathcal{U}$.

Theorem 1. $\|P\|_2 \leq \|P\|$.

Theorem 1 implies that if $f \in \mathcal{H}(B)$ then f depends only on a countable number of variables in B. See [6] for another example.

Proof of the Theorem. It is obvious that we may assume that $C_0(A) = C^\ell$ for some $\ell \in \mathbf{N}$ and C^ℓ is given the topology induced by C_0. For the proof of the Theorem we need the following:

Lemma 1. Let $a_j \in C$ and $P_j(z) = \sum_{k=1}^{\ell_j} P_{j,k} \cdot z^k$ for $1 \leq j \leq r$, where $\ell_j \in \mathbf{N}$ and $P_{j,k} \in C$. Then there is $\theta \in [0,2\pi]$ such that $\sum_{j=1}^{r} |a_j + P_j(e^{i\theta})|^2 \geq \sum_{j=1}^{r} |a_j|^2 + \sum_{j=1}^{r} \sum_{k=1}^{\ell_j} |P_{j,k}|^2$.

Proof: Let $\mathrm{Re}^+ z = \begin{cases} \mathrm{Re}\, z & \text{if } \mathrm{Re}\, z > 0 \\ 0 & \text{if } \mathrm{Re}\, z \leq 0 \end{cases}$. We may assume that $a_j = \mathrm{Re}^+ a_j$ for all j, since otherwise we consider r_j and $e^{-i\theta_j} P_j(z)$ where $a_j = r_j e^{i\theta_j}$. Now

$$\sum_{j=1}^{r} |a_j + P_j(e^{i\theta})|^2 = \sum_{j=1}^{r} \left| a_j + \sum_{k=1}^{\ell_j} \mathrm{Re}\, P_{j,k} e^{ik\theta} + i \sum_{k=1}^{\ell_j} \mathrm{Im}\, P_{j,k} e^{i\theta k} \right|^2$$

$$= \sum_{j=1}^{r} a_j^2 + \sum_{j=1}^{r} \sum_{k=1}^{\ell_j} (\mathrm{Re}\, P_{j,k} e^{ik\theta})^2 + \sum_{j=1}^{r} \sum_{k=1}^{\ell_j} (\mathrm{Im}\, P_{j,k} e^{ik\theta})^2$$

$$\underbrace{+ 2 \sum_{j=1}^{r} a_j \cdot \left(\sum_{k=1}^{\ell_j} \mathrm{Re}\, P_{j,k} e^{ik\theta} \right) + 2 \cdot \sum_{j=1}^{r} \sum_{s<k\leq \ell_j} (\mathrm{Re}\, P_{j,k} e^{ik\theta})(\mathrm{Re}\, P_{j,s} e^{is\theta}) + 2 \cdot \sum_{j=1}^{r} \sum_{s<k\leq \ell_j} (\mathrm{Im}\, P_{j,k} e^{ik\theta})(\mathrm{Im}\, P_{j,s} e^{is\theta})}_{A(\theta)}$$

$$= \sum_{j=1}^{r} |a_j|^2 + \sum_{j=1}^{r} \sum_{k=1}^{\ell_j} |P_{j,k}|^2 + A(\theta).$$

Also $\mathrm{Re}\, P_{j,k} e^{ik\theta} = \cos k\theta \cdot \mathrm{Re}\, P_{j,k} - \sin k\theta\, \mathrm{Im}\, P_{j,k}$ and $\mathrm{Im}\, P_{j,k} e^{ik\theta} = \sin k\theta\, \mathrm{Re}\, P_{j,k} + \cos k\theta\, \mathrm{Im}\, P_{j,k}$. Hence if $s < k$, $\mathrm{Re}\, P_{j,k} e^{ik\theta} \cdot \mathrm{Re}\, P_{j,s} e^{is\theta} = \cos k\theta \cdot \cos s\theta\, \mathrm{Re}\, P_{j,k} \mathrm{Re}\, P_{j,s} + \sin k\theta \cdot \sin s\theta\, \mathrm{Im}\, P_{j,k} \cdot \mathrm{Im}\, P_{j,s}$ $- \cos k\theta \sin s\theta \cdot \mathrm{Re}\, P_{j,k} \cdot \mathrm{Im}\, P_{j,s} - \sin k\theta \cdot \cos s\theta \cdot \mathrm{Im}\, P_{j,k} \cdot \mathrm{Re}\, P_{j,s} = \tfrac{1}{2}[\cos(k+s)\theta\, \mathrm{Re}\, P_{j,k} \cdot \mathrm{Re}\, P_{j,s}$ $+ \cos(k-s)\theta\, \mathrm{Re}\, P_{j,k} \cdot \mathrm{Re}\, P_{j,s} + \text{etc}]$. There are similar expressions for $a_j \cdot \mathrm{Re}\, P_{j,k} e^{ik\theta}$ and $\mathrm{Im}\, P_{j,k} e^{ik\theta} \cdot \mathrm{Im}\, P_{j,s} e^{is\theta}$. But since $\int_0^{2\pi} \sin p\theta d\theta = \int_0^{2\pi} \cos p\theta d\theta = 0$ if $P \in \mathbf{N}$ it follows that $\int_0^{2\pi} A(\theta) d\theta = 0$. Hence there is a $\theta_1 \in [0,2\pi]$ such that $A(\theta_1) \geq 0$. Therefore $\sum_{j=1}^{r} |a_j + P_j(e^{i\theta_1})|^2 \geq \sum_{j=1}^{r} |a_j|^2 + \sum_{j=1}^{r} \sum_{k=1}^{\ell_j} |P_{j,k}|^2$. Q.E.D.

Completion of the proof of Theorem 1.

Let $P(z) = \sum_{s=0}^{n} z_1^s \cdot P_s(z)$ where P_s does not depend on z_1 for any s. $\|P\|_2^2 = \sum_{s=0}^{n} \|P_s\|_2^2 = \|P_{(1)}\|_2^2$ where $P_{(1)}(z) = \sum_{s=0}^{n} P_s(z)$. Assume now that we have replaced z_r if $r \leqslant q \leqslant \ell$, by $e^{i\theta_r}$ where $\theta_r \in [0,2\pi]$ such that $P_{(q)}(z) = P([1, e^{i\theta_2}, \ldots, e^{i\theta_q}, z_{q+1}, \ldots, z_\ell])$ and $\|P_{(q)}\|_2 \geqslant \|P\|_2$. For q = 1 this follows from the above. Let $P_{(q)}(z) = \sum_\alpha (P_{q,\alpha}(z_{q+1}) + a_\alpha) \cdot P_\alpha(z)$ where $P_{q,\alpha}(0) = 0$, $a_\alpha \in C$ and $P_\alpha(z)$ is one term with coefficient 1 which does not depend on z_r if $1 \leqslant r \leqslant q+1$. $\|P_{(q)}\|_2^2 = \sum_\alpha (\|P_{q,\alpha}\|_2^2 + |a_\alpha|^2)$. Lemma 1 implies that there is $\theta_{q+1} \in [0,2\pi]$ such that $\sum_\alpha |P_{q,\alpha}(e^{i\theta_{q+1}}) + a_\alpha|^2 \geqslant \sum_\alpha (\|P_{q,\alpha}\|_2^2 + |a_\alpha|^2)$. Hence $\|P_{(q+1)}\|_2 \geqslant \|P_{(q)}\|_2 \geqslant \|P\|_2$ where $P_{(q+1)}(z) = \sum_\alpha (P_{q,\alpha}(e^{i\theta_{q+1}}) + a_\alpha) P_\alpha(z)$. Hence there is, for $1 \leqslant r \leqslant \ell$, $\theta_r \in [0,2\pi]$ such that $|P_{(\ell)}| = \|P_{(\ell)}\|_2 \geqslant \|P\|_2$ where $P_{(\ell)} = P([1, e^{i\theta_2}, \ldots, e^{i\theta_\ell}])$. Therefore $\|P\|_2 \leqslant |P_{(\ell)}| \leqslant \|P\|$. Q.E.D.

Example. There is $P \in \mathcal{P}(^2c_0, C)$ such that $\sum_{\substack{\alpha_j \in N \\ \alpha_1 \neq \alpha_2 \\ r_{\alpha_j} \in N \cup \{0\} \\ r_{\alpha_1} + r_{\alpha_2} = 2}} |a_{\alpha_1,\alpha_2}^{r_{\alpha_1}, r_{\alpha_2}}| = \|P\|_1 = \infty$ where

$$P(z) = \sum_{\substack{\alpha_j \in N \\ \alpha_1 \neq \alpha_2 \\ r_{\alpha_j} \in N \cup \{0\} \\ r_{\alpha_1} + r_{\alpha_2} = 2}} a_{\alpha_1,\alpha_2}^{r_{\alpha_1}, r_{\alpha_2}} \cdot z_{\alpha_1}^{r_{\alpha_1}} \cdot z_{\alpha_2}^{r_{\alpha_2}}.$$

It is enough to prove that, for every $\epsilon > 0$, there is $P_\epsilon \in \mathcal{P}(^2c_0, C)$ such that $\epsilon \cdot \|P_\epsilon\|_1 \geqslant \|P_\epsilon\|^2$ because then $\sum_{n=1}^{\infty} \frac{P_{2^{-n}}}{2^n \cdot \|P_{2^{-n}}\|^2} \in \mathcal{P}(^2c_0, C)$ and $\| \sum_{n=1}^{\infty} \frac{P_{2^{-n}}}{2^n \cdot \|P_{2^{-n}}\|^2} \|_1 = \infty$, if we have taken $P_{2^{-n}}$ such that $\{j \in N : P_{2^{-n_1}}$ depends on $z_j\} \cap \{j \in N : P_{2^{-n_2}}$ depends on $z_j\} = \phi$ if $n_1 \neq n_2$ which of course is possible.

Let $n \in N$ and let $\{X^{(\nu)}\}_{\nu \in U}$ be the set of all $X^{(\nu)} = \{X_1^{(\nu)}, X_2^{(\nu)}, \ldots, X_p^{(\nu)}, \ldots, X_{n^n}^{(\nu)}\}$ where $X_p^{(\nu)} = 1$ or -1 for all p and ν. There are 2^{n^n} elements in $\{X^{(\nu)}\}_{\nu \in U}$. Let $a = \{a_1, a_2, \ldots, a_{n^n}\}$ where $a_p \in R$ and $|a_p| \leqslant 1$. Divide $\{1, 2, \ldots, n^n\}$ into 2n disjoint parts $\mathcal{U}_p$ such that $\sup_{r,s \in \mathcal{U}_p} |a_r - a_s| \leqslant 1/n$ for all $p \in \{1, \ldots, 2n\}$. Put $a' = \{a'_1, \ldots, a'_{n^n}\}$ where $|a'_r - a'_s| = 0$ if $r, s \in \mathcal{U}_p$ and $\sup_{r \in \mathcal{U}_p} |a_r - a'_r| \leqslant 1/n$ if $\mathcal{U}_p$ contains more than n^4 elements and $a'_r = 0$ otherwise. Now the number of elements in $\{X^{(\nu)}\}_{\nu \in U}$ such that $|\sum_{r \in \mathcal{U}_p} X_r^{(\nu)}| \geqslant 2[\frac{g(p)}{2n}]$ is less than $2^{n^n - g(p)} \cdot \binom{g(p)}{[\frac{g(p)}{2}] + 1 - [\frac{g(p)}{2n}]} \cdot g(p)$ according to the binomial theorem where g(p) is the number of elements in $\mathcal{U}_p$ and [ ] denotes the integer part. From Stirlings formula it follows that

$$2^{n^n-g(p)} \cdot \left(\begin{matrix} g(p) \\ [\frac{g(p)}{2}] + 1 - [\frac{g(p)}{2n}] \end{matrix} \right) \cdot g(p) <$$

$$< (g(p))^5 \cdot 2^{n^n-g(p)} \cdot \frac{(g(p))^{g(p)+\frac{1}{2}}}{\left((1-\frac{1}{n})\frac{g(p)}{2}\right)^{\frac{(1-\frac{1}{n})g(p)+1}{2}} \cdot \left((1+\frac{1}{n})\frac{g(p)}{2}\right)^{\frac{(1+\frac{1}{n})g(p)+1}{2}}}$$

$$= \frac{(g(p))^5 \cdot 2^{n^n-g(p)} \cdot 2^{g(p)+1}}{(g(p))^{\frac{1}{2}} \cdot \left((1-\frac{1}{n})^{1-\frac{1}{n}} \cdot (1+\frac{1}{n})^{1+\frac{1}{n}}\right)^{\frac{g(p)}{2}} (1-\frac{1}{n^2})^{\frac{1}{2}}} < \frac{(g(p))^5\, 2^{n^n+1}\, (1-\frac{1}{n})^{\frac{g(p)}{2n}}}{(1-\frac{1}{n^2})^{\frac{g(p)}{2}}\, (1+\frac{1}{n})^{\frac{g(p)}{2n}}} < 2^{n^n}\, e^{-n}$$

if n is sufficiently large and $g(p) \geqslant n^4$ by an application of $\lim_{n\to\infty} (n \log(1-\frac{1}{n}) - n^2 \log(1-\frac{1}{n^2}) - n \log(1+\frac{1}{n})) = -1$.

We now have

$$\sum_{\nu\in U} \left| \sum_{r=1}^{n^n} x_r^{(\nu)} a_r \right| \leqslant \sum_{\nu\in U} \left| \sum_{r=1}^{n^n} x_r^{(\nu)} a_r' \right| + 2^{n^n} \cdot n^4 \cdot 2 \cdot n + \frac{1}{n} \cdot 2^{n^n} \cdot n^n \quad \text{since}$$

$$\sum_{\nu\in U} \sum_{r=1}^{n^n} |a_r' - a_r| \leqslant 2^{n^n} \cdot n^4 \cdot 2 \cdot n + \frac{1}{n} 2^{n^n} \cdot n^n. \text{ Hence}$$

$$\sum_{\nu\in U} \left| \sum_{r=1}^{n^n} x_r^{(\nu)} a_r \right| < 2^{n^n} e^{-n} \cdot n^n + 2^{n^n} \cdot n^{n-1} + 2^{n^n+1} \cdot n^5 + 2^{n^n} n^{n-1}.$$

Therefore $\sum_{\nu\in U} \left| \sum_{r=1}^{n^n} x_r^{(\nu)} a_\nu \right| \leqslant 4 \cdot 2^{n^n} n^{n-1}$. Thus

$$(1) \qquad \sum_{\nu\in U} \left| \sum_{r=1}^{n^n} z_r \cdot x_r^{(\nu)} \right| \leqslant 8 \cdot 2^{n^n} \cdot n^{n-1}$$

if $z_r \in \mathbf{C}$ and $|z_r| \leqslant 1$ for all r. We may assume that $U \subset \mathbf{N}$. Put $P(z) = \sum_{\nu\in U} z_{n^n+1+\nu} \left(\sum_{r=1}^{n^n} x_r^{(\nu)} z_r \right)$.

$P \in \mathcal{P}(^2C_0, \mathbf{C})$ and $\|P\|_1 = 2^{n^n} \cdot n^n$. But it follows from (1) that $\|P\|^2 \leqslant 8 \cdot n^{n-1} \cdot 2^{n^n}$. Hence the example follows if $\frac{8}{n} < \epsilon$. Q.E.D.

§2.

Let $U = \{ z \in C_0(A) : \forall \alpha \in A,\ \forall \beta \in A,\ \alpha \neq \beta \neq 1 \neq \alpha \Rightarrow |x_\alpha| + |x_\beta| < 1+2\epsilon \text{ and } |y_\alpha| < \frac{\epsilon}{3}\}$ where $0 < \epsilon < \frac{1}{6}$. U is an open and convex set in $C_0(A)$. Let

$\Omega = \{ z \in U : x_1 - \sup_{\alpha\neq 1} \varphi(x_\alpha) + \sup_{\alpha\neq 1} (y_\alpha + \epsilon x_\alpha)^2 < 0\}$ where $\varphi(t) = ((t-1)^+)^2$, $t \in \mathbf{R}$ and

$(t-1)^+ = \{\begin{smallmatrix} t-1 & \text{if } t-1>0 \\ 0 & \text{if } t-1\leq 0\end{smallmatrix}\}$. Ω is an open and connected set.

Lemma 2. Ω is pseudoconvex.

Proof. Since U is convex it is enough to check neighbourhoods of $z^o \in \delta\Omega \cap U$, where δ denotes the boundary of Ω. We shall prove that the boundary of Ω is given locally by a plurisubharmonic function.

1) If for every $\alpha \neq 1$, $|x_\alpha^o| < 1$, there is a neighbourhood ω of z^o, such that, for every $z \in \omega$ and every $\alpha \neq 1$, $\varphi(x_\alpha) = 0$. Hence $\omega \cap \Omega$ is defined by $x_1 + \sup_{\alpha\neq 1} (y_\alpha + \epsilon x_\alpha)^2 < 0$. But $x_1 + \sup_{\alpha \neq 1} (y_\alpha + \epsilon x_\alpha)^2$ is a convex function.

2) If there is an $\alpha \neq 1$ such that $|x_\alpha^o| \geq 1$, it follows that $|x_\beta^o| \leq 2\cdot\epsilon$ if $1 \neq \beta \neq \alpha$. Hence $|y_\beta^o + \epsilon x_\beta^o| \leq 2\epsilon^2 + \frac{\epsilon}{3} < \epsilon - \frac{\epsilon}{3} \leq |\epsilon x_\alpha^o + y_\alpha^o|$ if $\epsilon < \frac{1}{6}$. Therefore there is a neighbourhood ω of z^o such that $\sup_{1\neq\beta\neq\alpha} (y_\beta + \epsilon x_\beta)^2 < (y_\alpha + \epsilon x_\alpha)^2$ if $z \in \omega \cap \Omega$. Hence Ω is defined near z^o by $x_1 - \varphi(x_\alpha) + (y_\alpha + \epsilon x_\alpha)^2 < 0$, and since $x_1 - \varphi(x_\alpha) + (y_\alpha + \epsilon x_\alpha)^2$ is a plurisubharmonic function the lemma follows.

Q.E.D.

The proof of Lemma 2 is due to C. O. Kiselman. My original proof was much more complicated.

Lemma 3. There is an open set $\Omega_1 \subset C_o(A)$ such that $\Omega_1 \not\subset \Omega \subset \Omega_1$ and such that every $f \in \mathcal{H}(\Omega)$ may be extended to an $f_1 \in \mathcal{H}(\Omega_1)$.

Proof. (1) $-e_1 + \frac{K\cdot\epsilon^3}{2} e_\alpha + \epsilon^3 \cdot B \subset \Omega$ if $\alpha \neq 1$, where $\epsilon^3 B$ denotes the open ball with centre at 0 and radius ϵ^3 and $K \in \mathbb{N} \cup \{\frac{2}{\epsilon^3}(1+\frac{3}{2}\epsilon)\}$, $0 \leq K \leq \frac{2}{\epsilon^3}(1+\frac{3}{2}\epsilon)$. Let $f \in \mathcal{H}(\Omega)$. From Theorem 1 it follows that there is an $\alpha \neq 1$ such that $f(-e_1+z) \equiv f(-e_1 + \frac{\epsilon^3}{2} e_\alpha + z)$ if $\|z\| < \frac{\epsilon^3}{2}$. Hence it follows from (1) that this also is true if $\|z\| < \epsilon^3$. Assume that $f(-e_1+z) \equiv f(-e_1 + \frac{K\epsilon^3}{2} e_\alpha + z)$ if $K < t \in \mathbb{N} \cup \{\frac{2}{\epsilon^3}(1+\frac{3}{2}\epsilon)\}$ and $\|z\| < \epsilon^3$. Thus $f(-e_1+z) \equiv f(-e_1 + \frac{(t-1)\epsilon^3}{2} e_\alpha + z) \equiv f(-e_1 + \frac{t\epsilon^3}{2} e_\alpha + z)$ if $\|z\| < \frac{\epsilon^3}{2}$, and according to (1) the identity also holds if $\|z\| < \epsilon^3$. Therefore $f(-e_1+z) \equiv f(-e_1 + (1+\frac{3}{2}\epsilon)e_\alpha + z)$ if $\|z\| < \epsilon^3$. Also $(1+\frac{3}{2}\epsilon)e_\alpha + \epsilon^3 B \subset \Omega$ since $\epsilon^3 - (\frac{3\epsilon}{2} - \epsilon^3)^2 + (\epsilon + \epsilon^3 + \epsilon^4)^2 < 0$. Thus $f((1+\frac{3}{2}\epsilon)e_\alpha + z) \in \mathcal{H}(\epsilon^3 B)$. Let $\Omega_1 = \Omega \cup \epsilon^3 B$. $\Omega_1 \not\subset \Omega \subset \Omega_1$ since $0 \in \delta\Omega$. Let $f_1(z) = \begin{cases} f(z) & \text{if } z \in \Omega \\ f((1+\frac{3}{2}\epsilon)e_\alpha + z) & \text{if } z \in \epsilon^3 B. \end{cases}$ Now $f(z) \equiv f((1+\frac{3}{2}\epsilon)e_\alpha + z)$ if $z \in \Omega \cap \epsilon^3 B$ for the same reasons as above. Hence $f_1 \in \mathcal{H}(\Omega_1)$ and this completes the proof.

Q.E.D.

Remark. Lemma 3 is also true if $f \in \mathcal{H}(\Omega, F)$ where F is a Banach space.

Proposition 1. An open convex set $\mathcal{U} \subset \ell^\infty(A)$ is a domain of existence for analytic functions.

Proof. We may assume that there is an $\epsilon > 0$ such that $\epsilon B \subset \mathcal{U}$ where B is the closed unit ball in $\ell^{\infty}(A)$. To every $x \in \delta\mathcal{U}$, (the boundary of $\mathcal{U}$), there is a $\varphi_x \in (\ell^{\infty}(A))'$, the dual, such that $\|\varphi_x\| \leq \frac{2}{\epsilon}$, $\mathrm{Re}\,\varphi_x(x) = 1$ and $\mathrm{Re}\,\varphi_x(z) < 1$ when $z \in \mathcal{U}$. Let $f_x(z) = \dfrac{1}{(\varphi_x(z)-\varphi_x(x))^3}$. Then $f_x \in \mathcal{H}(\mathcal{U})$ since $\mathrm{Re}(\varphi_x(z)-\varphi_x(x)) \neq 0$ if $z \in \mathcal{U}$.

$$(1) \qquad \sup_{x\in\delta\mathcal{U}} \sup_{y\in\gamma B} |f_x(z+y)| < M_{z,\gamma} < \infty \text{ if } z+\gamma B \subset \mathcal{U}$$

because if not there are $x^{(1)} \in \delta\mathcal{U}$ and $z^{(1)} \in \mathcal{U}$ such that $|\varphi_{x^{(1)}}(x^{(1)})-\varphi_{x^{(1)}}(z^{(1)})| < \dfrac{\lambda}{(\|z^{(1)}\|+1)(2\lambda+2)} < \frac{1}{2}$ where $2\lambda = \inf_{\xi\in\complement\mathcal{U}} \|z^{(1)}-\xi\|$. ($\complement$ denotes the complement). But then it follows that

$\mathrm{Re}\,\varphi_{x^{(1)}}(z^{(1)} + \dfrac{2\lambda z^{(1)}}{(\|z^{(1)}\|+1)(2\lambda+2)}) > 1$ and this is a contradiction since $z^{(1)} + \dfrac{2\lambda z^{(1)}}{(\|z^{(1)}\|+1)(2\lambda+2)} \in \mathcal{U}$:

According to [9] there is a bounded linear mapping φ of $\ell^{\infty}(A)$ onto $\ell^2(\mathfrak{B})$ where card $\mathfrak{B} = 2^{\text{card } A}$. Let $e_\gamma = \{z_\beta\}_{\beta\in\mathfrak{B}} \in \ell^2(\mathfrak{B})$ where $z_\gamma = 1$ and $z_\beta = 0$ if $\beta \neq \gamma$. Let $e'_\beta \in \ell^{\infty}(A)$ be such that $\varphi(e'_\beta) = e_\beta$. Take a constant $C > 0$ and a subset $I \subset \mathfrak{B}$ such that card I = card $\mathfrak{B}$ and $\|e'_\beta\| \leq C$ if $\beta \in I$. That C and I exist follows from the fact that card $\mathfrak{B}$ > card **N**.

Let $\varphi'_\beta \in (\ell^2(\mathfrak{B}))'$, the dual, be such that $\varphi'_\beta(e_\beta) = 1$, $\varphi'_\beta(e_\gamma) = 0$ if $\gamma \neq \beta$, and $\|\varphi'_\beta\| = 1$ for all $\beta \in \mathfrak{B}$. Put $\psi_\beta = \varphi'_\beta \circ \varphi$. Then $\psi_\beta \in (\ell^{\infty}(A))'$ and

$$(2) \qquad \sum_{\beta\in\mathfrak{B}} \psi_\beta^2(z) \leq \|\varphi\|^2 \cdot \|z\|^2 \text{ for all } z \in \ell^{\infty}(A).$$

We may assume that I is well ordered and that card$\{\gamma \in I;\ \gamma < \beta\}$ < card I for every $\beta \in I$.

Put $\delta\mathcal{U} = \{x_\beta\}_{\beta\in I}$. This is possible because card$\{x \in \delta\mathcal{U}\}$ = card I. Now for every $\beta \in I$ let $\ell(\beta) \in I$ be such that $\ell(\beta_1) \neq \ell(\beta_2)$ if $\beta_1 \neq \beta_2$ and $\psi_{\ell(\beta)}(x_\gamma) = 0$ if $\gamma \leq \beta$. This is possible since supp $\varphi(x_\gamma)$ is countable for all $\gamma \in \mathfrak{B}$, (supp z = $\{\beta \in \mathfrak{B};\ z_\beta \neq 0\}$), and hence card$\{\bigcup_{\gamma\leq\beta} \text{supp}\,\varphi(x_\gamma)\}$ < card I.

$\psi^2_{\ell(\beta)}(z)\cdot f_{x_\beta}(z) \in \mathcal{H}(\mathcal{U})$ and it can not be continued over x_β because $\psi_{\ell(\beta)}(x_\beta) = 0$ according to the above. Hence

$$|\psi^2_{\ell(\beta)}((1-r)x_\beta+re'_{\ell(\beta)})\cdot f_{x_\beta}((1-r)x_\beta+re'_{\ell(\beta)})| \geq \frac{r^2\cdot\epsilon^3}{(2\cdot r\cdot c+2\cdot r\cdot\|x_\beta\|)^3} = \frac{\epsilon^3}{r\cdot 2^3(c+\|x_\beta\|)^3}$$

and this goes to infinity as $r \to 0$. Put $g(z) = \sum_{\beta\in I} C_\beta\,\psi^2_{\ell(\beta)}(z)\cdot f_{x_\beta}(z)$ where $C_\beta = 0$ if $\sum_{\gamma<\beta} C_\gamma\,\psi^2_{\ell(\gamma)}((1-r)x_\beta+re'_{\ell(\beta)})\cdot f_{x_\gamma}((1-r)x_\beta+re'_{\ell(\beta)}) \to \infty$ as $r \to \infty$ and otherwise $C_\beta = 1$. $g \in \mathcal{H}(\mathcal{U})$ since $\sup_{y\in r\cdot B} |g(z+y)| < M_{z,r}\cdot\|\varphi\|^2(\|z\|+r)^2$ according to (1) and (2). But g can not be continued over any boundary point because of the contruction and the fact that

$\sum_{\gamma>\beta} C_\gamma\,\psi^2_{\ell(\gamma)}(\xi\cdot x_\beta+ye'_{\ell(\beta)})f_{x_\gamma}(\xi\cdot x_\beta+ye'_{\ell(\beta)}) = 0$ for all ξ and $y \in \mathbb{C}$. Q.E.D.

Let $U = \{z \in \ell^\infty(A);\ \forall \alpha \in A,\ \forall \beta \in A,\ \alpha \neq \beta \neq 1 \neq \alpha \Rightarrow |x_\alpha| + |x_\beta| < 1+2\cdot\epsilon \text{ and } |y_\alpha| < \frac{\epsilon}{3}\}$ and let $\Omega = \{z \in U;\ x_1 - \sup_{\alpha \neq 1} \varphi(x_\alpha) + \sup_{\alpha \neq 1} (y_\alpha + \epsilon x_\alpha)^2 < 0\}$ where $0 < \epsilon < \frac{1}{6}$ and $\varphi(t) = ((t-1)^+)^2$. Ω is open.

Lemma 4. **Ω is pseudoconvex but not a domain of holomorphy in any sense.**

Proof. The proof is the same as in Lemma 2 and 3. Q.E.D.

§ 3.

Let Ω be the domain in Lemma 2. Lemma 3 implies that Ω is not a domain of holomorphy and the proof of Lemma 3 gives that there is an open connected set $\Omega_1 \subset C_0(A)$ such that $\Omega \subset \Omega_1$, $xe_1 \in \Omega_1$ if $x < \frac{32\cdot\epsilon^2}{9} - \frac{8}{3}\epsilon^3 - 4\epsilon^4$ and $\mathcal{H}(\Omega) = \mathcal{H}(\Omega_1)$ since $\inf_{z\in\Omega} (-\sup_{\alpha\neq 1} \varphi(x_\alpha) + \sup_{\alpha \neq 1} (y_\alpha + \epsilon\cdot x_\alpha)^2) = -4\cdot\epsilon^2 + (\epsilon(1+2\epsilon) - \frac{\epsilon}{3})^2 = -\frac{32}{9}\epsilon^2 + \frac{8}{3}\epsilon^3 + 4\cdot\epsilon^4$. That is to say (Ω, Ω_1) is an extension pair in the sense of Alexander. If $\mathcal{H}(\Omega)$ and $\mathcal{H}(\Omega_1)$ are given topologies T_Ω and T_{Ω_1} respectively such that the induced algebraic isomorphism $\mathcal{H}(\Omega) \to \mathcal{H}(\Omega_1)$ is a topological isomorphism we say that (Ω, Ω_1) is a normal extension pair for (T_Ω, T_{Ω_1}).

τ_1 will denote the Nachbin topology (the locally convex topology generated by the seminorms which are ported by compact sets. see [7]), τ_2 the locally convex topology generated by the seminorms which are ported by bounding sets (see [4]), and τ_3 will denote the bornological topology associated with the compact open topology τ_0 on $\mathcal{H}(\Omega)$, respectively $\mathcal{H}(\Omega_1)$. $\tau_0 < \tau_1 \leqslant \tau_2 \leqslant \tau_3$.
Let $\epsilon < \frac{1}{15}$. Then it follows that $\frac{30}{9}\epsilon^2\cdot e_1 \in \Omega_1$ since $-\frac{2}{9}\epsilon^2 + \frac{8}{3}\epsilon^3 + 4\epsilon^4 < 0$. Hence $P(f) = |f(\frac{30}{9}\epsilon^2 e_1)|$ is a continuous seminorm on $\mathcal{H}_{\tau_0}(\Omega_1)$ hence on $\mathcal{H}_{\tau_3}(\Omega_1)$ and therefore on $\mathcal{H}_{\tau_3}(\Omega)$ since every extension pair is normal for the associated bornological topology. Thus we have proved the first part of

Lemma 5. **P is a continuous seminorm on $\mathcal{H}_{\tau_3}(\Omega)$ but P is not ported by a bounding set in Ω.**

Proof. Assume that the Lemma is false. Then there is a bounding set $K \subset \Omega$ such that P is ported by K. Since $\mathcal{H}(\Omega)$ is an algebra it follows that $P(f) \leqslant \sup_{z\in K} |f(z)|$. But a bounding set for $\mathcal{H}(C_0(A))$ is a compact set in $C_0(A)$ hence it follows, since $\mathcal{H}(C_0(A)) \subset \mathcal{H}(\Omega)$, that $\overline{K}$ is compact and $\overline{K} \subset \overline{\Omega}$ where $\overline{}$ denotes the closure in $C_0(A)$. Since $\overline{K}$ is compact there are $r \in N$ and $\{\alpha_1, \ldots, \alpha_r\} \subset A$ such that, for every $z \in \overline{K}$, $|z_\alpha| < 1$ if $\alpha \notin \{\alpha_1, \ldots, \alpha_r\}$. Let $\Omega_{1,\alpha_1,\ldots,\alpha_r} = \{z \in \Omega;\ z_\alpha = 0 \text{ if } \alpha \notin \{\alpha_1, \ldots, \alpha_r, 1\}\}$. $\Omega_{1,\alpha_1,\ldots,\alpha_r}$ can be considered as an open set in C^{r+1} where we give C^{r+1} the topology induced by C_0. Let
$U_2 = \{z \in C^{r+1};\ \forall i, j \in \{1,2,\ldots,r+1\}\ i \neq 1 \neq j \neq i \Rightarrow |x_i| + |x_j| < 1+4\epsilon \text{ and } |y_i| < \frac{2\epsilon}{3}\}$ and let
$\Omega_2 = \{z \in U_2;\ x_1 - \sup_{i\neq 1} \varphi(x_i) + \sup_{i \neq 1} (y_i + \epsilon x_i) < \frac{32}{9}\epsilon^2 - \frac{8}{3}\epsilon^3 - 4\epsilon^4\}$. Ω_2 is an open connected set in C^{r+1} and $\Omega_{1,\alpha_1,\ldots,\alpha_r} \subset \Omega_2$. The proof of Lemma 2 gives, since $2\epsilon < \frac{1}{6}$, that Ω_2 is pseudoconvex hence Ω_2 is a domain of existence according to the Theorem of Oka-Norguet-Bremermann. Let $\overline{K}' = \mathrm{Proj}_{[1,\alpha_1,\ldots,\alpha_r]} \overline{K}$,

where $\mathrm{Proj}_{[1,\alpha_1,\dots,\alpha_r]}z = \{z'_\alpha\}_{\alpha\in A}$ where $z'_\alpha = z_\alpha$ if $\alpha \in \{1,\alpha_1,\dots,\alpha_r\}$ and $z'_\alpha = 0$ otherwise. $\overline{K}' \subset \subset \overline{\Omega}_{1,\alpha_1,\dots,\alpha_r}$ because if $z' \in \overline{K}'$ there is $z'' \in C_o(A)$ such that $z''_\alpha = 0$ if $\alpha \in \{1,\alpha_1,\dots,\alpha_r\}$ and $z' + z'' \in \overline{K}$. Therefore $\|z''\| < 1$ hence $z' \in \overline{\Omega}$ according to the construction of Ω. Now

(1) $$\inf_{z\in \complement\Omega_2} \|z - \frac{30}{9}\epsilon^2 e_1\| \leqslant \frac{2\epsilon^2}{9} \text{ since } \frac{32}{9}\epsilon^2 e_1 \notin \Omega_2.$$

(2) $$\inf_{\substack{z\in\complement\Omega_2\\ \xi\in\Omega_{1,\alpha_1,\dots,\alpha_r}}} \|z-\xi\| \geqslant \frac{\epsilon^2}{3} \text{ since } \epsilon < \epsilon^2 \text{ and}$$

$$\sup_{z\in\Omega} (x_1 + \frac{\epsilon^2}{3} - \sup_{\alpha\neq 1} \varphi(x_\alpha - \frac{\epsilon^2}{3}) + \sup_{\alpha\neq 1} (y_\alpha + \frac{\epsilon^2}{3} + \epsilon(x_\alpha + \frac{\epsilon^2}{3})^2)$$

$$< \frac{\epsilon^2}{3} + \frac{4\cdot\epsilon^3}{3} + (\frac{\epsilon^2}{3} + \frac{\epsilon^3}{3})^2 + 2(\frac{\epsilon}{3} + \epsilon + 2\epsilon^2)(\frac{\epsilon^2}{3} + \frac{\epsilon^3}{3}) = \frac{\epsilon^2}{3} + \frac{20\cdot\epsilon^3}{9} + \frac{21\cdot\epsilon^4}{9} + \frac{14\cdot\epsilon^5}{9} + \frac{\epsilon^6}{9} <$$

$< \frac{32}{9}\epsilon^2 - \frac{8}{3}\epsilon^3 - 4\epsilon^4$. Therefore $\inf_{\substack{z\in\complement\Omega_2\\ \xi\in\overline{K}'}} \|z-\xi\| \geqslant \frac{\epsilon^3}{3}$. Hence it follows that $\frac{30\cdot\epsilon^2}{9} e_1$ does not belong to the $\mathcal{H}(\Omega_2)$–hull of $\overline{K}'$, and therefore there is $f \in \mathcal{H}(\Omega_2)$ such that $f(\frac{30\cdot\epsilon^2}{9} e_1) = 1$ and $\sup_{z\in\overline{K}'} |f(z)| < 1$.

Put $\Omega_3 = \Omega_2 \times \prod_{\alpha\in A\setminus\{1,\alpha_1,\dots,\alpha_r\}} C_\alpha$ where $C_\alpha = \mathbf{C}$. $\Omega_3 \subset C_o(A)$ and $\Omega \subset \Omega_3$ since $(\frac{32\cdot\epsilon^2}{9} - \frac{8}{3}\epsilon^3 - 4\cdot\epsilon^4)e_1 \subseteq \overline{\Omega}_2$. Now f can trivially be extended to Ω_3. Hence $f' \in \mathcal{H}(\Omega)$ where f' is the extension of f. But $P(f') = f(\frac{30}{9}\epsilon^2 e_1) = 1 > \sup_{z\in\overline{K}'} |f(z)| = \sup_{z\in K} |f'(z)|$ which is a contradiction. Q.E.D.

Lemma 6. (Ω, Ω_1) is not a normal extension pair for τ_1 or τ_2.

Proof. P(f) is a continuous seminorm on $\mathcal{H}_{\tau_1}(\Omega_1)$ and $\mathcal{H}_{\tau_2}(\Omega_1)$ but not on $\mathcal{H}_{\tau_1}(\Omega)$ or $\mathcal{H}_{\tau_2}(\Omega)$ according to Lemma 5. Q.E.D.

§4.

Let $C_{o,\tau}(A)$ be the vectorspace $C_o(A)$ with the topology generated by the seminorms $\{P_{\mathcal{A}}\}_{\mathcal{A}\subset A}$ where $\mathcal{A} \subset A$ is countable and $P_{\mathcal{A}}(z) = \sup_{\alpha\in\mathcal{A}} |z_\alpha|$. It is easy to see that $C_{o,\tau}(A)$ is a complete locally convex topological vectorspace and that if A is uncountable, the norm topology is strictly finer.

Proposition. $\mathcal{H}(C_{o,\tau}(A)) = \mathcal{H}(C_o(A))$.

Proof. This follows easily from the fact that if $f \in \mathcal{H}(C_o(A))$ then there is $\mathcal{A} \subset A$ such that $\mathcal{A}$ is countable and f depends on z_α only if $\alpha \in \mathcal{A}$ (Theorem 1). Q.E.D.

This answers negatively the following question of P. Boland.

Let E be a complex vectorspace, τ_i a locally convex topology for E, i=1,2. Write $E_i = (E,\tau_i)$. Does $\tau_1 \neq \tau_2 \Rightarrow \mathcal{H}(E_1) \neq \mathcal{H}(E_2)$. If we assume E_1 and E_2 separable the question is still open.

Bibliography

1. H. Alexander, **Analytic functions on Banach spaces,** Thesis, University of California at Berkeley (1968).

2. G. Coeuré, **Fonctions plurisousharmonique sur les espaces vectoriels topologiques et applications à l'etude des fonctions analytiques,** Thesis, Université de Nancy (1969).

3. G. Coeuré, Book to appear. North Holland.

4. S. Dineen, **Bounding subsets of á Banach space,** Math. Ann. 192, 61–70 (1971).

5. L. Gruman and C. O. Kiselman, **Le problème de Levi dans les espaces de Banach à base,** C. R. Acad. Sc. Paris, 274, 1296–1299 (1972).

6. A. Hirschowitz, **Sur le non-plongement de variétés analytiques Banachiques réelles,** C. R. Acad. Sc. Paris, 269, 844–846 (1969).

7. L. Nachbin, **Topology on spaces of holomorphic mappings,** Ergebnisse der Mathematik und ihrer Grenzgebiete, Springer-Verlag, Germany 47(1969).

8. Ph. Noverraz, **Pseudo-convexité, convexité polynomiale et domaines d'holomorphie en dimension infinie,** Notas de Matematica, 4, North Holland (1973).

9. H. P. Rosenthal, **On quasicomplemented subspaces of Banach spaces,** Journal of Functional Analysis (1969), 176–214.

10. M. Schottenloher, **Analytische Fortsetzung in Banachräumen,** Mathematische Annalen, 199, 313–336 (1972).

Uppsala University
Sysslomansgatan 8
75223 Uppsala, Sweden

APPROXIMATION OF HOLOMORPHIC OR PLURISUBHARMONIC FUNCTIONS IN CERTAIN BANACH SPACES

Philippe Noverraz

In (2) we have proved some polynomial approximation theorems of the Runge and Oka-Weil type. Using recent results of (1), we shall prove more general theorems, the polynomials being replaced by the holomorphic functions in an open pseudo-convex set. The results are the following:

Let U be an open pseudo-convex subset of a Banach space with the bounded approximation property (BAP), then:

1) $\hat{K}_{H(U)} = \hat{K}_{P_c(U)}$, for every compact subset K of U.

2) Any holomorphic function in a neighborhood of a compact set K, such that $K = \hat{K}_{H(U)}$, can be approximated uniformly on K by elements of H(U).

3) If U and U' are two open pseudo-convex sets, $U \subset U'$, the following conditions are equivalent:

a- H(U') is dense in H(U) for the compact open topology,

b- $\hat{K}_{H(U')} = \hat{K}_{H(U)}$ for every compact subset K of U,

c- $\hat{K}_{P_c(U')}$ is a compact subset of U for every compact subset K of U.

4) If U is an open pseudo-convex set and F a complemented subspace, the restriction mapping $H(U) \to H(U \cap F)$ has a dense image.

5) The set of finite supremum of functions $a \cdot \log |f|$, with $a > 0$ and f in H(U), is dense in $P_c(U)$ for the compact open topology.

Let us recall that H(U) (resp. P(U), $P_c(U)$) denote the set of all holomorphic (resp. plurisubharmonic, plurisubharmonic and continuous) functions in an open set U and, if A(U) is a family of complex valued functions in U, the A(U)-convex hull of a compact subset K of U is defined by

$$\hat{K}_{H(U)} = \{z \in U, |f(z)| \leq |f|_K, \ \forall f \in A(U)\}.$$

For other definitions and properties of holomorphic and plurisubharmonic functions in Banach spaces, the reader is referred to (3).

Let E be a Banach space with a countable Schauder basis $(e_n)_{n \in N}$, we denote by E_n (resp. E^n) the closed subspace spanned by $e_1, \ldots, e_n$ (resp. $e_{n+1}, \ldots$) and by u_n the natural projection $x \to \sum_{i=1}^{n} x_i e_i$. We shall always suppose that $\|u_n\| \leq 1$ for all n.

The following result is due to Pelczynski (3):

Proposition 1: If E is a separable Banach space, the following conditions are equivalent:

a) The identity map in E is the pointwise limit of a sequence of finite dimensional operators.

b) There is a constant $C > 0$ such that for any compact subset K and any $\epsilon > 0$ there exists a finite dimensional operator u such that $\|u\| \leq C$ and $\|u(x) - x\| \leq \epsilon$ on K.

c) E is isomorphic with a complemented subspace of a Banach space with a basis.

In such a space, said to have the bounded (or Banach) approximation property (shortly BAP), the Levi problem has a positive answer, which means that every pseudo-convex open set is the domain of existence of a holomorphic function (3). Examples of such spaces:

1- The Banach spaces with a basis,

2- $\mathcal{L}^p_{(\mu)}$, $1 \leq p < +\infty$, with μ a positive Radon measure on a locally compact space,

3- the Hardy spaces H^p, $1 \leq p < +\infty$,

4- the space of continuous functions in the closed disk, holomorphic in the interior.

If U is an open pseudo-convex set of a Banach space with a basis, let us denote, as in (1):

$$U_n = \{x \in U, \|x\| \leq n,\ d(x, \complement U) \geq \frac{1}{n}\},$$

$$L_n = \{x \in U, \|u_n(x) - x\| \leq \frac{1}{2} d(x, \complement U)\}, \text{ and}$$

$$A_n = U_n \cap L_{n-1} \cap E_n = \{x \in U_n \cap E_n, -\log d(x, \complement U) + |x_n| \leq \log (2\|e_n\|)\}.$$

The set A_n is then a Runge compact of $E_n \cap U$; moreover, $u_n(L_n) = E_n \cap U$.

Lemma 1: Let U be an open pseudo-convex set in a Banach space with a basis and K a compact subset of U, then there is an integer n_0 such that:

$$\widehat{u_{n+1}(K)}_{H(U \cap E_{n+1})} \subset A_{n+1} \quad \text{for all} \quad n \geq n_0 .$$

As the identity map in E is the uniform limit on K of the sequence (u_n), there exists an integer n_0 such that for all $n \geq n_0$

$$u_n(K) \subset U_n \text{ and } \|u_n(x) - x\| \leq \frac{1}{5} d(K) \text{ for all x in K},$$

where $d(K) = \inf_{x \in U} d(x, \complement U)$.

Let $y = u_{n+1}(x)$ with x in K, it follows that

$$\|u_n(y) - y\| = \|u_n(x) - u_{n+1}(x)\| \leq \|u_n(x) - x\| + \|u_{n+1}(x) - x\| \leq \frac{2}{5} d(K).$$

As $|d(x) - d(y)| \leq \|x - y\| \leq \frac{1}{5} d(K)$ implies $d(y) \geq d(x) - \frac{1}{5} d(K) \geq \frac{4}{5} d(K)$, it follows that

$$\|u_n(y) - y\| \leq \frac{1}{2} d(y, \complement U),$$

which proves that $u_{n+1}(K) \subset L_n$. We then have

$$u_{n+1}(K) \subset L_n \cap U_{n+1} \cap E_{n+1} = A_{n+1}$$

which proves the lemma since $\widehat{A_{n+1}}_{H(U \cap E_{n+1})} = A_{n+1}$.

We shall use the following result of (1):

Lemma 2: **If E is a Banach space with a basis, for any $\epsilon > 0$ and any f_n in $H(U \cap E_n)$ there exists f_{n+1} in $H(U \cap E_{n+1})$ such that:**

1) $f_{n+1}\big|_{E_n} = f_n$

2) $\left|f_{n+1} - f_n \circ u_n\right|_{A_{n+1}} \leqslant \epsilon$.

Moreover, the domain of existence of the function f_{n+1} can be chosen to be equal to $U \cap E_{n+1}$.

It is now possible to prove:

Theorem 1: **Let U be an open pseudo-convex set of a Banach space with a basis and K a compact subset of U, then there is an integer n_K such that:**

$$\hat{K}_{H(U)} = \bigcap_{n \geqslant n_K} \widehat{u_n(K)}_{H(U \cap E_n)} \oplus E^n.$$

As we shall see further, the intersection is decreasing so it is always possible to neglect a finite number of indices.

Corollary: **Let U be an open pseudo-convex subset of a Banach space with BAP, then for every compact subset K of U, we have $\hat{K}_{H(U)} = \hat{K}_{P_c(U)}$.**

First, we shall prove:

Proposition 2: **Let U be an open pseudo-convex subset of a Banach space with a basis and let K be a compact subset of U, then there is an integer n_1 such that for any $n \geqslant n_1$, $\epsilon > 0$ and f_n in $H(U \cap E_n)$, there exists a g in H(U) such that:**

1) $g\big|_{E_n \cap U} = f_n$

2) $|g - f_n \circ u_n|_K < \epsilon$.

Let n_0 be defined as in lemma 1, $n_1 = n_0 + 1$, we shall apply lemma 2 by taking $2^{-(n+1)}$ instead of ϵ. For a given f_{n_1} in $H(U \cap E_{n_1})$, we shall construct a sequence (f_n) of elements of $H(U \cap E_n)$.

Let K_1 be an arbitrary compact subset of U and let $\epsilon > 0$; for any integer $n \geqslant n_0$ and $p \geqslant 0$, we have, if we write $\varphi_n = f_n \circ u_n$:

$$|\varphi_{n+p} - \varphi_n|_{K_1} \leqslant \sum_{i=0}^{p-1} |\varphi_{n+i+1} - \varphi_{n+i}|_{K_1}$$

$$\leqslant \sum_{i=0}^{p-1} |f_{n+i+1} - f_{n+i} \circ u_{n+i}|_{u_{n+i+1}(K_1)}$$

$$\leqslant \sum_{i=0}^{p-1} |f_{n+i+1} - f_{n+1} \circ u_{n+i}|_{A_{n+1}}$$

$$\leqslant \sum_{i=0}^{p-1} 2^{-(n+i+1)} \leqslant 2^{-n}.$$

The sequence (φ_n) is, then, a Cauchy sequence and converges uniformly on every compact subset of U to a function g which is analytic in U, because for any x in U the function φ_n is analytic in a neighborhood of x, from a certain index.

The proposition follows if we take $n_1 = \sup(n_0, n_0')$, where $2^{-n_0'} \leqslant \epsilon$.

Lemma 3: Let K be a compact subset of an open pseudo-convex subset of a Banach space with a basis, then the sequence of sets $\widehat{u_n(K)}_{H(U)} \oplus E^n$ is decreasing from a certain index.

If n_0 is defined as in lemma 1, we have $\widehat{u_{n+1}(K)} \subset A_{n+1}$ for $n \geqslant n_0$. To prove the lemma, it is sufficient to prove that, for $n \geqslant n_0$, we have:

$$\widehat{u_{n+1}(K)} \subset \widehat{u_n(K)} \oplus C \cdot e_{n+1}.$$

If x_0 is a point of the set $A_{n+1} - \widehat{u_n(K)} \oplus C\, e_{n+1}$, there exists a function f in $H(U \cap E_n)$ such that

$$a = |f \circ u_n(x_0)| - |f|_{u_n(K)} > 0.$$

Then lemma 2 implies that there is a g in $H(U \cap E_{n+1})$ such that

$$|g - f \circ u_n|_{A_{n+1}} < \frac{a}{3}.$$

It follows that

$$|g(x_0)| \geqslant |f \circ u_n(x_0)| - \frac{a}{3} = |f \circ u_n|_{u_{n+1}(K)} + \frac{2a}{3} \geqslant |g|_{u_{n+1}(K)} + \frac{a}{3},$$

which proves that $x_0 \notin \widehat{u_{n+1}(K)}$.

Proof of theorem 1: We only have to show the inclusion $\hat{K}_{H(U)} \subset \cap \widehat{u_n(K)} \oplus E^n$ since we know (see (2)) that the reverse is true for any family of continuous functions in U.

Let $x_0 \notin \cap \widehat{u_n(K)} \oplus E^n$ and n_0 such that $u_{n_0}(K \cup \{x_0\}) \subset A_n$; since the intersection is decreasing, there is an integer n such that $x_0 \notin \widehat{u_n(K)} \oplus E^n$ and $n \geqslant n_1$, where n_1 is the integer arising in proposition 2.

Using proposition as well as the same argument of the preceding lemma, we can show that $x_0 \notin \hat{K}_{H(U)}$.

The corollary follows since in a Banach space with a basis:

$$\hat{K}_{P_c(U)} \supset \cap \widehat{u_n(K)}_{P_c(U \cap E_n)} \oplus E^n = \cap \widehat{u_n(K)}_{H(U \cap E_n)} \oplus E^n = \hat{K}_{H(U)}$$

which shows that $\hat{K}_{P_c(U)} = \hat{K}_{H(U)}$. In order to obtain the result in the case of a Banach space with the BAP, we use the fact that if F is a complemented subspace of E, $E = F \oplus G$, and if K is a compact subset of an open set U in F, we then have: $\hat{K}_{A(U)} = \hat{K}_{A(U \oplus G)}$ for A equal to H or P_c.

We shall now prove an approximation theorem.

Theorem 2: If U is an open pseudo-convex subset of a Banach space with the bounded approximation property and if K is a compact subset of U such that $K = \hat{K}_{P_c(U)}$, then every holomorphic function in a neighborhood of K can be approximated uniformly on K by elements of H(U).

Corollary: If U and U' are two open pseudo-convex subsets of a Banach space with the BAP, $U \subset U'$, then the following conditions are equivalent:

a) H(U') is dense in H(U) for the compact open topology,

b) $\hat{K}_{H(U)} = \hat{K}_{H(U')}$ for every compact subset K of U,

c) $\hat{K}_{P_c(U')}$ is a compact subset of U for every compact subset K of U.

The corollary is an easy consequence of the theorem as b) ⇒ a) and a) ⇒ c) are obvious, and if $\hat{K}_{P_c(U)}$ is compact in U, every $f \in H(U)$ is holomorphic in a neighborhood of $\hat{K}_{P_c(U)}$.

We shall first prove the following lemma of the Arens-Calderon type:

Lemma 4: If E is a Banach space with a basis, U an open pseudo-convex subset of E and K a compact subset of U such that $K = \hat{K}_{H(U)}$, then for every neighborhood V of K there is an integer n_0 such that $\widehat{u_n(K)}_{H(U \cap E_n)} \subset V$, for every $n \geq n_0$.

Let $K_n = \widehat{u_n(K)}_{H(U \cap E_n)} \oplus E^n$, K_c be the closed convex hull of K, and $K'_n = \hat{K}_n \cap K_c$. Then, $u_n(\hat{K}_n) = u_n(K'_n)$: the inclusion $u_n(K'_n) \subset u_n(\hat{K}_n)$ is obvious; in the other direction, as $u_n(K_c)$ is convex, the inclusion $u_n(K) \subset u_n(K_c)$ implies $\widehat{u_n(K)} = u_n(\hat{K}_n) \subset u_n(K_c)$. If x belongs to $u_n(\hat{K}_n)$, there is a y in K_c with $u_n(y) = x$, which proves that x belongs to $u_n(K'_n)$.

Suppose now that the lemma is not true, then there would exist a neighborhood V of K and an infinite sequence (y_n) such that $y_n \in K'_n$ and $u_n(y_n) \notin V$. As the sequence (y_n) is contained in the compact K_c, there is a subsequence, again denoted by (y_n), which converges to a point y_0; the sequence $u_n(y_n)$ converges also to y_0 because

$$\|u_n(y_n) - y_0\| \leq \|u_n(y_n) - u_n(y_0)\| + \|u_n(y_0) - y_0\|$$
$$\leq \|y_n - y_0\| + \|u_n(y_0) - y_0\|.$$

As the sequence (K'_n) is decreasing, it follows that $y_0 \in \cap K'_n = (\cap K_n) \cap K_c = K$; hence, the contradiction since $y_0 \notin V$.

It is now easy to prove theorem 2 if E has a basis: let f be an holomorphic function in a neighborhood V of K and $\epsilon > 0$, there exists an integer n_0 such that $n \geq n_0$ implies $u_n(K) \subset V$ and $|f - f \circ u_n|_K < \epsilon$. Let n_1 be an integer such that $\widehat{u_n(K)}_{H(U \cap E_n)} \subset V$ for every $n \geq n_1$. If n' is an index such that $n' \geq \sup(n_0, n_1)$, we consider the restriction f_n of the function f to $V \cap E_n$. This function is holomorphic in $V \cap E_n$, which is a neighborhood of the compact $\widehat{u_n(K)}_{H(U \cap E_n)}$ so, by the finite dimensional theorem, there exists a function g_n in $H(U \cap E_n)$ such that $|g_n - f_n|_{u_n(K)} \leq \epsilon$.

By proposition 2, there is a function g in H(U) such that $g|_{U \cap E_n} = g_n$ and $|g - g_n \circ u_n|_K \leq \epsilon$. It follows that

$$
\begin{aligned}
|f - g|_K &\leq |f - f_n \circ u_n|_K + |f_n \circ u_n - g_n \circ u_n|_K + |g_n \circ u_n - g|_K \\
&\leq 3\epsilon .
\end{aligned}
$$

Now, if the space E has the bounded approximation property, there is a Banach space F such that $E \oplus F$ has a basis. We consider K as a compact subset of $U \oplus F$ and we have $\hat{K}_{H(U \oplus F)} = \hat{K}_{H(U)}$. The first part of the proof of the theorem shows, then, that there exists a function g in $H(U \oplus F)$ such that $|f - g|_K \leq \epsilon$. We finish the proof by taking the restriction of g to F.

If E is a Banach space with a basis, U an open pseudo-convex subset of E and F a finite dimensional subspace, it is not difficult to see, by choosing the first elements of the basis so that they generate F, that the restriction mapping $H(U) \to H(U \cap F)$ is surjective; it follows that, for every compact subset K of $U \cap F$, we have $\hat{K}_{H(U)} = \hat{K}_{H(U \cap F)}$. This is also true in a Banach space with the BAP.

It is quite natural to ask whether these results are still valid if we take a direct subspace instead of a finite dimensional subspace. We do not know if the surjectivity still holds, but it is possible to prove, as an easy consequence of theorem 2, what follows.

Proposition 2: If U is an open pseudo-convex subset of a Banach space with the BAP and F a complemented subspace of E then the space of the restrictions to F of holomorphic function in U is dense in $H(U \cap F)$ for the compact open topology and so, $\hat{K}_{H(U)} = \hat{K}_{H(U \cap F)}$ for every compact subset of $U \cap F$.

Proof: If f is an holomorphic function in $U \cap F$, K a compact subset of $U \cap F$ and $\epsilon > 0$, the function $f \circ u$, where u is a projection $E \to F$, is holomorphic in a neighborhood of $\hat{K}_{H(U)}$ and then, by theorem 2, there exists a function g holomorphic in U such that $|f - g| < \epsilon$, which proves the proposition.

Let us now give an approximation theorem of continuous plurisubharmonic functions by holomorphic functions, which was stated in the finite dimensional case by Bremerman; but the proof he gives is not correct; we shall here adapt an unpublished proof due to N. Sibony, to the infinite dimensional case.

Proposition 4: If U is an open pseudo-convex subset of a Banach space with the BAP, then for every compact subset K of U, $\epsilon > 0$ and continuous plurisubharmonic function f in U, there exist a finite number of f_i in H(U) and constants $a_i > 0$ such that

$$\left| v - \sup_{i=1,\ldots,p} (a_i \log |f_i|) \right|_K < \epsilon.$$

Let $U_1 = \{(z,w) \in EC,\ z \in U,\ |w| < e^{-v(z)}\}$. If z_0 is a point of U, the function h_0 defined by $h_0(w) = \Sigma\, e^{n\, v(z_0)}\, w^n$ is holomorphic and its domain of existence is $\{w \in \mathbf{C},\ |w| < e^{-v(z_0)}\}$. As the space E has the bounded approximation property, so has E X **C**; the pseudo-convexity of U_1 implies (see (3)) that there exists an holomorphic function f whose domain of existence is U_1 and such that $f(z_0,w) = h_0(w)$. Let $f(z,w) = \Sigma\, \ell_{0,n}(z)\, w^n$; the radius of convergence R is defined by

$$-\log R(z) = (\limsup_{n\to\infty} \frac{1}{n} \log |\ell_{0,n}(z)|)^* \leq v(z)$$

where * means the upper semi-continuous regularization of the function.

The Hartogs lemma, (3), implies the existence of an integer n_0 such that

$$\frac{1}{n} \log |\ell_{0,n}(z)| \leq v(z)+\epsilon$$

uniformly in K for every $n \geq n_0$.

The equality $\ell_{0,n}(z_0) = e^{-n\, v(z_0)}$ implies by continuity that

$$\frac{1}{n_0} \log |\ell_{0,n_0}(z)| \geq v(z)-\epsilon$$

in a neighborhood of z_0. If we take every point of K, then we can extract a finite covering of K with the neighborhoods of the points, which finish the proof of the proposition.

It also follows that any continuous plurisubharmonic function in an open pseudo-convex subset of a Banach space with the BAP is equal to the supremum of the functions $a \cdot \log |f|$, with $a > 0$ and f in H(U), which are smaller than v.

Bibliography

1. L. Gruman and C. O. Kiselman, Le problème de Levi dans les espaces de Banach à base, C. R. Acad. Sc. t 274, 1972, p. 1296.

2. Ph. Noverraz, Pseudo-convexité et convexité polynomiale en dimension infinie, Ann. Inst. Fourier, t 23, 1973.

3. Ph. Noverraz, Pseudo-convexité, convexité polynomiale et domaines d'holomorphie en dimension infinie, Notas de Matematica 3, North-Holland, 1973.

4. A. Pelczynski, Any separable space. . ., Studia Math., t 40, 1971, p. 239.

Universite de Nancy I
Case Officielle 140
54037 Nancy-Cedex
France

SOME RECENT RESULTS AND OPEN PROBLEMS IN INFINITE DIMENSIONAL ANALYTIC GEOMETRY

J. P. Ramis

Taking into account recent works in infinite dimensional analytic geometry (Y. Colombé [1], I. F. Donin [2], A. Douady [3], P. Mazet [8], [9], [10], J. P. Ramis [11], [12] and G. Ruget [12], [14]), one can make out two interesting types of problems:

a) Problems arising from the study of finite dimensional subspaces (or more generally of subspaces with a finite dimensional tangent space) of infinite dimensional analytic spaces.

b) Problems arising from the study of finite codimensional analytic subspaces of an analytic manifold.

In these two cases we will have finiteness hypothesis and we will deduce from these hypothesis some finiteness theorems. (There are interesting problems with subspaces of infinite dimensional spaces without the property of finiteness or cofiniteness; unfortunately I can't say anything about these; there are very pathological examples [4], [11], and one must study any particular case encountered in applications with original methods.)

The problems studied below are chiefly local ones, and the difficulties encountered to solve these problems are often of algebraic nature. However we will have functional analysis troubles with some problems of the type a). And we will need some global machinery in the applications.

As we will see below in some examples, one can find problems of the type a) or b) in studying various questions in classical analytic geometry (i.e. finite dimensional analytic geometry) or in topology; for example, modular spaces and classifying spaces. One will not encounter such situations only in the field of analytic geometry; the study of infinite dimensional manifolds has very interesting applications in differential topology.

We will study the case b) in part I of this paper, and the case a) in part II. In part I we will also give some applications of finite codimension study: as existence of a "verselle"(*) deformation for some germs of analytic spaces, of course of finite dimension (we will only outline the proof and give some conjectures: much work must still be done. . .), or, following G. Ruget an infinite dimensional interpretation of the Chern classes. In part II one will find applications to other modular space problems of results obtained in the case a): modular space of the analytic subspaces of a compact analytic space (following A. Douady) and modular space of the holomorphic fiber bundles over a compact analytic space (following I. F. Donin or R. Narasimhan). At last in part III we will discuss a problem of direct image.

(*) or semi universal

I. FINITE CODIMENSION.

A. Results and conjectures.

Let E be a topological vector space (over the field C). The ring of germs of scalar analytic functions at the origin of E will be denoted by O(E).

The notion of "germ of analytic subset at a point of E" is clear enough. We will say such a germ X is finitely defined if it is possible to find $f_1, \ldots, f_n \in O(E)$ with $X = V(f_1, \ldots, f_n)$. We will say such a germ X is nice when, modulo perhaps an analytic local diffeomorphism, it is possible to find a topological direct decomposition $E = E' \oplus E''$ $(x = (x', x''))$, with $\dim E'' < +\infty$, and germs $f_1, \ldots, f_n$ in O(E), independent of the variable x', with $X = V(f_1, \ldots, f_n)$ (i.e. X is a cylinder with a finite dimensional analytic set as basis). A nice germ of an analytic subset is finitely defined; however there are finitely defined but not nice germs (cf. [11]). We can associate to a germ X of an analytic subset an ideal I(X); on the contrary we can't always associate a germ of an analytic subset to an ideal I of O(E); although this can be done when I is finitely generated or when I is prime with finite height [11].

We have the following results (they are proved in [11] when E is a Banach space, and with slight modifications the proof extends to the case of more general topological spaces and to more general situations: as polynomial rings, formal power series rings...).

Theorem 1. - **The ring O(E) is a unique factorization domain. The proof is based on a "Weierstrass division theorem" (and E must be a locally convex space).**

Theorem 2. - **(Nullstellensatz for prime ideals with finite height).**

The following conditions are equivalent

(i) P is a prime ideal of O(E) and ht P is finite.

(ii) There is an irreducible finitely defined analytic germ X at the origin of E with P = I(X) (and X = V(P)).

The proof uses a local description of irreducible analytic germs.

Theorem 3. - **If $I \subset O(E)$ is finitely generated, there is a finite family of irreducible finitely defined analytic germs $\{X_i\}_{i \in I}$ with**

$$V(I) = \bigcup_{i \in I} X_i.$$

Modulo a usual supplementary assumption, this family is unique. Then the X_i are the irreducible components of V(I).

Theorem 4. - **(Nullstellensatz for finitely generated ideals: P. Mazet [8], [11].)**

If $I \subset O(E)$ is finitely generated, we have

$$I(V(I)) = \text{Rad } I.$$

Theorem 5. - If $P \subset O(E)$ is prime and if ht $P < +\infty$, $O(E)_P$ is a Noetherian regular ring.

We have also the following results (obtained by P. Mazet [9], [10]; Y. Colombe has proved some of them using localizations of categories)(*):

Definition: Let A be a ring and M an A-module. A prime ideal P of A is associate to M ($P \in \text{Ass } M$) if there is $x \in M$, with $P \in \text{Min}(\text{Ann } x)$. A prime ideal P is strongly associate to M if there is $x \in M$, with $P = \text{Ann } x$.

Proposition 6. - For a finitely generated O(E)-module M, and a prime ideal $P \subset O(E)$, with ht $P < +\infty$, the following conditions are equivalent

(i) P is associate to M.

(ii) P is strongly associate to M.

Definition: Let n be an integer. A ring A is n-Noetherian if the ascending chain condition is true for ideals I of A satisfying the following condition:

if $P \in \text{Ass } I$, then ht $P \leqslant n$.

Theorem 7. - For any integer n, O(E) is an n-Noetherian ring.

Theorem 8. - Let I be an ideal of O(E).

(i) For any $n \in N$, there is only a finite number of prime ideals P_i^n satisfying $P_i^n \in \text{Ass } I$ and ht $P_i^n = n$.

(ii) There is a decomposition

$$I = \left(\bigcap_{n \in N} \left(\bigcap_i Q_i^n\right)\right) \cap Q, \text{ with}$$

a) Q_i^n is a primary ideal, with Rad $Q_i^n = P_i^n$.

b) The associate ideals of Q have infinite height.

So we have a "primary decomposition" for ideals of O(E). The following corollary can also be proved using a different method (cf. [11]):

Corollary. - If $I \subset O(E)$, with I = Rad I:

(i) For any $n \in N$ there are only a finite number of prime isolated ideals P_i^n of I with ht $P_i^n = n$.

(ii) There is a decomposition

$$I = \left(\bigcap_{n \in N} \left(\bigcap_i P_i^n\right)\right) \cap P, \text{ with ht } P = +\infty.$$

Using these algebraic results, we obtain geometric results on germs of analytic subsets and then on analytic subsets: we have a codimension theory, and an analytic subset of a Banach manifold has analytic

(*) P. Mazet has also obtained a generalization of Cohen Macauley rings: CM rings, and has proved that O(E) is a CM ring [10].

irreducible components of finite codimension. Will only give the following result:

Theorem 9. - Any finite codimensional analytic irreducible component of an analytic subset of a projective space P(E) (E is a Banach space) is algebraic (i.e. defined by a finite number of scalar continuous polynomials on E).

This generalization of Chow's theorem is proved using a generalization of a theorem of R. Remmert and K. Stein (cf. [11]).

Remark: If every day life we will often find nice analytic subsets. For such subsets the algebraic local study is straightforward. We must however observe that the proof of theorem 9 for nice subsets needs local study of not necessarily nice analytic subsets.

Then some conjectures. (I think most of them are rather difficult to solve!.)

Conjecture 10. - Let P be a prime ideal of O(E), with ht $P < +\infty$. Then P is finitely generated.

Conjecture 11. - Let I be an ideal of O(E), with Ass O(E)/I finite and ht $P < +\infty$, for $P \in$ Ass O(E)/I. Then I is finitely generated.

Conjecture 12. - Let I be a finitely generated ideal of O(E). Then Ass O(E)/I is finite and, for $P \in$ Ass O(E)/I, ht $P < +\infty$.

Conjecture 13. - Let $I \subset O(E)$ be a finitely generated ideal and E' be a closed hyperplane of E. Then $I' = O(E') \cap I$ is a finitely generated ideal of O(E').

Conjecture 14. - Let $E = E' \oplus E''$ be a topological direct decomposition, with dim $E'' = 1$. Then $O(E') \to O(E)$ is flat.

Conjecture 15. - The ring O(E) is coherent (i.e., if $f_1, \ldots, f_n \in O(E)$, the module of relations between the f_1 is finitely generated).

Remark: Conjecture 15 ⇒ Conjectures 10, 11, 13, 14.

Conjecture 16. - Let I be a finitely generated ideal of O(E). Then we have a finite syzygie

$$0 \to O(E)^{m_k} \to \ldots \to O(E)^{m_0} \to O(E)/I \to 0.$$

Conjecture 17. - Let M be a finitely generated O(E)-module, with Ass M finite, and ht $P < +\infty$ for any $P \in$ Ass M. Then we have a finite syzygie

$$0 \to O(E)^{m_k} \to \ldots \to O(E)^{m_0} \to M \to 0.$$

Remark: Modulo the conjectures 12 and 15, conjectures 16 and 17 are equivalent.

Let P and Q be two prime ideals of O(E), with ht P and ht Q finite. Let R be the ideal of an irreducible component of $V(P) \cap V(Q)$, with ht R = ht P + ht Q. We denote by $A = O(E)_R$ the local ring of O(E) at R. The following formula gives the multiplicity of intersection of P and Q along R:

$$\chi^A(A/P', A/Q') = \sum_{i=0,\ldots,\mathrm{ht}\,R} (-1)^i \, \ell_A(\mathrm{Tor}_i^A(A/P', A/Q')). \quad \text{(cf. [17])}.$$

In this formula P' and Q' are the images of P and Q in A and $\ell_A(M)$ is the length of the A-module M. (We can use this formula because A is a Noetherian local ring.)

Conjecture 18. - This multiplicity and the multiplicity of [14] (defined geometrically by G. Ruget) are equal.

Remark: This result is true if $\dim E < +\infty$. However the proofs in the literature are not perfectly clear, and before any attempt to prove the conjecture above a thorough study of the finite dimensional case would be necessary.

Conjecture 18 bis. - Let I a finitely generated ideal of O(E). Then I is closed (in a reasonable sense).

Taking into account some results of G. Fischer [5], a positive answer to this conjecture would be very interesting for applications.

B. Deformations.

Definition: Let E be a DFN space(*), C{E} the ring of germs of scalar analytic functions at the origin of E (resp. C[[E]] the ring of formal powers series with "variable" in E and "values" in C), and I a finitely generated ideal of C{E} (resp. C[[E]]). We will say that A = C{E}/I is a **finitely defined DFN-analytic algebra** (resp. A = C[[E]]/I a finitely defined DFN-formal algebra).

A series $f \in F\{E\}$ (resp. F[[E]]) gives a morphism of analytic algebra (resp. formal algebra) $C\{F\} \to C\{E\}$ (resp. $C[[F]] \to C[[E]]$). Then the notion of morphism of finitely defined analytic (or formal) algebra is straightforward.

During 1972, there were important works on the problem of deformations of an analytic singularity [6], [18], [19]. This problem is thoroughly solved for a reduced singularity smooth outside the origin (isolated singularity): the proofs use the formal result of [16] and an analytization of this result(**); one obtains a "verselle" deformation for such singularities (i.e. a classifying deformation of the given singularity). But the problem remains open for more general singularities; in the general case we cannot hope to have a "verselle" deformation with a finite dimensional basis, even though I believe the problem very interesting when the expected basis is not too pathological. (In such cases we will associate integers with some geometric meaning to the given singularity: as minimal codimension for a smooth imbedding of the basis. . .). The study of deformations of rather general singularities can also give useful tools for the study of the deformations of

(*) DFN = dual of (nuclear Frechet space) = nuclear (dual of Frechet space).
(**) Another proof uses the method of Douady [3] : cf. [20].

isolated singularities: following some ideas of G. Ruget and using the verselle deformation of a complete intersection, one can prove that the "versality" is open for an isolated singularity(*).

When X is a germ of reduced isolated singularity, the Zariski tangent space of the basis of the "verselle" deformation of X is given by the fibre T^1 at the origin of the coherent sheave $\mathrm{Ext}^1_{0_X}(\Omega_X, 0_X)$ [15]; this fiber has finite dimension (the support of the sheave is the origin). In the general case, T^1 no longer has a finite dimension, however it has a natural structure of DFN-space [12]. We have another interesting space T^2 (cf. [15]); when X is a reduced singularity and I/I^2 the conormal sheave of a smooth inbedding of X, we have

$$T^2 = \text{fibre of } \mathrm{Ext}^1_{0_X}(I/I^2, 0_X)$$

at the origin. The space T^2 also has a natural structure of DFN-space.

If $T^2 = 0$ (this happens in the case of complete intersection and in other situations), we have the

Conjecture 19. - Let X be a germ of analytic singularity, with $T^2 = 0$. Then there is a finitely defined DFN-algebra A, and a "flat" (in a sense perhaps weaker than the classical one) morphism of analytic algebra $B = \mathbb{C}\{T^1\} \to A$, such that the deformation $B \to A$ is "verselle" for the family of flat deformations with a basis of finite dimension.

If $\dim T^2 < +\infty$, we have the

Conjecture 20. - Let X be a germ of analytic singularity, with $\dim T^2 < +\infty$. Then there are two finitely defined DFN-algebras A and B ($B = \mathbb{C}\{T^1\}/J$, with J finitely generated), and a "flat" morphism of analytic algebra $B \to A$, such that the deformation $B \to A$ is "verselle" for the family of flat deformations with a basis of finite dimension.

We can hope the situation of conjecture 20 will be generic in a very strong sense: the subset $\{\dim T^2 = +\infty\}$ will be of "infinite codimension" in the "space" (sic) of all the germs of singularities.

If $\dim T^2 = +\infty$, I think the situation can be very pathological.

The proofs of 19 and 20 will follow the paths of [19]: first, as in [16], we will have a formal result (i.e. A and B finitely defined formal DFN-algebras, and $B \to A$ "verselle" for formal deformations), then we will analyticize the result (using a generalization of the theorem of M. Artin [0]) (One can also use the methods of [20]). However to complete this program a lot of difficulties must be overcome.(**)

C. An interpretation of Chern classes.

Let X be an analytic manifold modelled on a Banach space, and Y an analytic subset of finite co-dimension p in X. Following G. Ruget (cf. [14]), we can associate to Y a cohomology class in $H^{2p}(X;\mathbb{Z})$. We know that the space Ω of Fredholm operators of a separable Hilbert space (of infinite dimension) is a

(*) Cf. also [20] for another proof.

(**) One can also try to use the methods of [20].

classifying space for $GU(\infty)$ (Ω is a Banach space). In [14] G. Ruget builds some nice analytic subsets of Ω; he denotes by γ_p the associated classes in $H^{2p}(\Omega;Z)$, and gives the following interpretation of the Chern classes:

Let U be a connected paracompact topological space; we have a map $K(U) \to [U,\Omega]$ and for α fixed in $K(U)$, a map $H^{2p}(\Omega;Z) \to H^{2p}(U;Z)$: the image of γ_p is the p-th Chern class of α.

II. FINITE DIMENSION.

A. The first finiteness criterion (following A. Douady).

Let X, X' be Banach analytic spaces, h: X → X' a morphism (h(a) = a'). If we can find models U, U', f, for small neighborhoods of a, a', and the corresponding restriction of h, with Tf compact at the image of a, we will say the morphism h is compact at a.

Proposition 21. - (A. Douady) **Let X be a Banach analytic space, and $a \in X$. If Id_X is compact at a, $\dim_a X < +\infty$.**

A Douady encounters such a situation in [3], solving the "modular space problem" for the analytic subspaces of a compact analytic space. He built a Banach analytic space as a solution, then he proved that this space is of finite dimension using the criterion above.

B. The second finiteness criterion (following G. Ruget).

Definition: Let X be an analytic Banach manifold and let A be a closed subset of X. If we can find locally an analytic submanifold Y of X, with $A \subset Y$ and $\dim Y < +\infty$, we will say that A is a **finite dimensional analytic subset of X.**

Theorem 22. - (G. Ruget) **Let X be a Hilbert analytic manifold. Let Y be a finite dimensional analytic space. Let f: Y → X be an analytic proper map. Then f(Y) is a finite dimensional analytic subset of X.**

This theorem is proved in [14]. If X is only a Banach analytic manifold, the theorem remains true if f has finite fibers; for f proper, the problem is open (and more or less related to approximation questions). The problem is also open for X an open subset of a topological vector space E: we will encounter such a situation in every day life with E = space of holomorphic functions, space of currents,. . .

C. The third finiteness criterion (following I. F. Donin).

One can find the definitions below and a sketch of proof for the following theorems in [2]. (Independantly R. Narasimhan has communicated similar ideas to me.)

Definition: Let X be a analytic Banach space, let a be a point of X, let H be a finite dimensional subspace of $T_a(X)$ (Zariski tangent space to X at a). A finite dimensional subset for X in a neighbourhood of a is called maximal relative to H if its tangent space is H and if it is not contained in any finite dimensional analytic subset of X having this property.

Definition: An analytic Banach space is said to be complete at one of its points a if in a neighbourhood of a there exists a model (U,f,F) of this space with df of closed range at the point corresponding to a.

Theorem 23. - (I.F. Donin) **Let X be an analytic Banach space complete at the point $a \in X$, H a finite dimensional subspace of the tangent space $T_a(X)$. Then in some neighborhood of a there exists a maximal analytic subset relative to H.**

Theorem 24. - (I.F. Donin) **Let X be an analytic Banach space with finite dimensional tangent space $T_a(X)$ at $a \in X$, and complete at a. Then in a neighborhood of the point a there exists a unique finite dimensional analytic subset A containing all other finite dimensional analytic subsets in X.**

Remark: We can have $T_a(X) = (0)$ and a not isolated in X. Then $A = \{a\}$. (Cf. an example in [11].)

I.F. Donin uses these results to prove the following modular space theorem:

Theorem 25. - **Let X be a compact complex space, G a complex Lie group, P a holomorphic fibering over X with structure group G. Then P has a complete finite dimensional family of deformations with tangent space of dimension dim $H^1(X;\mathrm{Ad}\,P)$.**

D. A conjecture.

Conjecture 25 bis. - (P. Mazet) Let J be a prime ideal of O(E), with finite coheight, then there is an irreducible finite dimensional germ of analytic subset X such that J = I(X).

III. DIRECT IMAGES.

Definition: Let X, Y be analytic Banach manifolds. An analytic map f: $X \to Y$ is said to be **Fredholm** if for any $x \in X$, the tangent map df_x is Fredholm.

Theorem 26. - (J.P. Ramis and G. Ruget) **Let X, Y be analytic Banach manifolds, Z a finitely defined analytic subset of X, f: $X \to Y$ an analytic Fredholm map, with $f|_Z$ proper. Then f(Z) is a finitely defined analytic subset of Y.**

This theorem is proved in [14] (cf. also [12]).

Theorem 27. - (C. Houzel [7]) **Let X be an analytic family of finite dimensional analytic spaces parametrised by an analytic Banach manifold S (for a more precise definition, see [7]). Let F be a "coherent" sheave on X. If we denote by f the natural map $X \to S$, then the sheaves $R^k f_* F$ are coherent.**

Modulo some conjectures given in I, theorem 26 is a particular case of theorem 27.

Bibliography

0. M. Artin, **On the solutions of Analytic Equations**, Inventiones Math. 5(1968), 277–292.

1. Y. Colombé, Secret papers.

2. I.F. Donin, **On analytic Banach spaces and on the space of modules of holomorphic fiberings**, Dokl. Akad. Nauk SSSR (Soviet. Math. Dokl. Vol. 11(1970), no. 6.

3. A. Douady, Thèse, Ann. Inst. Fourier, Grenoble, 16, 1, (1966), 1–98.

4. A. Douady, **A remark on Banach analytic spaces**, Stanford University.

5. G. Fischer, Book to appear, Ergebnisse (Springer-Verlag).

6. H. Grauert, **Uber die Deformation isolierte Singularitäten analytischer Mengen.**

7. G. Houzel, Thèse, Université de Grenoble, to appear.

8. P. Mazet, **Le Nullstellensatz pour un germe analytique banachique**, C.R. Acad. Sci., Paris 269, Sér. A, 235–237.

9. P. Mazet, **Propriétés noethériennes relatives** (1971), Université de Tunis.

10. P. Mazet, **Généralisation des notions d'anneau noethérien et d'anneau de Cohen-Macauley.** (Cf. also the Mazet's lecture for the "Colloque International du C.N.R.S.: Analyse Complexe" (Lecture Notes, Springer Verlag).)

11. J. P. Ramis, **Sous-ensembles analytiques d'une variété banachique complexe**, Ergebnisse Bd. 53 (Springer-Verlag).

12. J. P. Ramis and G. Ruget, **Un théorème sur les images directes analytiques banachiques**, C.R. Acad. Sc. Paris 267, Sér. A, 995–996.

13. J. P. Ramis and G. Ruget, **Complexe dualisant et théorèmes de dualité en géométrie analytique**, Publ. Math. I.H.E.S. 38(1971), 77–91.

14. G. Ruget, **A propos des cycles analytiques de dimension infinie**, Inventiones Math. 8(1969), 267–312.

15. G. Ruget, **Deformations des germes d'espace analytique I, II**, Séminaire de l'Ecole Normale Supérieure (Paris), 1972.

16. M. Schlessinger, **Functors of Artin Rings.**

17. J. P. Serre, **Algèbre locale et multiplicités**, Lecture Notes (Springer-Verlag).

18. Tiourina, **Locally semi-universal flat deformations of isolated singularities of complex spaces**, Math. of the USSR Izvestija, vol. 3; no. 5, 967–999.

19. J. L. Verdier, **Séminaire de l'Ecole Normale Supérieure**, (Paris), 1972.

20. G. Pourcin, **Déformation de singularitiés isolées**, C.R. Acad. Sc. Paris 1276(1973), 1217–1219.

Departement de Mathématiques
Université Louis Pasteur
67–Strasbourg
France

RIEMANN DOMAINS: BASIC RESULTS AND OPEN PROBLEMS

Martin Schottenloher

Introduction.

This paper is divided into two parts. The first part, sections 1–3, is an elementary and nearly self-contained introduction to the notion of a Riemann domain spread over a locally convex complex Hausdorff space. In section 1 we first consider Riemann domains spread over arbitrary Hausdorff spaces and prove the intersection theorem. In section 2 we discuss the topological and geometric structure of a domain spread over a locally convex space. Finally, in section 3 the notion of analyticity on such domains is introduced as well as the fundamental concepts of a domain of holomorphy, an envelope of holomorphy and a domain of existence.

The second part of this paper, consisting only of section 4, deals with a number of open problems concerning analytic continuation and domains of holomorphy which arise immediately after having established some basic results, as for example those in section 3. Besides formulating the problems, we discuss to which extent answers are already known at the present stage and where to find these partial answers.

§ 1. Domains and Intersection

Definition 1.1. Let E be a Hausdorff space. A pair (X,p) is called a **manifold (spread) over** E if $X \neq \phi$ is a Hausdorff space and if $p: X \to E$ is a local homeomorphism. p is called the projection map. (Ω,p) is said to be a **domain (or a Riemann domain) spread over** E if (Ω,p) is a manifold over E and if Ω is connected. (The Greek letters Ω, Ω', Σ will be reserved for domains.)

For short we often write X instead of (X,p) although the map p is the essential part of the manifold (X,p) over E. In fact, a Hausdorff space X which admits a local homeomorphism $p: X \to E$ can be the underlying space for many different manifolds (X,q) over E; e.g. $q = \varphi \circ p$ where $\varphi: E \to E$ is a local homeomorphism.

A standard example of a domain over $E = \mathbf{C}$ is the Riemann domain of the function log. A trivial example of a manifold over E is (E, id_E). A manifold (X,p) is said to be **schlicht** if p is injective. The class of the schlicht manifolds over E can be identified with the class of the open sets of E with respect to the following notion of isomorphy.

Definition 1.2. A map $j: X \to Y$ between manifolds (X,p), (Y,q) over E is called a **morphism (or a weak inclusion)** if is continuous and satisfies $p = q \circ j$. Notation: $j: (X,p) \to (Y,q)$.

Theorem 1.3. Let $(j_k)_{k\in I}$ be a family of morphisms of domains $j_k\colon (\Omega,p) \to (\Omega_k,p_k)$ over a locally connected Hausdorff space E. Then there exists a morphism $\underline{j}\colon (\Omega,p) \to (\underline{\Omega},\underline{p})$ of domains such that

1^0 $j_k = \underline{j}_k \circ \underline{j}$ for a suitable family $(\underline{j}_k)_{k\in I}$ of morphisms $\underline{j}_k\colon (\underline{\Omega},\underline{p}) \to (\Omega_k,p_k)$.

2^0 $\underline{j}$ is maximal in the following sense:

If $j'\colon (\Omega,p) \to (\Omega',p')$ is a morphism of domains with $j_k = j_k' \circ j'$ for a suitable family $(j_k')_{k\in I}$ of morphisms then there is a unique morphism $\underline{j}'\colon (\Omega',p') \to (\underline{\Omega},\underline{p})$ such that $\underline{j} = \underline{j}' \circ j'$ and $j_k' = \underline{j}_k \circ \underline{j}'$. Hence the following diagram is commutative:

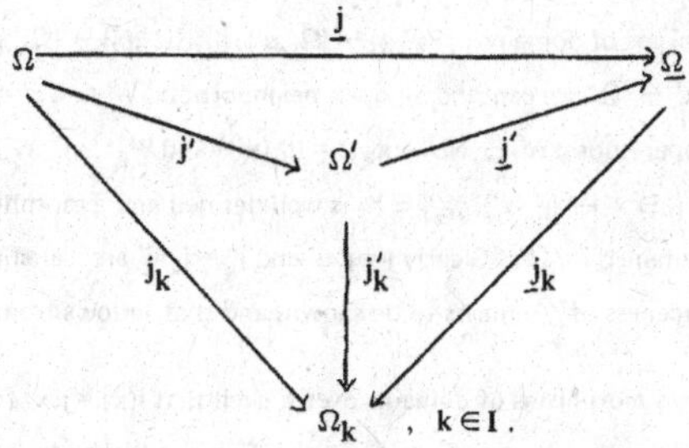

Remarks. 1^0 $(\underline{\Omega},\underline{p})$ is unique up to isomorphisms of domains (cf. 1.6).

2^0 $(\underline{\Omega},\underline{p})$ is the maximal domain over E such that all j_k, $k \in I$, can be extended similtaneously to $(\underline{\Omega},\underline{p})$ as morphisms.

3^0 For example, let $\Omega \subset E$ be a domain and $(\Omega_k)_{k\in I}$ be a family of domains in E containing Ω. Then $\underline{\Omega}$ is that component of $\widehat{\bigcap_{k\in I} \Omega_k}$ which contains Ω.

Definition 1.4. $\underline{j}\colon (\Omega,p) \to (\underline{\Omega},\underline{p})$ as in the theorem (resp: $\underline{j}$, or $(\underline{\Omega},\underline{p})$, or $\underline{\Omega}$) is called the **intersection** of the $j_k\colon(\Omega,p) \to (\Omega_k,p_k)$ (resp. of the j_k, or (Ω_k,p_k), or Ω_k).

1.5. The intersection is nothing else than the product of objects in a suitable category; namely the category $\Omega/D(E)$ of domains under Ω: The objects of $\Omega/D(E)$ are the morphisms $j\colon (\Omega,p) \to (\Sigma,q)$ over E. The morphisms from $j\colon (\Omega,p) \to (\Sigma,q)$ to $j'\colon (\Omega,p) \to (\Sigma',q')$ are the morphisms $\varphi\colon (\Sigma,q) \to (\Sigma',q')$ which satisfy $j' = \varphi \circ j$. Theorem 1.3 now asserts that $\Omega/D(E)$ has arbitrary products and it follows that $\Omega/D(E)$ has arbitrary projective limits.

Proof of Theorem 1.3. Let $X \subset \prod_{k\in I} \Omega_k$ be the set of all $(x_k)_{k\in I}$ such that

i) $p_k(x_k) = p_\ell(x_\ell)$ for all $k, \ell \in I$.

ii) There is a family $(W_k)_{k\in I}$ of open sets $W_k \subset \Omega_k$ such that $x_k \in W_k$, $p_k|W_k\colon W_k \to p_k(W_k)$ is a homeomorphism and $p_k(W_k) = p_\ell(W_\ell)$ for all $k,\ell \in I$.

A topology on X is given by defining the neighborhoods: $W \subset X$ is said to be a neighborhood of $(x_k)_{k\in I}$ if there exists a family $(W_k)_{k\in I}$ satisfying ii) such that $X \cap \prod_{k\in I} W_k \subset W$. The topology defined in this manner is finer than the induced product topology. Thus, X is a Hausdorff space. The map $\tilde{p}: X \ni (x_k)_{k\in I} \mapsto p_\ell(x_\ell) \in E$, $\ell \in I$, is well-defined according to i) and locally a homeomorphism according to ii). Hence, $(X,\tilde{p})$ is a manifold over E and $\underline{j}: \Omega \ni x \mapsto (j_k(x))_{k\in I} \in X$ is a morphism, in particular continuous. Now let $\underline{\Omega}$ be that component of X which contains $\underline{j}(\Omega)$ and let $\underline{p} := \tilde{p}|\underline{\Omega}$. Then $\underline{j}: (\Omega,p) \to (\underline{\Omega},\underline{p})$ has the required properties:

1° Define $\underline{j}_\ell$, $\ell \in I$, by $\underline{j}_\ell: \underline{\Omega} \ni (x_k)_{k\in I} \mapsto j_\ell(x_\ell) \in \Omega_\ell$. Then $\underline{j}_\ell$ is a morphism and satisfies $j_\ell = \underline{j}_\ell \circ \underline{j}$ for all $\ell \in I$.

2° Assume there are morphisms of domains $j': (\Omega,p) \to (\Omega',p')$, $j'_k: (\Omega',p') \to (\Omega_k,p_k)$ so that $j_k = j'_k \circ j'$ for all $k \in I$. For each $x' \in \Omega'$ we can find an open neighborhood $W' \subset \Omega'$ of x' which is mapped homeomorphically by p onto an open subset of E. Now $(x_k) := (j'_k(x'))$ and $W_k := j'_k(W') \subset \Omega_k$ satisfy i) and ii). Therefore, the map $\underline{j}': \Omega' \ni x' \mapsto (j'_k(x'))_{k\in I} \in X$ is well-defined and a morphism. It follows that $\underline{j}'(\Omega')$ is connected and, hence, contained in $\underline{\Omega}$. Clearly $\underline{j} = \underline{j}' \circ j'$ and $j'_k = \underline{j}_k \circ \underline{j}'$ are satisfied. To complete the proof of the theorem the uniqueness of $\underline{j}'$ remains to be shown, and that follows from the next lemma.

Lemma 1.6. Let $i,j: \Omega \to \Sigma$ be two morphisms of domains over E such that $i(x) = j(x)$ for at least one point $x \in \Omega$. Then $i = j$. In particular, every morphism $j: \Omega \to \Omega$ with a fixpoint is the identity.

Proof. The set $Z := \{x \in \Omega \mid i(x) = j(x)\}$ is closed since i,j are continuous maps into a Hausdorff space. Z is open because i,j are morphisms. Thus, $Z = \Omega$.

Similarly one can prove the following useful result

Lemma 1.7. Let (Ω,p) be a domain over E and let $U,V \subset \Omega$ be open sets such that $U \cap V \neq \phi$, $p(U) \cap p(V)$ is connected and $p|U: U \to p(U)$, $p|V: V \to p(V)$ are homeomorphisms. Then $p|U \cup V: U \cup V \to p(U \cup V)$ is a homeomorphism.

Proof. It is enough to show that $p|U \cup V$ is injective. Denote $q_U := (p|U)^{-1}|p(U) \cap p(V): p(U) \cap p(V) \to U$ and analogously q_V. By the continuity of q_U, q_V the set $Z := \{z \in p(U) \cap p(V) \mid q_U(z) = q_V(z)\}$ is closed. Z is non-void and open since $Z = p(U \cap V)$ and because of the fact that p is an open map. Now, for $x \in U$, $y \in V$ with $p(x) = p(y) =: z$ it follows that $x = q_U(z) = q_V(z) = y$.

Final remark: The notion of a domain spread over C^n and the intersection theorem for domains over C^n can be found in [7].

§2. Riemann Domains Spread Over Locally Convex Spaces

For the remainder of this article, let E be a locally convex Hausdorff space (lcs) over the field of the complex numbers **C**. A Riemann domain Ω spread over E has some of the topological properties of E; e.g. Ω

is pathwise connected and locally pathwise connected. Moreover, Ω carries in a natural way a part of the linear and metric structure of E; e.g. in Ω there are real and complex lines, and one can define a boundary distance of a point $x \in \Omega$ with respect to a continuous seminorm on E. The object of this section is to compile these elementary properties of a Riemann domain over a lcs E.

It is easy to see that E is pathwise connected (i.e. for each pair (x_0, x_1) of points of E there is a continuous map $\varphi : [0,1] \to E$ so that $\varphi(0) = x_0$ and $\varphi(1) = 1$) and locally pathwise connected (i.e. each point of E has a neighborhood base of pathwise connected sets). It follows that every manifold over E is locally pathwise connected and, moreover, that every domain over E is pathwise connected. From this we deduce the following

Lemma 2.1. An open set Z in a domain Ω over E is closed if and only if it is sequentially closed.

Proof. We can assume $Z \neq \phi$. Let $y \in \Omega$. Then there is a continuous map $\varphi : [0,1] \to \Omega$ so that $\varphi(0) \in Z$ and $\varphi(1) = y$. $\varphi^{-1}(Z) \neq \phi$ is open in $[0,1]$. $\varphi^{-1}(Z)$ is also closed because Z is sequentially closed. Hence, $y \in Z$ which implies $Z = \Omega$.

Let cs(E) denote the family of all continuous seminorms $\alpha: E \to [0,\infty[$, $\alpha \neq 0$, on the locally convex space E. For $\alpha \in cs(E)$, $z \in E$, $r > 0$ the "α–ball" $B^{\alpha}_{E}(z,r)$ is defined by $B^{\alpha}_{E}(z,r) := \{y \in E \mid \alpha(y-z) < r\}$.

Definition 2.2. Let (X,p) be a manifold over E and let $\alpha \in cs(E)$. The **α-boundary distance** at a point $y \in X$ is defined by

$d^{\alpha}_{X}(y) := \sup\{r > 0 \mid$ there is an open neighborhood U of y so that $p|U: U \to B^{\alpha}_{E}(py,r)$ is a homeomorphism$\} \cup \{0\}$. For $r \in \,]0, d^{\alpha}_{X}(y)]$ let $B^{\alpha}_{X}(y,r)$ denote that connected component of $p^{-1}(B^{\alpha}_{E}(py,r))$ which contains the point y (cf. [34], [39]).

Remark. With the notation of 2.2 the construction on the proof of 1.3 can now be described in the following way: $X = \{(x_k) \in \prod_{k \in I} \Omega_k \mid p_k(x_k) = p_\ell(x_\ell)$ for all $\ell, k \in I$ and there is $\alpha \in cs(E)$ such that $\inf\{d^{\alpha}_{\Omega_k}(x_k) \mid k \in I\} > 0\}$. $W \subset X$ is a neighborhood of $(x_k) \in X$ iff there are $\alpha \in cs(E)$ and $r > 0$, $r < \inf\{d^{\alpha}_{\Omega_k}(x_k) \mid k \in I\}$, such that $X \cap \prod_{k \in I} B^{\alpha}_{\Omega_k}(x_k, r) \subset W$.

Proposition 2.3. Let (X,p) be a manifold over E.

1^0 For each $x \in X$ there is $\alpha \in cs(E)$ so that $d^{\alpha}_{X}(x) > 0$.

2^0 $d^{\alpha}_{X}: X \to [0,\infty]$ is continuous for every $\alpha \in cs(E)$: For all $y \in B^{\alpha}_{X}(x, d^{\alpha}_{X}(x))$ one has $|d^{\alpha}_{X}(x) - d^{\alpha}_{X}(y)| \leq \alpha(px-py)$.

3^0 Each component of $\{x \in X \mid d^{\alpha}_{X}(x) = \infty\}$ is isomorphic to E.

Proof. 1^0 and 3^0 are obvious.

2^0 Let $d_X^\alpha(x) > 0$ and $y \in B_X^\alpha(x, d_X^\alpha(x))$. Then $B_E^\alpha(py, d_X^\alpha(x) - \alpha(px-py)) \subset B_E^\alpha(px, d_X^\alpha(x))$. Hence, $d_X^\alpha(y) \geqslant d_X^\alpha(x) - \alpha(px-py)$. Similarly, $d_X^\alpha(x) \geqslant d_X^\alpha(y) - \alpha(px-py)$, from which follows that $|d_X^\alpha(x) - d_X^\alpha(y)| \leqslant \alpha(px-py)$, and that d_X^α is continuous at x. Now let $d_X^\alpha(x) = 0$ and $\epsilon > 0$. According to 1^0 there is $\beta \in cs(E)$ with $\alpha \leqslant \beta$ and $d_X^\beta(x) = \frac{1}{2}\epsilon$. Thus we can find a neighborhood U of x such that $|d_X^\beta(x) - d_X^\beta(y)| < \frac{1}{2}\epsilon$ for all $y \in U$. It follows that $d_X^\alpha(y) \leqslant d_X^\beta(y) < \epsilon$ for all $y \in U$. This completes the proof.

2.4. Given $a, z \in E$ and $r > 0$ put $D_a(z,r) := \{z + \lambda a \mid \lambda \in C, |\lambda| < r\}$. For a point x in a manifold (X,p) over E, $d_X(x,a)$ is defined by $d_X(x,a) := \sup\{r > 0 \mid$ there is a connected set $D \subset X$, $x \in D$, such that $p|D : D \to D_a(px,r)$ is a homeomorphism$\}$. $d_X : X \times E \to]0,\infty]$ is lower semicontinuous but in general not continuous. The connection with d_X^α is given by:

Proposition 2.5. Let X be a manifold over E, $\alpha \in cs(E)$ and $x \in X$ with $d_X^\alpha(x) > 0$. Then $d_X^\alpha(x) = \inf\{d_X(x,a) \mid \alpha(a) = 1, a \in E\}$.

The proof of 2.5, which is similar to the normed case [8, p.394], can be found in [39]. Note that the above formula is in general not true if $d_X^\alpha(x) = 0$.

To conclude this section we introduce the notion of a boundary sequence. Let us recall that for a domain Ω in C^n the following holds:

2.6. Ω is holomorphically convex if and only if for each sequence (x_n) in Ω without accumulation point there is a holomorphic function f on Ω satisfying $\sup|f(x_n)| = \infty$.

For domains in infinite dimensional spaces E this property is too restrictive, since E is not locally compact and there are bounded sequences without an accumulation point. In particular, $\Omega = E = \ell_\infty$ would not be holomorphically convex in this sense. So, if one wants to work with a convexity property similar to 2.6 it is reasonable to require that the sequences in 2.6 satisfy additional conditions. A possible condition is that they are boundary sequences in the following sense (see definition 4.2):

Definition 2.7. A sequence (x_n) in the domain (Ω,p) over E is called a **boundary sequence (relative to p)**, if (x_n) has no accumulation point (in Ω), (px_n) converges to a point $a \in E$ and if for every neighborhood V of a there exists $N \in \mathbf{N}$ such that the remainder $\{x_n \mid n \geqslant N\}$ is contained in a connected component of $p^{-1}(V)$. (Compare with the notion of a "Randpunkt" in [3, p.13]).

Remarks. 1^0 A boundary sequence satisfies $d_\Omega^\alpha(x_n) \to 0$ for all $\alpha \in cs(E)$.

2^0 If $j : (\Omega,p) \to (\Sigma,q)$ is a morphism of domains and if (x_n) is a boundary sequence (relative to p), then either $(j(x_n))$ is a boundary sequence (relative to q) or there is $\alpha \in cs(E)$ with $d_\Sigma^\alpha(j(x_n)) \not\to 0$. In the case of $d_\Sigma^\alpha(j(x_n)) \not\to 0$ it follows that $(j(x_n))$ is convergent in Σ.

3° It can happen that there is no boundary sequence in the α-ball $B^{\alpha}_{\Omega}(x, d^{\alpha}_{\Omega}(x))$ [39]. But nevertheless in every domain there exist a lot of boundary sequences.

Proposition 2.8. **Let (Ω, p) be a domain spread over E. Given $x \in \Omega$, $\alpha \in cs(E)$ and $\epsilon > 0$ there is a boundary sequence (x_n) in Ω satisfying $\alpha(px_n - px) < d^{\alpha}_{\Omega}(x) + \epsilon$.**

Proof. For $s \in]0, d_{\Omega}(x,a)]$ let $D_a(x,s)$ denote that connected component of $p^{-1}(D_a(px,s))$ which contains x. There is a vector $a \in E$ such that $\alpha(a) < d^{\alpha}_{\Omega}(x) + \epsilon$ and $d_{\Omega}(x,a) \leq 1$. Otherwise, by virtue of Lemma 1.7 it can be shown that $Z := \cup\{D_a(x,1) \mid a \in E,\ \alpha(a) < d^{\alpha}_{\Omega}(x)+\epsilon\}$ is open and that

$$p|Z: Z \to B^{\alpha}_{E}(px, d^{\alpha}_{\Omega}(x)+\epsilon)$$

is a homeomorphism; a contradiction to the definition of $d^{\alpha}_{\Omega}(x)$. By a compactness argument we see that there is $\theta \in [0, 2\pi]$ such that for each sequence (t_n) in $]0, d_{\Omega}(x,a)[$, $t_n \to d_{\Omega}(x,a) \leq 1$, the sequence $x_n := (p|D_a(x, d_{\Omega}(x,a)))^{-1}(px + t_n e^{i\theta} a)$ is a boundary sequence in Ω with $\alpha(px_n - px) = \alpha(t_n e^{i\theta} a) \leq \alpha(a) < d^{\alpha}_{\Omega}(x) + \epsilon$.

§3. Envelope of Holomorphy and Simultaneous Analytic Continuation

In this section we prove in different ways the uniqueness and the existence of the envelope of holomorphy of a given domain Ω over a lcs E. First, let us introduce the notion of an analytic map on Ω. We assume that the reader is acquainted with the concept of an analytic map on an open set of E (e.g. [31]).

Definition 3.1. Let (X,p) be a manifold spread over the lcs E and let F be another lcs. A map $f: X \to F$ is called **analytic (weakly analytic, G-analytic, LF-analytic)** if for all $x \in X$ there are $\alpha \in cs(E)$ and $r > 0$, $r < d^{\alpha}_{X}(x)$, such that with $B := B^{\alpha}_{X}(x,r)$ the map

$$f \circ (p|B)^{-1} : B^{\alpha}_{E}(px, r) \to F$$

is analytic (weakly analytic, G-analytic, LF-analytic [35]). A map $f: X \to Y$ into a manifold (Y,q) over F is called **analytic** if it is continuous and if $q \circ f$ is analytic. $\mathcal{H}(X,Y)$ is the set of analytic mappings from X to Y; $\mathcal{H}(X) := \mathcal{H}(X, \mathbb{C})$.

3.2. It follows from elementary properties of analytic maps on open sets of E that a map $f: X \to F$ is analytic if and only if for all $x \in X$ there exists a sequence $(\hat{d}^n f(x))_{n \geq 0}$ of continuous homogeneous polynomials $\hat{d}^n f(x) : E \to F$ of degree n such that for each $\beta \in cs(F)$ there are $\alpha \in cs(E)$ and $r > 0$ satisfying $r < d^{\alpha}_{X}(x)$ and

$$\lim_{N \to \infty} \sup_{y \in B} \beta\Big(f(y) - \sum_{n=0}^{N} \frac{1}{n!} \hat{d}^n f(x)(py - px)\Big) = 0,$$

where $B := B^{\alpha}_{X}(x,r)$ (cf. [31], [34]).

Definition 3.3. Let A be a set of analytic mappings defined on the domain (Ω,p) over E.

1° A morphism $j: \Omega \to \underline{\Omega}$ of domains is called a **simultaneous analytic continuation** (s.a.c.) of A if each $f \in A$ factors analytically through j.

2° A s.a.c. $j: \Omega \to \underline{\Omega}$ of A is called **maximal** if j factors through each s.a.c. $j': \Omega \to \Omega'$ of A as a morphism (i.e. there is a morphism $\underline{j}': \Omega' \to \underline{\Omega}$ such that $j = \underline{j}' \circ j'$). If $j: \Omega \to \&(\Omega)$ is a maximal s.a.c. of $\mathcal{H}(\Omega)$, then $\&(\Omega)$ is called the **envelope of holomorphy** of Ω.

3° Ω is called an **A-domain of holomorphy (domain of holomorphy** if $A = \mathcal{H}(\Omega)$) if each s.a.c. $j: \Omega \to \underline{\Omega}$ of A is an isomorphism of domains. Ω is the **domain of existence of the analytic map** $f \in \mathcal{H}(\Omega,F)$ if Ω is an $\{f\}$-domain of holomorphy.

Remark 3.4. The above definition make sense also for weakly analytic, G-analytic or LF-analytic maps with values in locally convex spaces, in particular in **C**. The existence and uniqueness results of the rest of this section remain valid for these more general situations.

First of all, a maximal s.a.c. $j: \Omega \to \underline{\Omega}$ of A is unique (i.e. $\underline{\Omega}$ is unique up to isomorphisms of domains, which is the same as to say that j is unique up to isomorphisms in $\Omega/D(E)$, (see 1.5)). The next proposition shows that the maximal s.a.c. can be constructed step by step.

Proposition 3.5. Let $A = \cup\{ A_k \mid k \in I\}$ be a set of analytic mappings on the domain Ω over E and let $j_k: \Omega \to \Omega_k$ be a maximal s.a.c. of A_k for each $k \in I$. Then the intersection (1.4) $\underline{j}: \Omega \to \underline{\Omega}$ of the j_k is a maximal s.a.c. of A.

Proof. Evidently $\underline{j}: \Omega \to \underline{\Omega}$ is a s.a.c. of A. Let $j': \Omega \to \Omega'$ be another s.a.c. of A. Since $j_k: \Omega \to \Omega_k$ is a maximal s.a.c. of A_k and j is in particular a s.a.c. of $A_k \subset A$ there is a morphism $j_k': \Omega' \to \Omega_k$ such that $j_k = j_k' \circ j'$. Now the maximality of the intersection $\underline{j}$ implies that $\underline{j}$ factors through j' as a morphism.

Theorem 3.6. Each set A of analytic mappings on the domain Ω over E has a maximal s.a.c.

Proof. According to the foregoing proposition it is enough to prove the theorem for a singleton $A = \{f\}$. To simplify the notations we assume further that the range of f is a lcs F.

Let $F\{E\}$ be the set of all convergent power series at $0 \in E$ with values in F, i.e. the set of all sequences $P = (P_n)_{n \geq 0}$ of polynomials $P_n: E \to F$ such that

(1) There exist a balanced convex neighborhood U of $0 \in E$ and a $g \in \mathcal{H}(U,F)$ such that $P_n := \frac{1}{n!} \hat{d}^n g(0)$.

Let $\mathcal{O} := E \times F\{E\}$. A topology on $\mathcal{O}$ is determined by defining the neighborhoods of a point $(a,P) \in \mathcal{O}$: $W \subset \mathcal{O}$ is a neighborhood of (a,P) if there are U and g as in (1) so that $\{(a+z, (\frac{1}{n!}\hat{d}^n g(z))_{n\geq 0}) \mid z \in U\} =: U(a,P) \subset W$. The map $\pi: \mathcal{O} \ni (b,Q) \mapsto b \in E$ is a local homeomorphism, since for U and g as in (1) and $W := U(a,P)$ the restriction $\pi \mid W: W \to a+U$ is a homeomorphism. By virtue of the identity theorem, $\mathcal{O}$ is a

Hausdorff space. Hence, $(\mathcal{O},\pi)$ is a manifold over E. Now the map

$$j: \Omega \ni x \mapsto (px, (\frac{1}{n!}\hat{d}^n f(x))_{n\geq 0}) \in \mathcal{O}$$

is a morphism: Clearly $p = \pi \circ j$, and for $B := B^{\alpha}_{\Omega}(x,r)$, $\alpha \in cs(E)$, $r > 0$ as in 3.2 it follows that $j(B) = U(px, (\frac{1}{n!}d^n f(x))_{n\geq 0})$ where $U = B^{\alpha}_{E}(0,r)$. Let Ω_f be that connected component of $\mathcal{O}$ which contains the connected set $j(\Omega)$. f factors through the analytic map $\underline{f}$: $\Omega_f \ni (a,P) \mapsto P_0 \in F$. Thus, $j: \Omega \to \Omega_f$ is a s.a.c. of $\{f\}$.

It remains to show that $j: \Omega \to \Omega_f$ is maximal. Let $j': \Omega \to \Omega'$ be a morphism of domains and $f' \in \mathcal{H}(\Omega',F)$ so that $f = f' \circ j'$. The same construction as above for Ω' and f' gives a connected component $\Omega_{f'}$ of $\mathcal{O}$. From $\phi \neq \Omega_f \cap \Omega_{f'} \supset j(\Omega)$ it follows that $\Omega_f = \Omega_{f'}$ and that j factors through j' as a morphism.

Remarks 3.7. 1° The domain Ω_f is obviously the domain of existence of the analytic map $\underline{f}: \Omega_f \to F$. Thus, each connected component Σ of $\mathcal{O}$ is the domain of existence of an analytic map $\Sigma \to F$. Conversely, each domain of existence of an analytic F-valued map is isomorphic as a domain to a connected component of $\mathcal{O}$.

2° The construction of Ω_f depends, of course, on Ω and f, and it depends on the projection map $p: \Omega \to E$ (cf. problem 1 in the next section).

3° The construction of $\mathcal{O}$ in the above proof depends on E and F and should better be denoted by $\mathcal{O}_E(F)$. For an open set $U \subset E$, the set $\mathcal{H}(U,F)$ can be identified with the set $\Gamma(U,\mathcal{O}_E(F))$ of all sections $U \to \mathcal{O}_E(F)$, i.e. with the set of continuous maps $s: U \to \mathcal{O}_E(F)$ with $\pi \circ s = id_U$. Hence, $(\mathcal{O}_E(F),\pi)$ is the **sheaf of germs of F-valued analytic maps on E.**

4° The above proof for domains over C^n is given in [3], [7] and [40], an abstract version in terms of sheaf theory in [6], [28]. In the infinite dimensional case a proof using sheaf theory and not using the intersection theorem can be found in [22], [34] and [37]. Therefore, we have presented here the older proof, in which one needs not explicitly the concept of a sheaf and in which one directly sees how to construct the maximal s.a.c. with the aid of power series expansion. But the sheaf theoretic viewpoint is more general [9]:

5° Let g be a sheaf of germs of maps on a Hausdorff space E (not necessarily a lcs) with values in a Hausdorff space Y which satisfy the identity theorem (i.e. two sections over an open connected set $U \subset E$ are equal if they are equal on a non-void open subset of U). On each domain (Ω,p) over E a sheaf g_Ω is induced by g and p. Then, one obtains as above the domain of existence of a global section $f \in \Gamma(\Omega,g_\Omega)$ as a suitable component Ω_f of g. Thus, the existence of the maximal extension of a set $A \subset \Gamma(\Omega,g_\Omega)$ with respect to g can be proved using 3.5. As a result, we obtain the existence of the envelope of "weak holomorphy", or "G-holomorphy" or "LF-holomorphy" (cf. remark 3.4), and also the existence of the envelope of $\mathfrak{A}$ –holomorphy in the sense of Rickart [36].

6⁰ Theorem 3.7 implies in particular that the envelope of holomorphy $\mathcal{E}(\Omega)$ of a given domain Ω exists. Following the method of [18, p.49] we obtain a different existence proof using the spectrum $\mathcal{S}(\mathcal{H}(\Omega))$ of $\mathcal{H}(\Omega)$ ($\mathcal{S}(\mathcal{H}(\Omega))$ is the set of non-zero (algebra-) homomorphisms $\mathcal{H}(\Omega) \to \mathbf{C}$).

3.8. Alternative construction of the envelope of holomorphy.

Let (Ω,p) be a domain spread over E. With the notation of 3.2 for every $f \in \mathcal{H}(\Omega)$ each of the functions

$$\hat{d}_a^n f: \Omega \ni x \mapsto \hat{d}^n f(x).a \in \mathbf{C}, \quad n \in \mathbf{N}, \ a \in E,$$

is analytic. Let $\mathcal{M}(\Omega) \subset \mathcal{S}(\mathcal{H}(\Omega))$ denote the set of all $h \in \mathcal{S}(\mathcal{H}(\Omega))$ satisfying:

- * There is an open neighborhood V of $0 \in E$ such that $h_a(f) := \Sigma \frac{1}{n!} h(\hat{d}_a^n f)$ converges for all $a \in V$ and $f \in \mathcal{H}(\Omega)$ absolutely and defines a homomorphism $h_a \in \mathcal{S}(\mathcal{H}(\Omega))$.
- ** For all $f \in \mathcal{H}(\Omega)$ and $a_0 \in V$ there exists a neighborhood U of a_0, $U \subset V$, such that $\sup\{|h_a(f)| \mid a \in U\} < \infty$.
- *** There is a point $q(h) \in E$ such that $h(\mu \circ p) = \mu(q(h))$ for all $\mu \in E'$, i.e. $h \circ p^*: E' \ni \mu \mapsto h(\mu \circ p) \in \mathbf{C}$ is weakly analytic.

Let $h \in \mathcal{M}(\Omega)$ and V as above. For $a,b \in V$, $a+b \in V$ we have

$$(1)\quad (h_a)_b(f) = \Sigma \frac{1}{n!} h_a(\hat{d}_b^n f) = \Sigma\Sigma\, h(\frac{1}{m!}\hat{d}_a^m \frac{1}{n!}\hat{d}_b^n f) = \sum_{i=0}^{\infty} \sum_{n+m=i} h(\frac{1}{m!}\hat{d}_a^m f \frac{1}{n!}\hat{d}_b^n f) = \sum_{i=0}^{\infty} \frac{1}{i!} h(\hat{d}_{a+b}^i f) = h_{a+b}(f).$$

$$(2)\qquad h_a(\mu \circ p) = h(\mu \circ p) + h(\mu(a)) = \mu(q(h)+a) \ \text{ for all } \mu \in E'.$$

Hence, $\{h_a \mid a \in V\} \subset \mathcal{M}(\Omega)$; and the sets $\{\{h_a \mid a \in V\} \mid V \text{ with } *\text{—}**\}$ are the neighborhood base of a Hausdorff topology on $\mathcal{M}(\Omega)$ (use (1)). Evidently, $q: \mathcal{M}(\Omega) \to E$ is a local homeomorphism (by (2)). Moreover, $j: \Omega \ni x \mapsto \hat{x} \in \mathcal{M}(\Omega)$, $\hat{x}: \mathcal{H}(\Omega) \ni f \mapsto f(x) \in \mathbf{C}$, is a morphism of manifolds: Let $f \in \Omega$ and $\alpha \in cs(E)$ with $d_\Omega^\alpha(x) > 0$. For $y \in B_\Omega^\alpha(x, d_\Omega^\alpha(x))$, $a = py - px$, it follows that $\hat{y}(f) = \Sigma \frac{1}{n!}\hat{d}^n f(x)a = \Sigma \frac{1}{n!}\hat{x}(\hat{d}_a^n f) = \hat{x}_a(f)$. Therefore, j is continuous and $p(y) = px + a = q(\hat{x}) + a = q(\hat{x}_a) = q(\hat{y}) = q \circ j(y)$. For every $f \in \mathcal{H}(\Omega)$ the function

$$\hat{f}: \mathcal{M}(\Omega) \ni h \mapsto h(f) \in \mathbf{C}$$

is analytic by virtue of * and **: For $\lambda \in \mathbf{C}$ with small absolute value, * implies $\hat{f}(h_{\lambda a}) = \Sigma[\frac{1}{n!} h(\hat{d}_a^n f)]\lambda^n$; hence f is G-analytic. Condition ** shows that $\hat{f}$ is locally bounded.

Now, let $\mathcal{E}(\Omega)$ denote that component of $\mathcal{M}(\Omega)$ which contains the connected set $j(\Omega)$. Until now we have established that $j: \Omega \to \mathcal{E}(\Omega)$ is a s.a.c. of $\mathcal{H}(\Omega)$. To show that j is maximal, let $j': \Omega \to \Omega'$ be a s.a.c. of $\mathcal{H}(\Omega)$. Each point $x \in \Omega'$ defines a homomorphism

$$\hat{x}: \mathcal{H}(\Omega) \ni f \mapsto f'(x) \in \mathbf{C}$$

if $f' \in \mathcal{H}(\Omega)$ denotes the analytic extension of f. $\hat{x}$ satisfies *-***, hence $\hat{x} \in \mathcal{M}(\Omega)$, and, as above, we see that $\underline{j}'$: $\Omega' \ni x \mapsto \hat{x} \in \mathcal{M}(\Omega)$ is a morphism. Thus $\underline{j}'(\Omega')$ is connected and contains $j(\Omega)$ which implies $\underline{j}'(\Omega') \subset \mathcal{E}(\Omega)$ and $j = \underline{j}' \circ j'$. This completes the proof.

Let us summarize what we have proved in this section.

Theorem 3.9. The envelope of holomorphy $\mathcal{E}(\Omega)$ of a domain Ω exists and is unique up to isomorphisms of domains. Further, $\mathcal{H}(\mathcal{E}(\Omega))$ separates the points of $\mathcal{E}(\Omega)$.

The separation property follows immediately from the construction $\mathcal{E}(\Omega) \subset \mathcal{S}(\mathcal{H}(\Omega))$. In the other construction (3.6) it is a consequence of the identity theorem.

Remark 3.10. The first attempt to construct the envelope of holomorphy of a domain Ω over a Banach space via the spectrum was made by Alexander [1]. He started with the set of all homomorphism $\mathcal{H}(\Omega) \to \mathbb{C}$ **which are continuous in the compact open topology**, but he was not able to prove the maximality of his constructed s.a.c. of $\mathcal{H}(\Omega)$ (cf. the remarks before problem 5 in the next section). The same holds for Coeuré's construction [8] in which the **bounded** (i.e. bounded on bounded sets of $\mathcal{H}(\Omega)$) homomorphisms replace the continuous homomorphisms. A construction starting with the bounded homomorphisms, too, and actually leading to the envelope of holomorphy is finally given in [37]. Let us note that the viewpoint in 3.8, and hence the result, are different from [37], since in [37] the properties *-*** are derived from the boundedness of the homomorphism (see also the remarks after problem 6 in the next section).

§4. Open Problems

As already mentioned in the introduction we will now discuss some open problems concerning analytic continuation. We are also interested in giving a short survey of some known results.

Let (Ω,p) be a domain over the lcs E. Then all the constructions of the envelope $\mathcal{E}(\Omega)$ of Ω, as discussed in the third section, depend on the projection map $p: \Omega \to E$. A natural question is to ask whether or not the analytic structure of $\mathcal{E}(\Omega)$ is independent of p.

Problem 1. Let Ω, Σ be domains over E and let $\varphi : \Omega \to \Sigma$ be a bianalytic map. Does there exist a bianalytic map $\mathcal{E}(\varphi)$: $\mathcal{E}(\Omega) \to \mathcal{E}(\Sigma)$?

In [22] and [37] it is proved that the answer to the above question is yes, provided that Ω is a domain over a **Banach** space. The proofs need the implicit function theorem which does not hold for arbitrary E.

Problem 2. Let $j: \Omega \to \underline{\Omega}$ be a s.a.c. of $\mathcal{H}(\Omega)$ over E and let F be a complete, locally convex space. Is the restriction map j_F^*: $\mathcal{H}(\underline{\Omega},F) \ni g \mapsto g \circ j \in \mathcal{H}(\Omega,F)$ surjective?

This problem has been investigated by many authors; e.g. in [1], [8], [22], [29], [37] for the Banach space case, in [15], [25], [26], [30], [34], [38] for more general situations. A very useful lemma is proved in [4] from which follows that every $f \in \mathcal{H}(\Omega,F)$ has at least a G-analytic continuation $\underline{f}$: $\Omega \to F$

which is also weakly analytic and LF-analytic. Consequently, problem 2 can be solved if one can prove that "weak plus slight" holomorphy (in the sense of Nachbin [30], [31]) implies holomorphy. Corresponding to problem 2 one can consider a **similar problem** for open analytic maps $\varphi: \Omega \to \underline{\Omega}$ between analytic manifolds modelled on E for which $\varphi^*: \mathcal{H}(\underline{\Omega}) \to \mathcal{H}(\Omega)$ is surjective (e.g. [9], [22], [34]).

The answer to problem 2 is affirmative if E is a **metrizable** or a **Zorn space** (i.e. a space E with the following property: A G-analytic map $f: \Omega \to F$ on a domain $\Omega \subset E$ is analytic if and only if f is continuous at at least one point of Ω). Moreover, j_F^* is always surjective if E is an **arbitrary product** of metrizable or Zorn spaces, or, more generally, if E is an **open surjective limit** (in the sense of [15]) of such spaces. This is a direct consequence of the theorem of Liouville.

The next proposition shows that problem 2 reduces itself to the question of whether or not the analytic maps with values in the Banach spaces $\ell_\infty(N)$, N a set, can be extended analytically. Furthermore, this proposition links the problem to a certain isomorphism result. To explain this let $K(\Omega)$ be the set of all **absolutely convex, pointwise bounded and equicontinuous sets of** $\mathcal{H}(\Omega)$. Each $N \in K(\Omega)$ is endowed with the topology of **pointwise convergence**, or, equivalently, with the compact open topology. $K(\Omega)$ is then a **convex compactology** on $\mathcal{H}(\Omega)$ (cf. [5] for the definition of a compactology).

Proposition 4.1. For a s.a.c. $j: \Omega \to \underline{\Omega}$ the following properties are equivalent:

1° j_F^* is surjective for all complete F.

2° j_F^* is surjective for all Banach spaces of the form $F = \ell_\infty(N)$ where N is an index set.

3° $j^*: \mathcal{H}(\underline{\Omega}) \to \mathcal{H}(\Omega)$ is an isomorphism of the compactologies $K(\underline{\Omega})$ and $K(\Omega)$.

Proof. Obviously, 1° implies 2°.

2° ⇒ 3°. Clearly, for each $\underline{N} \in K(\underline{\Omega})$ we have $j^*(\underline{N}) \in K(\Omega)$ and $j^*|\underline{N}: \underline{N} \to j^*(\underline{N})$ is continuous. Let $N \in K(\Omega)$. Then the analytic map $\varphi_N: \Omega \ni x \mapsto (g(x))_{g\in N} \in \ell_\infty(N)$ has an analytic continuation $\underline{\varphi_N}: \underline{\Omega} \to \ell_\infty(N)$ with $\underline{\varphi_N}(x) = (\underline{g}(x))_{g\in N}$. Hence, $\underline{N} = \{\underline{g} \in \mathcal{H}(\underline{\Omega}) \mid g \in N\} = (j^*)^{-1}(N)$ is equicontinuous and pointwise bounded. It follows that $\underline{N} \in K(\underline{\Omega})$, and that $(j^*)^{-1}|N: N \to \underline{N}$ is a homeomorphism which completes the proof of 2° ⇒ 3°.

3° ⇒ 1°. Let $f \in \mathcal{H}(\Omega,F)$. f has a weakly analytic continuation $\underline{f}: \underline{\Omega} \to F$ according to the result mentioned above [4] or applying the completeness theorem of Grothendieck (cf. [38]). To prove the continuity of $\underline{f}$, let $x \in \underline{\Omega}$, $\beta \in cs(F)$ and $\epsilon > 0$. Then $\{\nu\circ f \mid \nu \in F', |\nu| \leq \beta\} \in K(\Omega)$, and $\{\nu\circ\underline{f} \mid \nu \in F', |\nu| \leq \beta\}$ must be equicontinuous. Hence, there is a neighborhood U of x such that $|\nu\circ\underline{f}(x) - \nu\circ\underline{f}(y)| < \epsilon$ for all $y \in U$ and all $\nu \in F'$. It follows that $\beta(\underline{f}(x) - \underline{f}(y)) \leq \epsilon$ for all $y \in U$; hence $\underline{f}$ is continuous at x.

For arbitrary, non-complete F, problem 2 has a negative answer. Hirschowitz gives in [22, p.269] an example of a s.a.c., $j: \Omega \hookrightarrow \underline{\Omega}$ of $\mathcal{H}(\Omega)$ over C^2 and a normed space F such that j_F^* is not surjective. But under certain completeness conditions, weaker than the ordinary completeness, j_F^* is surjective whenever $j_{\hat{F}}^*$

is surjective.

Problem 3. Assume that $j^*_{\hat{F}}$ is surjective. Under what conditions on F does it follow that j^*_F is also surjective?

It is easy to see that j^*_F is surjective (provided $j^*_{\hat{F}}$ is surjective) if F is **sequentially complete**. The same is true for **confined** [30] spaces F. It isn't known whether the same holds for arbitrary holomorphically complete spaces (cf. [22], [11] for the definition), but j^*_F is surjective if F is **holomorphically complete** and e.g. a **metrizable** [39], resp. a **Zorn space** [15].

One is also interested in extending analytic maps with values in a domain.

Problem 4. Let $\varphi: \Omega \to \Sigma$ be an analytic map into a domain of holomorphy Σ. Does there exist an analytic extension $\mathcal{E}(\varphi): \mathcal{E}(\Omega) \to \Sigma$ of φ to $\mathcal{E}(\Omega)$?

For example, φ can be extended analytically to $\mathcal{E}(\Omega)$ if Ω is **metrizable** and Σ is a domain over a **Fréchet space which is sequentially holomorphically convex** [39] (cf. definition 4.2 below). A similar result was obtained in [22]. Every metrizable domain of existence is sequentially holomorphically convex; therefore, problem 4 is related to the Levi problem which we consider later. Note that a positive answer to problem 4 helps to solve problem 1.

Let $j: \Omega \to \underline{\Omega}$ again be a s.a.c. of $\mathcal{H}(\Omega)$ over E. Of course, $j^*: \mathcal{H}(\underline{\Omega}) \to \mathcal{H}(\Omega)$ is an algebraic isomorphism and a continuous map in the compact open topology. j^* is in general not open in this topology as Josefson [24] has shown. It follows that Alexander's construction [1] of a s.a.c. of $\mathcal{H}(\Omega)$ via the continuous spectrum is in general not the envelope of holomorphy of Ω. But if E is **metrizable**, then j^* is an isomorphism of the **associated bornological topologies** [39] (cf. [8], [22], [36] for the Banach space case). This leads to

Problem 5. For which natural structures on $\mathcal{H}(\Omega)$, $\mathcal{H}(\underline{\Omega})$ is j^* an isomorphism?

We have already seen in 4.1 that problem 5 is related to problem 2. The **equicontinuous compactology** is a suitable structure for those domains for which problem 2 has a positive answer.

Let A be a C-algebra with a structure s (i.e. with a locally convex topology, or a bornology, or a compactology). Then $\mathcal{S}(A,s)$ denotes the set of all non-zero homomorphisms $A \to C$ which are s-morphisms (i.e. continuous, resp. bounded on bounded sets, resp. continuous on the compact sets of the compactology). To improve the construction 3.8 we pose

Problem 6. Is $\mathcal{E}(\Omega) = \mathcal{S}(\mathcal{H}(\Omega),s)$ for a suitable structure s on $\mathcal{H}(\Omega)$?

In the finite dimensional case, every homomorphism $\mathcal{H}(\Omega) \to C$ is continuous for the compact open topology [23] and $\mathcal{E}(\Omega) = \mathcal{S}(\mathcal{H}(\Omega))$ is true [18, p.61]. In the case of $E = C^\Lambda$, Λ a set, we have $\mathcal{E}(\Omega) = \mathcal{S}(\mathcal{H}(\Omega),b)$, where b is the bornological topology associated with the compact open topology [2]. If Ω is **metrizable** it is possible to construct $\mathcal{E}(\Omega)$ as a certain subset of $\mathcal{S}(\mathcal{H}(\Omega),b)$ [39], but we don't know whether $\mathcal{E}(\Omega) = \mathcal{S}(\mathcal{H}(\Omega),b)$ holds. Note, that the construction of $\mathcal{E}(\Omega)$ via $\mathcal{S}(\mathcal{H}(\Omega),b)$ gives more information about the envelope of holomorphy than the construction in 3.8; in particular, it enables us to prove the results on analytic continuation of domain-valued analytic maps mentioned above.

In the definition 3.3 of a simultaneous analytic continuation of a set A of holomorphic maps on the domain (Ω,p) over E one could allow (Ω,p) to be a domain spread over a space E_1 satisfying $E \subset E_1 \subset \hat{E}$. Of course, the notion of a morphism must be slightly modified too. For example, if $\Omega = E$ and $A = \{f\}$ is a singleton, then f can be extended analytically to a domain $\hat{\Omega}$ in $\hat{E}$. The maximal s.a.c. of $\mathcal{H}(E)$ with this more general definition is exactly the holomorphic completion E_θ of E (cf. [15] and [22]). Therefore, it is reasonable to require $E_1 \subset E_\theta$. We are led to two immediate questions.

Problem 7. Does the generalized maximal s.a.c. of $\mathcal{H}(\Omega)$ always exist? Is it a domain over E_θ?

To conclude this section and the paper we want to compare some convexity properties of domains and to present the Levi problem. First a definition

Definition 4.2. Let (Ω,p) be a domain over E. Ω is called

1^0 holomorphically convex if for each compact $K \subset \Omega$ the set $\hat{K} := \{x \in \Omega \mid |f(x)| \leq \|f\|_K$ for all $f \in \mathcal{H}(\Omega)\}$ is precompact. $\|f\|_K := \sup\{|f(x)| \mid x \in K\}$.

2^0 metrically holomorphically convex if for each compact $K \subset \Omega$ and for all $\alpha \in cs(E)$: $d^\alpha_\Omega(K) = d^\alpha_\Omega(\hat{K})$.

3^0 sequentially holomorphically convex if for each sequence (x_n) in Ω, which is a boundary sequence (2.7) relative to every locally bianalytic $q: \Omega \to E$, there exists an analytic function $f \in \mathcal{H}(\Omega)$ such that $\sup |f(x_n)| = \infty$.

4^0 pseudoconvex if $-\log d_\Omega$ is plurisubharmonic on $X \times E\setminus\{0\}$ (d_Ω as in 2.4), or, equivalently, if for each compact $K \subset \Omega$ there is $\alpha \in cs(E)$ such that $d^\alpha_\Omega(\hat{K}_P) > 0$, whereby $\hat{K}_P := \{x \in \Omega \mid v(x) \leq \sup_K v$ for all plurisubharmonic functions v on $\Omega\}$. (cf. [34, p.55]).

Theorem 4.3. Let Ω be a domain spread over a separable Banach space with a basis ([17], [19]) or over C^Λ, Λ a set [2]. Then the following statements are equivalent.

1^0 Ω is the domain of existence of an analytic function $f: \Omega \to C$.

2^0 Ω is the domain of existence of an analytic map $f: \Omega \to F$ with values in a Banach space F.

3^0 Ω is a domain of holomorphy.

4^0 Ω is holomorphically convex.

5^0 Ω is metrically holomorphically convex.

6^0 Ω is sequentially holomorphically convex.

7^0 Ω is pseudoconvex.

The equivalence of 1°–7° is also true if Ω is a domain in a separable Banach space with the bounded approximation property [34].

The question of whether or not 7° implies 3° or 1° is called the Levi problem. The example of Josefson [24] shows that $7^o \Rightarrow 3^o$ is not in general true even for domains over Banach spaces. By an example of Hirschowitz [21] it is also known that $3^o \Rightarrow 1^o$ and $2^o \Rightarrow 1^o$ fail to be true in general.

Problem 8. Which of the implications in theorem 4.3 hold for domains over arbitrary locally convex Hausdorff spaces?

The implications $1^o \Rightarrow 2^o$, $1^o \Rightarrow 3^o$, $4^o \Rightarrow 5^o$, $5^o \Rightarrow 7^o$, and $6^o \Rightarrow 3^o$ are trivial. $3^o \Rightarrow 5^o$ can be proved in general [34, p.62]. We don't know whether $1^o \Rightarrow 6^o$, but a positive answer to this question would imply that in problem 3 the restriction map j_F^* is surjective if F is holomorphically complete (and if $i_{\hat{F}}^*$ is surjective). For metrizable domains we have $2^o \Rightarrow 3^o$ and $2^o \Rightarrow 6^o$ ([37], [39]) but none of the converse implications is known. $2^o \Rightarrow 3^o$ for arbitrary domains is connected with problem 2.

Problem 9. For which classes of spaces E are the properties 1°–7° in 4.3 equivalent?

Partial answers (besides theorem 4.3) to problem 9 can be found in [12] and [32]. In particular, the class of Silva spaces seems to be a good candidate for proving the equivalence of 1°–7° [32]. Note that the equivalences 1°–7° are preserved under open surjective limits (cf. [12], there called N-projective limits).

Results similar to theorem 4.3 can be obtained under several aspects which are different from the above consideration:

1° With the viewpoint of Cartan-Thullen [7] one can consider, instead of $\mathcal{H}(\Omega)$, certain subalgebras of $\mathcal{H}(\Omega)$ and prove results similar to the equivalence of 1°–6° (cf. [10], [27], [37], [39], [9]).

2° Dineen [12], [13] and Noverraz [33], [34] consider polynomially convex and finitely polynomially convex domains in a lcs E and deduce, under additional assumptions on E, some of the properties 1°–6°.

3° A number of results similar to $1^o \Leftrightarrow 3^o \Leftrightarrow 7^o$ are obtained in [14] and [16] for G-analytic instead of analytic functions.

Bibliography

1. H. Alexander, **Analytic functions on Banach spaces,** Thesis, University of California, Berkeley (1968).

2. V. Aurich, The spectrum as the envelope of holomorphy of a domain over an arbitrary product of complex lines, this conference.

3. H. Behnke and P. Thullen, **Theorie der Funktionen mehrerer komplexer Veränderlichen,** Erg. d. Math. 51(1970); 1. edition 1932.

4. W. Bogdanowicz, Analytic continuation of holomorphic functions with values in a locally convex space, Proc. Amer. Math. Soc. 22(1969) 660–666.

5. H. Buchwalter, Topologies et compactologies, Publ. Dép. Math. Lyon, vol. 6, n° 2(1969) 1–74.

6. H. Cartan, Séminaire E. N. S. 1950/51, New York: Benjamin (reprint 1970).

7. H. Cartan and P. Thullen, Zur Theorie der Singularitäten der Funktionen mehrerer komplexen Veränderlichen, Math. Ann. 106(1932) 617–647.

8. G. Coeuré, Fonctions plurisousharmoniques sur les espaces vectoriels topologiques et applications à l'étude des fonctions analytiques, Ann. Inst. Fourier 20(1970) 361–432.

9. G. Coeuré, **Forthcoming book on manifolds spread over normed spaces,** To appear at North-Holland.

10. S. Dineen, The Cartan-Thullen Theorem for Banach spaces, Ann. d. Sc. Norm. Sup. d. Pisa 24(1970) 883–886.

11. S. Dineen, Holomorphically complete locally convex topological vector spaces, To appear in Séminaire Lelong.

12. S. Dineen, Holomorphic functions on locally convex topological vector spaces II, Ann. Inst. Fourier (to appear).

13. S. Dineen, Runge domains in Banach spaces, Proc. R. Irish Acad. 71(1971) 85–89.

14. S. Dineen, Sheaves of holomorphic functions on infinite dimensional vector spaces, Math. Ann. (to appear).

15. S. Dineen, Holomorphic functions and surjective limits, Preprint.

16. L. Gruman, The Levi problem in certain infinite dimensional vector spaces, Ill. J. of Math. (to appear).

17. L. Gruman and C. Kiselman, Le problème de Levi dans les espaces de Banach à base, C. R. Acad. Sci. 274(1972) 1296–1299.

18. R. C. Gunning and H. Rossi, **Analytic functions of several complex variables,** Englewood Cliffs, N.J.: Prentice-Hall 1965.

19. Y. Hervier, Sur le problème de Levi pour les espaces étalés banachiques, C. R. Acad. Sci. 275(1972) 821–824.

20. A. Hirschowitz, Bornologie des espaces de fonctions analytiques en dimension infinie, Séminaire Lelong 1970, Lecture Notes in Math. 205(1971) 21–33.

21. A. Hirschowitz, Diverses notions d'ouverts d'analyticité en dimension infinie, Séminaire Lelong 1970, Springer Lecture Notes in Math. 205(1971) 11–20.

22. A. Hirschowitz, Prolongement analytique en dimension infinie, Ann. Inst. Fourier 22(1972) 255–292.

23. J. Igusa, On a property of the domain of regularity, Mem. Coll. Sci. Univ. Kyoto, Ser. A, 27(1952) 95–97.

24. B. Josefson, A Counterexample in the Levi problem, This conference.

25. E. Ligocka, A local factorization of analytic functions and its applications, Stud. Math. (to appear).

26. E. Ligocka and J. Siciak, Weak analytic continuation, Bull. de l'Acad. Pol. de Sci. 20(1972).

27. M. C. Matos, On the Cartan-Thullen theorem for some subalgebras of holomorphic functions in a locally convex space, J. für die Reine und Angew. Math. (to appear).

28. B. Malgrange, Lectures on the theory of functions of several complex variables, Bombay: Tata Inst. of Fund. Research (1958).

29. L. Nachbin, Concerning spaces of holomorphic mappings, Rutgers University, USA (1970).

30. L. Nachbin, On vector-valued versus scalar-valued analytic continuation, Indag. Math. (to appear).

31. L. Nachbin, Holomorphy between locally convex spaces, This conference.

32. Ph. Noverraz, Sur le théorème de Cartan-Thullen-Oka en dimension infinie, An. Acad. brasil. Ciênc. (to appear).

33. Ph. Noverraz, Sur la pseudo-convexité et la convexité polynomiale en dimension infinie, Ann. Inst. Fourier (to appear).

34. Ph. Noverraz, Pseudo-convexité, convexité polynomiale et domaines d'holomorphie en dimension infinie, Amsterdam: North-Holland 1973.

35. D. Pisanelli, Applications analytique en dimension infinie, Bull. Sc. Math., 2e série, 96(1972) 181–191.

36. C. E. Rickart, Holomorphic extensions and domains of holomorphy for general function algebras, This conference.

37. M. Schottenloher, Über analytische Fortsetzung in Banachräumen, Math. Ann. 199(1972) 313–336.

38. M. Schottenloher, **ϵ-product and continuation of analytic mappings**, Colloque d'Analyse, Rio de Janeiro 1972, Hermann (to appear).

39. M. Schottenloher, **Analytic continuation and regular classes in locally convex Hausdorff spaces**, Port. Math. (to appear).

40. K. Stein, **Einführung in die Funktionentheorie mehrerer Veränderlichen**, Lecture notes, Univ. München (1962).

Added in proof: New results concerning the Levi problem in Silva spaces (cf. Problem 9 ff) can be found in:

Ph. Noverraz, **Le problème de Levi dans certains espaces de Silva** (to appear).

N. Popa, **Sur le problème de Levi dans les espaces de Silva à base**, C. R. Acad. Sci. 277(1973) 211–214.

Mathematisches Institut
der Universität München
Theresienstrasse 39
D-8 München 2
Germany

Vol. 215: P. Antonelli, D. Burghelea and P. J. Kahn, The Concordance-Homotopy Groups of Geometric Automorphism Groups. X, 140 pages. 1971. DM 16,-

Vol. 216: H. Maaß, Siegel's Modular Forms and Dirichlet Series. VII, 328 pages. 1971. DM 20,-

Vol. 217: T. J. Jech, Lectures in Set Theory with Particular Emphasis on the Method of Forcing. V, 137 pages. 1971. DM 16,-

Vol. 218: C. P. Schnorr, Zufälligkeit und Wahrscheinlichkeit. IV, 212 Seiten. 1971. DM 20,-

Vol. 219: N. L. Alling and N. Greenleaf, Foundations of the Theory of Klein Surfaces. IX, 117 pages. 1971. DM 16,-

Vol. 220: W. A. Coppel, Disconjugacy. V, 148 pages. 1971. DM 16,-

Vol. 221: P. Gabriel und F. Ulmer, Lokal präsentierbare Kategorien. V, 200 Seiten. 1971. DM 18,-

Vol. 222: C. Meghea, Compactification des Espaces Harmoniques. III, 108 pages. 1971. DM 16,-

Vol. 223: U. Felgner, Models of ZF-Set Theory. VI, 173 pages. 1971. DM 16,-

Vol. 224: Revêtements Etales et Groupe Fondamental. (SGA 1). Dirigé par A. Grothendieck XXII, 447 pages. 1971. DM 30,-

Vol. 225: Théorie des Intersections et Théorème de Riemann-Roch. (SGA 6). Dirigé par P. Berthelot, A. Grothendieck et L. Illusie. XII, 700 pages. 1971. DM 40,-

Vol. 226: Seminar on Potential Theory, II. Edited by H. Bauer. IV, 170 pages. 1971. DM 18,-

Vol. 227: H. L. Montgomery, Topics in Multiplicative Number Theory. IX, 178 pages. 1971. DM 18,-

Vol. 228: Conference on Applications of Numerical Analysis. Edited by J. Ll. Morris. X, 358 pages. 1971. DM 26,-

Vol. 229: J. Väisälä, Lectures on n-Dimensional Quasiconformal Mappings. XIV, 144 pages. 1971. DM 16,-

Vol. 230: L. Waelbroeck, Topological Vector Spaces and Algebras. VII, 158 pages. 1971. DM 16,-

Vol. 231: H. Reiter, L^1-Algebras and Segal Algebras. XI, 113 pages. 1971. DM 16,-

Vol. 232: T. H. Ganelius, Tauberian Remainder Theorems. VI, 75 pages. 1971. DM 16,-

Vol. 233: C. P. Tsokos and W. J. Padgett. Random Integral Equations with Applications to stochastic Systems. VII, 174 pages. 1971. DM 18,-

Vol. 234: A. Andreotti and W. Stoll. Analytic and Algebraic Dependence of Meromorphic Functions. III, 390 pages. 1971. DM 26,-

Vol. 235: Global Differentiable Dynamics. Edited by O. Hájek, A. J. Lohwater, and R. McCann. X, 140 pages. 1971. DM 16,-

Vol. 236: M. Barr, P. A. Grillet, and D. H. van Osdol. Exact Categories and Categories of Sheaves. VII, 239 pages. 1971. DM 20,-

Vol. 237: B. Stenström, Rings and Modules of Quotients. VII, 136 pages. 1971. DM 16,-

Vol. 238: Der kanonische Modul eines Cohen-Macaulay-Rings. Herausgegeben von Jürgen Herzog und Ernst Kunz. VI, 103 Seiten. 1971. DM 16,-

Vol. 239: L. Illusie, Complexe Cotangent et Déformations I. XV, 355 pages. 1971. DM 26,-

Vol. 240: A. Kerber, Representations of Permutation Groups I. VII, 192 pages. 1971. DM 18,-

Vol. 241: S. Kaneyuki, Homogeneous Bounded Domains and Siegel Domains. V, 89 pages. 1971. DM 16,-

Vol. 242: R. R. Coifman et G. Weiss, Analyse Harmonique Non-Commutative sur Certains Espaces. V, 160 pages. 1971. DM 16,-

Vol. 243: Japan-United States Seminar on Ordinary Differential and Functional Equations. Edited by M. Urabe. VIII, 332 pages. 1971. DM 26,-

Vol. 244: Séminaire Bourbaki - vol. 1970/71. Exposés 382-399. IV, 356 pages. 1971. DM 26,-

Vol. 245: D. E. Cohen, Groups of Cohomological Dimension One. V, 99 pages. 1972. DM 16,-

Vol. 246: Lectures on Rings and Modules. Tulane University Ring and Operator Theory Year, 1970-1971. Volume I. X, 661 pages. 1972. DM 40,-

Vol. 247: Lectures on Operator Algebras. Tulane University Ring and Operator Theory Year, 1970-1971. Volume II. XI, 786 pages. 1972. DM 40,-

Vol. 248: Lectures on the Applications of Sheaves to Ring Theory. Tulane University Ring and Operator Theory Year, 1970-1971. Volume III. VIII, 315 pages. 1971. DM 26,-

Vol. 249: Symposium on Algebraic Topology. Edited by P. J. Hilton. VII, 111 pages. 1971. DM 16,-

Vol. 250: B. Jónsson, Topics in Universal Algebra. VI, 220 pages. 1972. DM 20,-

Vol. 251: The Theory of Arithmetic Functions. Edited by A. A. Gioia and D. L. Goldsmith VI, 287 pages. 1972. DM 24,-

Vol. 252: D. A. Stone, Stratified Polyhedra. IX, 193 pages. 1972. DM 18,-

Vol. 253: V. Komkov, Optimal Control Theory for the Damping of Vibrations of Simple Elastic Systems. V, 240 pages. 1972. DM 20,-

Vol. 254: C. U. Jensen, Les Foncteurs Dérivés de $\varprojlim$ et leurs Applications en Théorie des Modules. V, 103 pages. 1972. DM 16,-

Vol. 255: Conference in Mathematical Logic - London '70. Edited by W. Hodges. VIII, 351 pages. 1972. DM 26,-

Vol. 256: C. A. Berenstein and M. A. Dostal, Analytically Uniform Spaces and their Applications to Convolution Equations. VII, 130 pages. 1972. DM 16,-

Vol. 257: R. B. Holmes, A Course on Optimization and Best Approximation. VIII, 233 pages. 1972. DM 20,-

Vol. 258: Séminaire de Probabilités VI. Edited by P. A. Meyer. VI, 253 pages. 1972. DM 22,-

Vol. 259: N. Moulis, Structures de Fredholm sur les Variétés Hilbertiennes. V, 123 pages. 1972. DM 16,-

Vol. 260: R. Godement and H. Jacquet, Zeta Functions of Simple Algebras. IX, 188 pages. 1972. DM 18,-

Vol. 261: A. Guichardet, Symmetric Hilbert Spaces and Related Topics. V, 197 pages. 1972. DM 18,-

Vol. 262: H. G. Zimmer, Computational Problems, Methods, and Results in Algebraic Number Theory. V, 103 pages. 1972. DM 16,-

Vol. 263: T. Parthasarathy, Selection Theorems and their Applications. VII, 101 pages. 1972. DM 16,-

Vol. 264: W. Messing, The Crystals Associated to Barsotti-Tate Groups: With Applications to Abelian Schemes. III, 190 pages. 1972. DM 18,-

Vol. 265: N. Saavedra Rivano, Catégories Tannakiennes. II, 418 pages. 1972. DM 26,-

Vol. 266: Conference on Harmonic Analysis. Edited by D. Gulick and R. L. Lipsman. VI, 323 pages. 1972. DM 24,-

Vol. 267: Numerische Lösung nichtlinearer partieller Differential- und Integro-Differentialgleichungen. Herausgegeben von R. Ansorge und W. Törnig. VI, 339 Seiten. 1972. DM 26,-

Vol. 268: C. G. Simader, On Dirichlet's Boundary Value Problem. IV, 238 pages. 1972. DM 20,-

Vol. 269: Théorie des Topos et Cohomologie Etale des Schémas. (SGA 4). Dirigé par M. Artin, A. Grothendieck et J. L. Verdier. XIX, 525 pages. 1972. DM 50,-

Vol. 270: Théorie des Topos et Cohomologie Etale des Schémas. Tome 2. (SGA 4). Dirigé par M. Artin, A. Grothendieck et J. L. Verdier. V, 418 pages. 1972. DM 50,-

Vol. 271: J. P. May, The Geometry of Iterated Loop Spaces. IX, 175 pages. 1972. DM 18,-

Vol. 272: K. R. Parthasarathy and K. Schmidt, Positive Definite Kernels, Continuous Tensor Products, and Central Limit Theorems of Probability Theory. VI, 107 pages. 1972. DM 16,-

Vol. 273: U. Seip, Kompakt erzeugte Vektorräume und Analysis. IX, 119 Seiten. 1972. DM 16,-

Vol. 274: Toposes, Algebraic Geometry and Logic. Edited by. F. W. Lawvere. VI, 189 pages. 1972. DM 18,-

Vol. 275: Séminaire Pierre Lelong (Analyse) Année 1970-1971. VI, 181 pages. 1972. DM 18,-

Vol. 276: A. Borel, Représentations de Groupes Localement Compacts. V, 98 pages. 1972. DM 16,-

Vol. 277: Séminaire Banach. Edité par C. Houzel. VII, 229 pages. 1972. DM 20,-

Vol. 278: H. Jacquet, Automorphic Forms on GL(2). Part II. XIII, 142 pages. 1972. DM 16,–

Vol. 279: R. Bott, S. Gitler and I. M. James, Lectures on Algebraic and Differential Topology. V, 174 pages. 1972. DM 18,–

Vol. 280: Conference on the Theory of Ordinary and Partial Differential Equations. Edited by W. N. Everitt and B. D. Sleeman. XV, 367 pages. 1972. DM 26,–

Vol. 281: Coherence in Categories. Edited by S. Mac Lane. VII, 235 pages. 1972. DM 20,–

Vol. 282: W. Klingenberg und P. Flaschel, Riemannsche Hilbertmannigfaltigkeiten. Periodische Geodätische. VII, 211 Seiten. 1972. DM 20,–

Vol. 283: L. Illusie, Complexe Cotangent et Déformations II. VII, 304 pages. 1972. DM 24,–

Vol. 284: P. A. Meyer, Martingales and Stochastic Integrals I. VI, 89 pages. 1972. DM 16,–

Vol. 285: P. de la Harpe, Classical Banach-Lie Algebras and Banach-Lie Groups of Operators in Hilbert Space. III, 160 pages. 1972. DM 16,–

Vol. 286: S. Murakami, On Automorphisms of Siegel Domains. V, 95 pages. 1972. DM 16,–

Vol. 287: Hyperfunctions and Pseudo-Differential Equations. Edited by H. Komatsu. VII, 529 pages. 1973. DM 36,–

Vol. 288: Groupes de Monodromie en Géométrie Algébrique. (SGA 7 I). Dirigé par A. Grothendieck. IX, 523 pages. 1972. DM 50,–

Vol. 289: B. Fuglede, Finely Harmonic Functions. III, 188. 1972. DM 18,–

Vol. 290: D. B. Zagier, Equivariant Pontrjagin Classes and Applications to Orbit Spaces. IX, 130 pages. 1972. DM 16,–

Vol. 291: P. Orlik, Seifert Manifolds. VIII, 155 pages. 1972. DM 16,–

Vol. 292: W. D. Wallis, A. P. Street and J. S. Wallis, Combinatorics: Room Squares, Sum-Free Sets, Hadamard Matrices. V, 508 pages. 1972. DM 50,–

Vol. 293: R. A. DeVore, The Approximation of Continuous Functions by Positive Linear Operators. VIII, 289 pages. 1972. DM 24,–

Vol. 294: Stability of Stochastic Dynamical Systems. Edited by R. F. Curtain. IX, 332 pages. 1972. DM 26,–

Vol. 295: C. Dellacherie, Ensembles Analytiques, Capacités, Mesures de Hausdorff. XII, 123 pages. 1972. DM 16,–

Vol. 296: Probability and Information Theory II. Edited by M. Behara, K. Krickeberg and J. Wolfowitz. V, 223 pages. 1973. DM 20,–

Vol. 297: J. Garnett, Analytic Capacity and Measure. IV, 138 pages. 1972. DM 16,–

Vol. 298: Proceedings of the Second Conference on Compact Transformation Groups. Part 1. XIII, 453 pages. 1972. DM 32,–

Vol. 299: Proceedings of the Second Conference on Compact Transformation Groups. Part 2. XIV, 327 pages. 1972. DM 26,–

Vol. 300: P. Eymard, Moyennes Invariantes et Représentations Unitaires. II. 113 pages. 1972. DM 16,–

Vol. 301: F. Pittnauer, Vorlesungen über asymptotische Reihen. VI, 186 Seiten. 1972. DM 18,–

Vol. 302: M. Demazure, Lectures on p-Divisible Groups. V, 98 pages. 1972. DM 16,–

Vol. 303: Graph Theory and Applications. Edited by Y. Alavi, D. R. Lick and A. T. White. IX, 329 pages. 1972. DM 26,–

Vol. 304: A. K. Bousfield and D. M. Kan, Homotopy Limits, Completions and Localizations. V, 348 pages. 1972. DM 26,–

Vol. 305: Théorie des Topos et Cohomologie Etale des Schémas. Tome 3. (SGA 4). Dirigé par M. Artin, A. Grothendieck et J. L. Verdier. VI, 640 pages. 1973. DM 50,–

Vol. 306: H. Luckhardt, Extensional Gödel Functional Interpretation. VI, 161 pages. 1973. DM 18,–

Vol. 307: J. L. Bretagnolle, S. D. Chatterji et P.-A. Meyer, Ecole d'été de Probabilités: Processus Stochastiques. VI, 198 pages. 1973. DM 20,–

Vol. 308: D. Knutson, λ-Rings and the Representation Theory of the Symmetric Group. IV, 203 pages. 1973. DM 20,–

Vol. 309: D. H. Sattinger, Topics in Stability and Bifurcation Theory. VI, 190 pages. 1973. DM 18,–

Vol. 310: B. Iversen, Generic Local Structure of the Morphisms in Commutative Algebra. IV, 108 pages. 1973. DM 16,–

Vol. 311: Conference on Commutative Algebra. Edited by J. W. Brewer and E. A. Rutter. VII, 251 pages. 1973. DM 22,–

Vol. 312: Symposium on Ordinary Differential Equations. Edited by W. A. Harris, Jr. and Y. Sibuya. VIII, 204 pages. 1973. DM 22,–

Vol. 313: K. Jörgens and J. Weidmann, Spectral Properties of Hamiltonian Operators. III, 140 pages. 1973. DM 16,–

Vol. 314: M. Deuring, Lectures on the Theory of Algebraic Functions of One Variable. VI, 151 pages. 1973. DM 16,–

Vol. 315: K. Bichteler, Integration Theory (with Special Attention to Vector Measures). VI, 357 pages. 1973. DM 26,–

Vol. 316: Symposium on Non-Well-Posed Problems and Logarithmic Convexity. Edited by R. J. Knops. V, 176 pages. 1973. DM 18,–

Vol. 317: Séminaire Bourbaki – vol. 1971/72. Exposés 400–417. IV, 361 pages. 1973. DM 26,–

Vol. 318: Recent Advances in Topological Dynamics. Edited by A. Beck, VIII, 285 pages. 1973. DM 24,–

Vol. 319: Conference on Group Theory. Edited by R. W. Gatterdam and K. W. Weston. V, 188 pages. 1973. DM 18,–

Vol. 320: Modular Functions of One Variable I. Edited by W. Kuyk. V, 195 pages. 1973. DM 18,–

Vol. 321: Séminaire de Probabilités VII. Edité par P. A. Meyer. VI, 322 pages. 1973. DM 26,–

Vol. 322: Nonlinear Problems in the Physical Sciences and Biology. Edited by I. Stakgold, D. D. Joseph and D. H. Sattinger. VIII, 357 pages. 1973. DM 26,–

Vol. 323: J. L. Lions, Perturbations Singulières dans les Problèmes aux Limites et en Contrôle Optimal. XII, 645 pages. 1973. DM 42,–

Vol. 324: K. Kreith, Oscillation Theory. VI, 109 pages. 1973. DM 16,–

Vol. 325: Ch.-Ch. Chou, La Transformation de Fourier Complexe et L'Equation de Convolution. IX, 137 pages. 1973. DM 16,–

Vol. 326: A. Robert, Elliptic Curves. VIII, 264 pages. 1973. DM 22,–

Vol. 327: E. Matlis, 1-Dimensional Cohen-Macaulay Rings. XII, 157 pages. 1973. DM 18,–

Vol. 328: J. R. Büchi and D. Siefkes, The Monadic Second Order Theory of All Countable Ordinals. VI, 217 pages. 1973. DM 20,–

Vol. 329: W. Trebels, Multipliers for (C, α)-Bounded Fourier Expansions in Banach Spaces and Approximation Theory. VII, 103 pages. 1973. DM 16,–

Vol. 330: Proceedings of the Second Japan-USSR Symposium on Probability Theory. Edited by G. Maruyama and Yu. V. Prokhorov. VI, 550 pages. 1973. DM 36,–

Vol. 331: Summer School on Topological Vector Spaces. Edited by L. Waelbroeck. VI, 226 pages. 1973. DM 20,–

Vol. 332: Séminaire Pierre Lelong (Analyse) Année 1971-1972. V, 131 pages. 1973. DM 16,–

Vol. 333: Numerische, insbesondere approximationstheoretische Behandlung von Funktionalgleichungen. Herausgegeben von R. Ansorge und W. Törnig. VI, 296 Seiten. 1973. DM 24,–

Vol. 334: F. Schweiger, The Metrical Theory of Jacobi-Perron Algorithm. V, 111 pages. 1973. DM 16,–

Vol. 335: H. Huck, R. Roitzsch, U. Simon, W. Vortisch, R. Walden, B. Wegner und W. Wendland, Beweismethoden der Differentialgeometrie im Großen. IX, 159 Seiten. 1973. DM 18,–

Vol. 336: L'Analyse Harmonique dans le Domaine Complexe. Edité par E. J. Akutowicz. VIII, 169 pages. 1973. DM 18,–

Vol. 337: Cambridge Summer School in Mathematical Logic. Edited by A. R. D. Mathias and H. Rogers. IX, 660 pages. 1973. DM 42,–

Vol. 338: J. Lindenstrauss and L. Tzafriri, Classical Banach Spaces. IX, 243 pages. 1973. DM 22,–

Vol. 339: G. Kempf, F. Knudsen, D. Mumford and B. Saint-Donat, Toroidal Embeddings I. VIII, 209 pages. 1973. DM 20,–

Vol. 340: Groupes de Monodromie en Géométrie Algébrique. (SGA 7 II). Par P. Deligne et N. Katz. X, 438 pages. 1973. DM 40,–

Vol. 341: Algebraic K-Theory I, Higher K-Theories. Edited by H. Bass. XV, 335 pages. 1973. DM 26,–

Vol. 342: Algebraic K-Theory II, "Classical" Algebraic K-Theory, and Connections with Arithmetic. Edited by H. Bass. XV, 527 pages. 1973. DM 36,–